Bioactive and Therapeutic Dental Materials

Bioactive and Therapeutic Dental Materials

Special Issue Editor

Salvatore Sauro

MDPI • Basel • Beijing • Wuhan • Barcelona • Belgrade

MDPI

Special Issue Editor
Salvatore Sauro
CEU Cardenal Herrera University,
Spain
King's College London Dental Institute,
UK

Editorial Office
MDPI
St. Alban-Anlage 66
4052 Basel, Switzerland

This is a reprint of articles from the Special Issue published online in the open access journal *Materials* (ISSN 1996-1944) from 2018 to 2019 (available at: http://www.mdpi.com/journal/materials/special_issues/bioactive_and_therapeutic_dental_materials)

For citation purposes, cite each article independently as indicated on the article page online and as indicated below:

LastName, A.A.; LastName, B.B.; LastName, C.C. Article Title. *Journal Name* **Year**, *Article Number*, Page Range.

ISBN 978-3-03921-419-8 (PDF)
ISBN 978-3-03921-420-4 (Pbk)

Cover image courtesy of Mary Anne Melo

Contents

About the Special Issue Editor

Sauro Salvatore (Orcid number: 0000-0002-2527-8776) obtained his Ph.D. in Dental Biomaterials Research Pre-clinical Dentistry, and postdoctorate in Dental Biomaterials/Pre-clinical Dentistry at King's College London Dental Institute, London.

Currently full professor in dental biomaterials and minimally invasive dentistry at the Departamento de Odontología, Facultad de Ciencias de la Salud, Universidad CEU-Cardenal Herrera, coordinator of the Dental Research, and principal investigator of the research group In Situ Dental Tissues Engineering and Minimally Invasive Therapeutic Adhesive Rehabilitation.

Professor Sauro is honorary senior lecturer, at the Faculty of Dentistry, Oral & Craniofacial Sciences, King's College London Dental Institute (KCLDI), and Visiting Professor at the School of Dentistry, Sechenov University of Moscow, Russia.

Professor Sauro has been working in dental research for more 15 years (JCR—h-index: 28) and in collaboration with internationally renowned researchers, has published more than 100 articles in international peer-review journals with high impact in the dental field.

Preface to "Bioactive and Therapeutic Dental Materials"

A new generation of dental materials has been developed during the last ten years. These are identified as bioactive dental materials, which are able to release calcium, phosphate and other specific ions to help rebuild demineralized dentin and enamel. Such a phenomenon is known as biomineralization, which refers to the exchange of therapeutic ions with the dental substrates forming new apatite or, in many cases, repairing existing demineralized apatite. Moreover, smart materials can react to pH changes in the oral environment, as well as elicit reparative processes within the bonding interface in the presence of body fluids such as saliva, crevicular fluid and blood.

This book focus principally on ion-releasing and other smart dental materials for application in preventive and restorative dentistry, as well as in endodontics in the form of adhesives, resin-based composites, pastes, varnishes, liners, and dental cements. Special attention has been given to bioactive materials developed to induce cells differentiation/stimulation, hard tissue formation, and exert antimicrobial actions. New innovations are necessary to continue to help reinforce existing technologies and to introduce new paradigms for treating dental disease and restoring teeth seriously compromised by caries lesions via biomimetic and more biological operative approaches. Dental bioactive materials is arguably the latest research area in dentistry and, thus, the amount of new research is overwhelming. However, in this day and age of evidence-based practice, it important for this new information to be distilled into a practical and understandable format. I would like to thank all the authors who took the time to contribute to this volume. Moreover, special thanks go to Professor Irina Makeeva (Sechenov University Russia, Moscow, Russia) and Dr. Massimo Giovarruscio (Sechenov University Russia, Moscow, Russia) for their scientific collaboration and contribution towards organizing and managing this Special Issue. I am sure you will find the material presented in this book enlightening and informative.

Salvatore Sauro
Special Issue Editor

materials

MDPI

Article

Effects of Ions-Releasing Restorative Materials on the Dentine Bonding Longevity of Modern Universal Adhesives after Load-Cycle and Prolonged Artificial Saliva Aging

Salvatore Sauro [1,2,*], **Irina Makeeva** [2], **Vicente Faus-Matoses** [3], **Federico Foschi** [2,4], **Massimo Giovarruscio** [2,4], **Paula Maciel Pires** [1,5], **Maria Elisa Martins Moura** [1,6], **Aline Almeida Neves** [5] and **Vicente Faus-Llácer** [3]

[1] Departamento de Odontologia, Facultad de Ciencias de la Salud, Universidad CEU Cardenal Herrera, 46115 Valencia, SPAIN
[2] Institute of Dentistry, I. M. Sechenov First Moscow State Medical University, 119146 Moscow, Russia; irina_makeeva@inbox.ru
[3] Departamento de Estomatología, Facultad de Medicina y Odontología, Universitat de Valencia, 46010 Valencia, SPAIN; fausvj@uv.es (V.F.-M.); Vicente.J.Faus@uv.es (V.F.-L.)
[4] Department of Restorative Dentistry, Faculty of Dentistry, Oral & Craniofacial Sciences at King's College London, Tower Wing, Guy's Hospital, Great Maze Pond, London SE1 9RT, UK; federico.foschi@kcl.ac.uk (F.F.); giovarruscio@me.com (M.G.)
[5] Department of Pediatric Dentistry, Federal University of Rio de Janeiro, 21941-617 Rio de Janeiro, Brazil; paulinha_pmp@hotmail.com (P.M.P.); aline.neves@odonto.ufrj.br (A.A.N.)
[6] Materiais Dentários, Universidade Federal do Ceará, Fortaleza, 60430-355 Ceará, Brazil; mariaelisa_martins@hotmail.com
* Correspondence: salvatore.sauro@uchceu.es

Received: 28 January 2019; Accepted: 25 February 2019; Published: 1 March 2019

Abstract: This study aimed at evaluating the microtensile bond strength (MTBS) and fractographic features of dentine-bonded specimens created using universal adhesives applied in etch-and-rinse (ER) or self-etching (SE) mode in combination with modern ion-releasing resin-modified glass-ionomer cement (RMGIC)-based materials after load cycling and artificial saliva aging. Two universal adhesives (FTB: Futurabond M+, VOCO, Germany; SCU: Scotchbond Universal, 3M Oral Care, USA) were used. Composite build-ups were made with conventional nano-filled composite (AURA, SDI, Australia), conventional resin-modified glass ionomer cement (Ionolux VOCO, Germany), or a (RMGIC)-based composite (ACTIVA, Pulpdent, USA). The specimens were divided in three groups and immersed in deionized water for 24 h, load-cycled (350,000 cycles; 3 Hz; 70 N), or load-cycled and cut into matchsticks and finally immersed for 8 months in artificial saliva (AS). The specimens were cut into matchsticks and tested for microtensile bond strength. The results were analyzed statistically using three-way ANOVA and Fisher's LSD post hoc test ($p < 0.05$). Fractographic analysis was performed through stereomicroscope and FE-SEM. FTB showed no significant drop in bond strength after aging. Unlike the conventional composite, the two RMGIC-based materials caused no bond strength reduction in SCU after load-cycle aging and after prolonged aging (8 months). The SEM fractographic analysis showed severe degradation, especially with composite applied on dentine bonded with SCU in ER mode; such degradation was less evident with the two GIC-based materials. The dentine-bond longevity may be influenced by the composition rather than the mode of application (ER vs. SE) of the universal adhesives. Moreover, the choice of the restorative material may play an important role on the longevity of the finalrestoration. Indeed, bioactive GIC-based materials may contribute to maintain the bonding performance of simplified universal adhesives over time, especially when these bonding systems are applied in ER mode.

Keywords: adhesion; cycling mechanical stress; dentine; longevity; glass-ionomer cements; universal adhesives

1. Introduction

Direct restorations in modern operative dentistry are frequently accomplished using conventional resin composites due to their excellent mechanical and aesthetic properties [1,2]. Nevertheless, such restorative materials are still characterized by important downsides associated to polymerization shrinkage; a phenomenon that may induce stress at resin–dentine interfaces during the light-curing procedures and jeopardize their longevity [3–5]. Indeed, it has been widely demonstrated that the volumetric contraction of conventional resin composites can transfer polymerization stress directly to the adhesive-bonded interface, causing its innermost deformation due to a lack of proper bonding performance of some adhesive systems [3,6,7]. Consequently, the sealing between composite and dental hard tissues (i.e., dentine and enamel) can be seriously compromised. This will result in gaps and marginal leakage formation, which are pathways for microleakage of oral fluids, bacteria, and enzymes penetration [3,8–10]. Such issues may translate into important clinical problems such as post-operative sensitivity, marginal discoloration, recurrent caries, and advanced pulp pathology in all those cases that are seriously compromised by the caries process [11,12].

The recently introduced universal adhesive systems are currently very popular in general dental practices, as well as in dental hospitals, due to the fact that they can be applied both in self-etching (SE) and etch-and-rinse (ER) modes. Considering their compositions, universal adhesives can be also classified as simplified systems because all ingredients, including acidic functional monomers and solvents, are incorporated into one bottle. They are similar to one-step self-etching systems, so that they might still present issues related to bonding performance, degradation, and longevity [9,13]. However, application in self-etching mode minimizes recontamination of the dentine by blood and saliva during etch washing and drying. This makes SE a less technique-sensitive procedure compared to ER application mode. Moreover, SE systems present further benefits such as less post-operative sensitivity due to residual smear plugs, which are usually only partially removed from inside the dentinal tubules because of the mild acidic nature of SE systems. Indeed, the tubules remain occluded and the dentinal fluid movement is less evident compared to that usually experienced with ER systems [9,11].

On the other hand, great attention has been given to improve the effectiveness and longevity of resin–dentine bonds through several clinical strategies that may abate stress concentration at the resin–dentine interface during polymerization [14]. For instance, the use of flowable composites or resin-modified glass-ionomer cements (RMGIC) as liners or as dentine substitute materials may represent a suitable method to provide a sort of "stress-absorption" effect at the bonding interface [15,16]. This has been advocated to prevent stress development at the dentine-bonded interface and reduce gap formation, microleakage, and degradation over time [14,17,18]. Although RMGIC are self-adhesive materials, they are also often applied in dentine after etching and adhesive application, especially in those situations where the structure of the dental crown is highly compromised and a lack of mechanical retention is encountered [19–21].

It is also important to consider that occlusal stress during mastication, swallowing, as well as in cases of parafunctional habits, can affect the integrity of the bonding interface, making such a structure more susceptible to "quicker" degradation in the oral environment [22]. This seems to be of particular interest in modern, minimally invasive therapeutic restorative dentistry since it has been demonstrated that cyclic mechanical stress can promote gap formation at the margins along the composite restorations; bacteria penetration into narrow marginal gaps might ultimately promote secondary caries formation [23]. Recently, it has been advocated that ion-releasing resin-based restorative materials can reduce such biofilm penetration into marginal gaps of simulated tooth restorations; the risk for development and propagation of secondary caries is also reduced [24].

It is widely accepted that glass-ionomer cement (GIC)-based materials have a bioactive ability to release therapeutic ions such as fluoride. The presence of such ions has been associated with long-term caries inhibition when GIC-based materials are applied as a dentine substitute [25–27]. Moreover, GIC-based materials are an ideal dentine substitute as their physical properties, such as the coefficient of thermal expansion, dimensional stability, optical properties (i.e., opacity), and microhardness, are very close to that of dentine [28]. ACTIVA BioActive Restorative is a new type of restorative, bioactive, flowable, resin-based composite comparable to RMGICs. It contains fluoro-aluminum silicate particles and polyacid components of glass ionomer that undergo the acid-base setting reaction. Moreover, a bioactive ionic resin matrix is also contained in ACTIVA, which confers both light and chemical polymerization. According to the manufacturer, ACTIVA release calcium, phosphate, and fluoride when in contact with saliva. It has been advocated that restorative materials able to release specific "therapeutic" ions (e.g., calcium, phosphates, fluoride, strontium, and other minerals) into the dental hard tissues may buffer the constant assault of day-to-day ingestion of acidic food and beverages and encourage remineralization along the margins of the restoration with the tooth [29]. However, it is of great relevance that the use of ion-releasing materials in restorative dentistry may contribute to the reduced activity of proteases such as metalloproteinases (MMPs) and cathepsins involved in collagen degradation. Such enzymes are considered one of the main causes for reduction of bonding longevity when simplified bonding systems are applied in dentine with self-etching or etch-and-rinse protocols [30,31]. Moreover, there is a lack of knowledge about the effects of modern ion-releasing materials based on glass ionomer cements on resin–dentine interfaces created using current universal adhesives after mechanical load cycling and prolonged storage in artificial saliva.

Thus, the aim of this study was to evaluate, after short-term load-cycle aging or after load-cycle stress followed by prolonged aging (8 months) in artificial saliva (AS), the microtensile bond strength (MTBS) of resin–dentine bonded specimens created using universal adhesives applied in an etch-and-rinse or self-etching mode in combination with modern ion-releasing RMGIC-based materials. Fractographic analysis was also performed using field-emission scanning electron microscopy (FE-SEM).

The hypothesis tested was that compared to conventional resin composite, the use of modern ion-releasing materials would preserve the bonding performance of modern universal adhesives, applied in etch-and-rinse or self-etching, after mechanical load cycling and/or prolonged storage in artificial saliva (8 months).

2. Materials and Methods

2.1. Preparation of Dentine Specimens and Experimental Design

Sound human molars were extracted for periodontal or orthodontic reasons (ethical approval number: LEC № 11.18, 05/12/2018) and stored in distilled water at 5 °C for no longer than 3 months. The roots were removed 1 mm beneath the cemento–enamel junction using a diamond blade (XL 12205; Benetec, London, UK) mounted on a low-speed microtome (Remet evolution, REMET, Bologna, Italy). A second parallel cut was made to remove the occlusal enamel and expose mid-coronal dentine.

Three main groups (n = 72 specimens/group) were created based on the restorative materials used in this study: (i) RC: Resin composite (Aura SDI, Bayswater Victoria, Australia), applied in 2 mm increment layers up to 6 mm, and light-cured as per manufacturer's instructions; (ii) RMGIC: Resin-modified glass-ionomer cement (Ionolux; VOCO GmbH, Cuxhaven, Germany) mixed for 10 s in a trituration unit and applied in bulk. Two capsules of RMGIC were used and each one was light-cured as per manufacturer's instructions to obtain 6 mm build-ups; (iii) ACTIVA (ACTIVA BioActive Restorative, PULPDENT, Watertown, MA, USA) applied in 2 mm increment layers up to 6 mm and light-cured as per manufacturer's instructions. Light-curing was performed using an light-emitting diode (LED) light source (>1000 mW/cm^2) (Radii plus, SDI Ltd., Bayswater Victoria, Australia). The experimental design of this study required that the specimens in each main group were subsequently subdivided into four sub-groups (n = 18 specimens/group) based on the protocol employed for bonding procedures. Two modern universal adhesives were employed in this study: SCU (Scotchbond Universal, 3M Oral Care, St. Paul, MN, USA); FTB: (Futurabond M+, VOCO, Cuxhaven, Germany). These adhesives were applied as per manufacturer's instructions in self-etching (SE) or in etch-and-rinse (ER) mode (Table 1). In groups SCU–ER and FTB–ER, dentine was etched with 37% orthophosphoric acid for 15 s and subsequently rinsed with distilled water (15 s) and blotted, leaving the substrate moist. Adhesives were light-cured for 10 s. In groups SCU–SE and FTB–SE, the adhesives were applied with a microbrush for 20 s and air dried for 5 s to evaporate the solvent. These were finally light-cured for 10 s using am LED light source (>1000 mW/cm^2) (Radii plus, SID Ltd., Bayswater VIC, Australia). The specimens were finally restored with the selected restorative materials as aforementioned in the main groups. At this point, the specimens in each sub-group were furtherly divided into three groups (n = 6 specimens/group) based on the aging protocol: CTR: no aging (control, 24 h in deionized water); LC: Load cycling (350,000 cycles in artificial saliva); LC–AS: Load cycling (350,000 cycles in artificial saliva), followed by prolonged water storage (8 months in artificial saliva). A detailed description of the test groups can be found in Table 2 (Experimental design). The composition of the artificial saliva was AS: 0.103 g L^{-1} of CaCl$_2$, 0.019 g L^{-1} of MgCl$_2$·6H$_2$O, 0.544 g L^{-1} of KH$_2$PO$_4$, 30 g L^{-1} of KCl, and 4.77 g L^{-1} HEPES (acid) buffer, pH 7.4] [32]. The specimens in the subgroup LC and LC–AS were mounted in plastic rings with acrylic resin for load cycle testing. A compressive load was applied to the flat surface (3 Hz; 70 N) using a 5 mm diameter spherical stainless-steel plunger attached to a cyclic-load machine (model S-MMT-250NB; Shimadzu, Tokyo, Japan) while immersed in AS [18,33].

Table 1. Adhesive system, composition, and application procedures.

Name	Composition	Application
Scotchbond Universal, 3M Oral Care, USA (lot: 627524)	10-MDP, HEMA, silane, dimethacrylate resins, Vitrebond™ copolymer, filler, ethanol, water, initiators, and catalysts (pH 2.7)	1. Apply the adhesive on the surface and rub it for 20 s. 2. Gently air-dry the adhesive for approximately 5 s for the solvent to evaporate. 3. Light cure for 10 s (>500 mW/cm^2).
FuturaBond M+, VOCO, Germany (lot: 1742551)	HEMA, BIS-GMA, ethanol, Acidic adhesive monomer (10-MDP), UDMA, catalyst ethanol, water, initiators, and catalysts (pH 2.8)	1. Apply the adhesive homogenously to the surface. 2. Rub for 20 s. 3. Dry off the adhesive layer with dry, oil-free air for at least 5 s. 4. Light cure for 10 s (>500 mW/cm^2).

Abbreviations: 10-MDP 10-methacryloxydecyl dihydrogen phosphate, Bis-GMA bisphenol A diglycidyl methacrylate, HEMA 2-hydroxyethyl methacrylate, UDMA urethane dimethacrylate.

2.2. Micro-Tensile Bond Strength and Failure/Fractographic Analysis

The specimens were cut after the aging period using a hard-tissue microtome (Remet evolution, REMET, Bologna, Italy) across the resin–dentine interface, obtaining approximately 15–18 matchstick-shaped specimens from each tooth (Ø 0.9 mm^2). These were submitted to microtensile bond strength tests using a device with a stroke length of 50 mm, peak force of 500 N, and a displacement resolution of 0.5 mm. Modes of failure were evaluated at 50× magnification using stereoscopic microscopy and conveyed in a percentage of adhesive (A), mixed (M), or cohesive (C) bonding fracture. Five representative fractured specimens from each sub-group were mounted on aluminum stubs with carbon glue after the critical-point drying process. The specimens were gold-sputter-coated and analyzed using field-emission scanning electron microscopy (FE-SEM S-4100; Hitachi, Wokingham, UK) at 10 kV and a working distance of 15 mm.

Bond strength values in MPa were initially assessed for normality distribution and variances homogeneity using Kolmogorov–Smirnov and Levene's tests, respectively. Data were then analyzed using a three-way Analysis of Variance (ANOVA Factors: restorative material, adhesive, and aging protocol) and Newman–Keuls multiple-comparison test ($\alpha = 0.05$). SPSS V16 for Windows (SPSS Inc., Chicago, IL, USA) was used.

Table 2. Experimental design. Distribution of specimens in groups and sub-groups for evaluation via microtensile bond strength (MTBS), interface confocal microscopy, and SEM fractographic analysis. CTR = control, no aging; LC = load-cycling; AS = artificial saliva.

Total Number of Specimens in Main Groups	RESIN COMPOSITE (72 Specimens)			RMGIC (72 Specimens)			ACTIVA (72 Specimens)		
Number of specimens in sub-groups (18/group)	Number of specimens in aging sub-groups (6/ group)								
SCU–ER: Scotchbond Etch and rinse	CTR 6 spec	LC 6 spec	LC+AS 6 spec	CTR 6 spec	LC 6 spec	LC+AS 6 spec	CTR 6 spec	LC 6 spec	LC+AS 6 spec
FTB–ER Futurabond M+ Etch and rinse	CTR 6 spec	LC 6 spec	LC+AS 6 spec	CTR 6 spec	LC 6 spec	LC+AS 6 spec	CTR 6 spec	LC 6 spec	LC+AS 6 spec
SCU–SE: Scotchbond Self-etch	CTR 6 spec	LC 6 spec	LC+AS 6 spec	CTR 6 spec	LC 6 spec	LC+AS 6 spec	CTR 6 spec	LC 6 spec	LC+AS 6 spec
FTB–SE: Futurabond M+ Self-etch	CTR 6 spec	LC 6 spec	LC+AS 6 spec	CTR 6 spec	LC 6 spec	LC+AS 6 spec	CTR 6 spec	LC 6 spec	LC+AS 6 spec

3. Results

Micro-Tensile Bond Strength (MTBS) and Failure Mode Analysis

There were no pre-test failures before the microtensile bond strength assessment. Three-way ANOVA revealed a significant effect of adhesive (F = 28.75, $p < 0.001$) and restorative material (F = 6.68, $p < 0.001$) on the bond strength, whereas the aging protocol was not statistically significant (F = 8.17; $p = 0.125$). The interactions between the three variables were significant ($p < 0.001$).

The results of the microtensile bond strength test (mean and $\pm$ SD) are depicted in Table 3. It was observed that there was no significant difference ($p > 0.05$) at 24 h testing between the two adhesives when applied in etch-and-rinse (ER) or self-etching (SE) mode and then restored using the conventional RC or the two RMGIC-based materials (IONOLUX and ACTIVA). Conversely, the specimens created with the conventional RMGIC presented no significant differences ($p > 0.05$) when bonded using the two adhesives applied in ER or SE mode. However, all the specimens created with the conventional RMGIC showed a significant lower bond strength compared to those created with RC or ACTIVA. The failure mode showed that all the specimens restored with the RMGIC failed mainly in the cohesive mode, leaving a clear presence of the material still bonded to the dentine. The specimens created with RC or ACTIVA failed mainly in the cohesive in composite and mixed mode, leaving part of the dentine still covered by the restorative material and the other part exposed.

The fractographic analysis showed that the restorative materials employed in this study had no influence on the outcomes in the control storage period (24 h), but all those specimens created with SCU in ER mode presented less resin infiltration within exposed acid-etched dentine collagen fibrils (Figure 1A,B), while the specimens bonded using the FTB applied in ER mode presented fractures mainly underneath the hybrid layer (Figure 1C). Moreover, in this latter case, there was mineralized peri-tubular dentine around the lumen of the dentine tubules and no demineralized and exposed collagen fibrils (Figure 1D). Conversely, all the specimens bonded with the two adhesives applied in SE or ER mode and then restored with RMGIC showed a surface still covered by the restorative material (cohesive mode within RMGIC) with no exposure of the dentine (Figure 1E,F) after microtensile bond strength testing. Furthermore, the fractographic analysis showed that the specimens created both with SCU (Figure 1F) and FTB (Figure 1G) applied in SE, and that failed in mixed or adhesive mode, presented a dentine surface still covered by a smear layer with no presence of collagen fibrils and/or exposed dentinal tubules (Figure 1I).

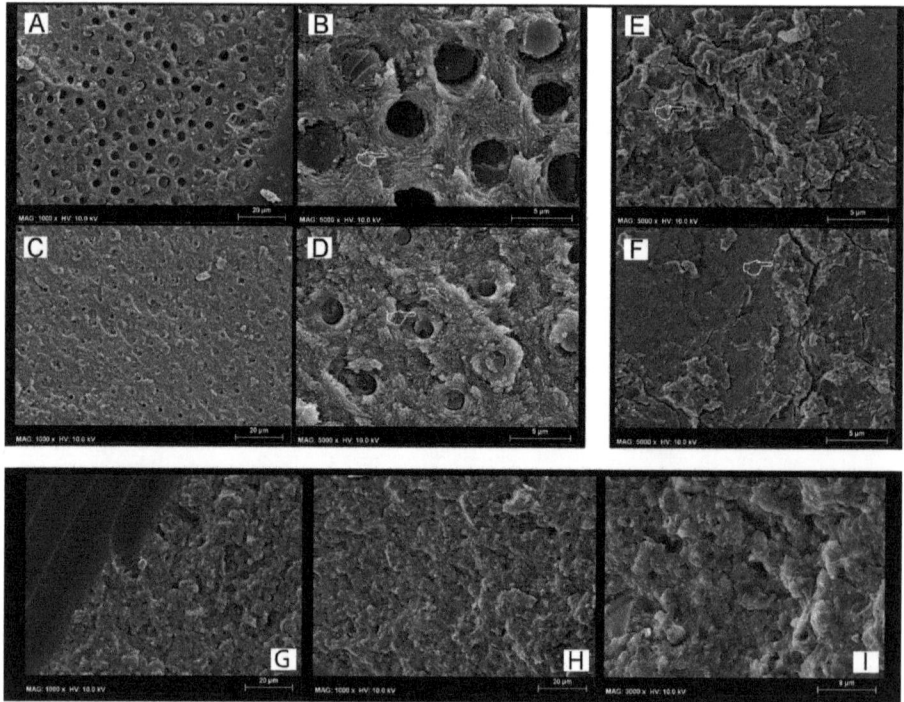

Figure 1. SEM fractographic analysis of the control specimens. (**A**) SEM fractography of a specimen created with SCU applied in ER mode and restored with resin composite (RC) showing the presence of exposed dentine and several resin tags still in the dentinal tubules. (**B**) At higher magnification, it is possible to note the presence of resin tags inside demineralized dentine tubules and collagen fibrils not well infiltrated by the SCU adhesive (pointer). This latter morphological characteristic may indicate that such resin–dentine interface would be affected by degradation over time and would drop in bond strength. (**C**) SEM fractography of a specimen created with FTB applied in ER mode and restored with ACTIVA showing the presence of exposed dentine and several resin tags still inside the small lumen of the dentinal tubules. (**D**) At higher magnification it is possible to observe a typical failure occurred at the bottom of the hybrid layer (HL) characterized by the presence of mineralized peritubular dentine (pointer), with tubules totally obliterated by resin tags and with no presence of demineralized exposed collagen fibrils. Conversely, the dentine specimens bonded with SCU (**E**) and FTB (**F**) applied in ER mode and restored with the RMGIC show the presence of the remaining RMGIC that totally covered the dentine surface. (**G**) SEM fractography of a specimen created with SCU applied in SE mode and restored with ACTIVA and (**H**) FTB applied in SE mode and restored with RC showing a characteristic failure in mixed mode. Note the presence of the remaining resin (**G**) and smear layer on the dentine surface; the latter was even more evident at higher magnification (**I**).

Table 3. The results show the mean (± SD) of the MTBS (MPa) to dentine and the percentage (%) of the failure mode analysis.

	RESIN COMPOSITE			RMGIC			ACTIVA		
	CTR	LC	LC+AS	CTR	LC	LC+AS	CTR	LC	LC+AS
SCU–ER:	48.9 (7.6)	33.5 (5.6)	28.1 (5.7)	35.1 (7.1)	33.4 (7.8)	31.1 (8.8)	55.3 (6.1)	53.1 (7.1)	50.1 (6.8)
Scotchbond	A1	B2	B2	B1	B1	B1	A1	A1	A1
Etch and rinse	50/45/5	15/55/30	10/50/40	80/20/0	75/20/5	50/35/15	45/55/0	55/40/5	30/50/20
FTB–ER	51.2 (5.9)	58.1 (7.3)	55.3 (6.5)	31.3 (6.7)	32.1 (6.6)	32.1 (7.1)	54.2 (5.7)	52.7 (6.2)	52.1 (5.6)
Futurabond M+	A1	A1	A1	B1	B1	B1	A1	A1	A1
Etch and rinse	55/40/5	45/50/5	20/65/15	70/30/0	65/35/0	60/30/5	45/55/0	55/40/5	30/50/20
SCU–SE:	45.1 (5.2)	44.4 (6.2)	34.1 (5.9)	32.3 (7.4)	34.4 (7.2)	29.6 (7.9)	46.1 (6.2)	49.8 (7.4)	49.5 (6.9)
Scotchbond	A1	A1	B1	B1	B1	B1	A1	A1	A1
Self-etch	45/50/5	40/50/10	10/55/35	70/30/0	65/30/5	50/45/5	40/55/5	30/65/5	45/50/5
FTB–SE:	49.2 (4.9)	48.3 (9.3)	45.6 (7.5)	34.1 (6.2)	31.5 (7.7)	30.5 (7.5)	48.1 (6.2)	51.1 (7.4)	50.5 (7.4)
Futurabond M+	A1	A1	A1	B1	B1	B1	A1	A1	A1
Self-etch	40/50/10	45/50/5	25/60/15	75/25/0	70/30/50	60/35/5	40/55/5	45/50/5	45/50/5

Failure mode [Cohesive/Mixed/Adhesive]. The same number indicates no significance in column, while the same letter indicates no significance in row ($p > 0.05$).

After submitting the specimens to load-cycle aging, the only group that showed a significant bond strength drop ($p < 0.05$) was that created with the SCU applied in ER mode and restored using the conventional RC. In this group, an important change in the failure mode was also observed; only 15% of the specimens failed in cohesive mode, while failure in mixed and adhesive modes were 55% and 30%, respectively (Table 3). This situation was not evident in the specimens bonded with the same adhesive but restored using IONOLUX (RMGIC) or ACTIVA; no significant bond strength drop ($p > 0.05$) and no radical change in failure mode was observed. The SEM fractography showed no important ultra-morphological changes in most of the fractured resin–dentine interfaces of these groups compared to the control group. Conversely, the specimens created with the SCU applied in ER mode (Figure 2A) and restored with the conventional RC, which failed prevalently in mixed and adhesive mode, showed that the fracture occurred underneath the hybrid layer with no sign of demineralized and/or poorly infiltrated collagen fibrils (Figure 2B,C).

Figure 2. SEM Fractographic analysis after load-cycle aging. (**A**) SEM fractography of a specimen created with SCU applied in ER mode and restored with RC showing a characteristic failure in mixed mode. The finger pointer indicates a brighter area of greater and more evident aging (pointer), which was probably induced by the cycling load. However, when we observed that specific area at higher magnifications (**B**), it was possible to observe that the fracture occurred underneath the hybrid layer (pointer), which is characterized by the presence of mineralized dentine (pointer), with tubules totally obliterated by resin tags and with no presence of demineralized exposed collagen fibrils (**C**).

The prolonged aging in artificial saliva performed subsequent load-cycling stress induced important changes on microtensile bond strength as well as on the ultramorphology of the fracture of some specific groups. In particular, the specimens bonded with SCU applied both in ER and SE mode and then restored with the conventional composite had a significant drop in bond strength compared to the specimens in the groups CTR and LC ($p < 0.05$). Moreover, the number of failures in mixed and adhesive modes increased in the aforementioned groups compared to the control

(CTR) group. The SEM fractography showed evident signs of dentine degradation in the group of specimens created with SCU applied in ER mode and then restored with the conventional composite (Figure 3(A1,A2)). The SEM fractography showed that specimens created with SCU applied in SE mode and then restored with the RC presented degradation both of the adhesive (Figure 3B) and dentine hybrid layers (Figure 3C). Conversely, the same specimens restored with the RMGIC or ACTIVA presented a stable bond strength with no significant drop ($p < 0.05$), and the type of failure remained quite similar to the control group. The SEM fractography showed no drastic changes in all those groups for the ultramorphology of fractured resin–dentine interfaces compared to the control group (Figure 3D,E). In particular, the SEM fractography of a specimen created with SCU applied in ER mode and restored with ACTIVA and RMGIC showed the presence of dentine that was well mineralized with no sign of demineralized collagen fibrils, but with the presence of mineral debrides as a possible result of the bioactivity of such GIC-based materials (Figure 3D).

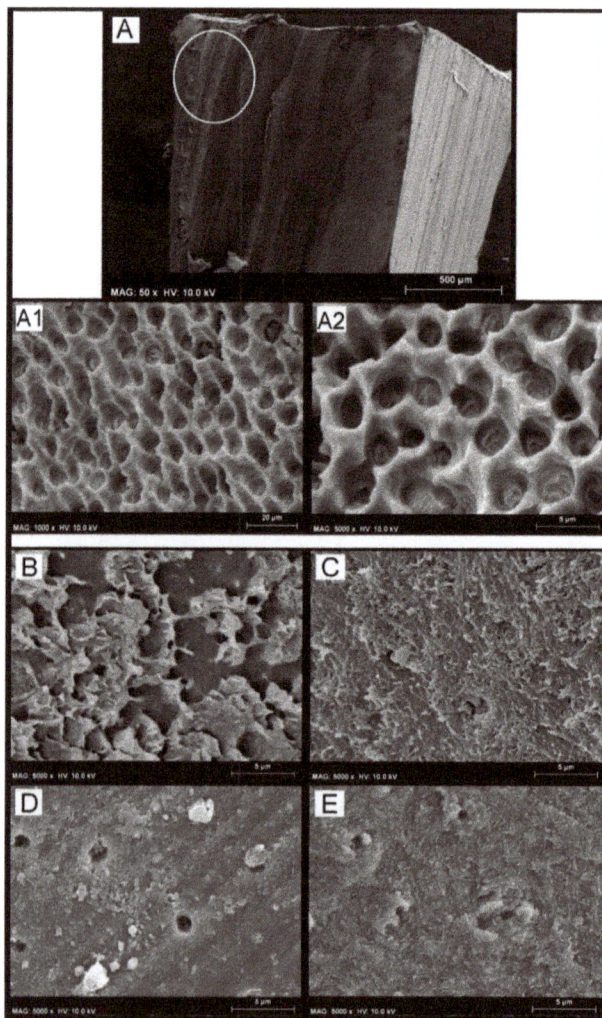

Figure 3. SEM Fractographic analysis after load cycling and aging in artificial saliva. (**A**) SEM fractography of a specimen created with SCU applied in ER mode and restored with RC showing

a characteristic failure in adhesive mode. Note that the white circle indicates no physical difference in the material; it was added to show the reader that images (**A1,A2**) were obtained by higher magnification in that zone. Indeed, in (**A1,A2**) it is possible to see severe collagen degradation without the presence of any resin residual. (**B**) SEM fractography of a specimen created with SCU applied in SE mode where it is possible to see a failure between composite and adhesive, probably due to degradation induced by excessive water sorption upon mechanical stress and prolonged AS storage. However, it was also possible to see, in those specimens that failed in mixed mode, signs of degradation of the collagen fibrils underneath the hybrid/interdiffusion layer (**C**). (**D**) SEM fractography of a specimen created with SCU applied in ER mode and restored with ACTIVA showing that the failure occurred underneath the hybrid layer, but the exposed dentine is well mineralized with no sign of exposed demineralized collagen fibrils. Note also the presence of mineral debrides that are a possible result of the bioactivity of ACTIVA, which released ions and diffused through the resin-bonded dentine. (**E**) SEM fractography of a specimen created with FTB applied in SE mode and restored with RMGIC. The specimens of this group failed mainly in cohesive and mixed mode; this latter zone is characterized by a fracture occurring underneath the hybrid layer, leaving behind a dentine surface completely mineralized with no sign of exposed, denatured, or demineralized collagen fibrils. Please note the presence of a well mineralized intratubular dentine inside the lumen of the dentine tubules.

4. Discussion

This study showed that the use of modern ion-releasing materials such as conventional RMGIC or RMGIC-based composite (ACTIVA) preserved the bonding performance of only one (SCU) of the two modern universal adhesives bonded to dentine in etch-and-rinse or self-etching mode, after the two aging protocols employed in the experimental design. Conversely, the dentine-bonded specimens created with the FTB universal bonding system applied in etch-and-rinse or self-etching showed no significant drop in bonding performance after aging, regardless the restorative material employed or the aging protocol. Hence, the hypothesis tested in this study needs to be partially accepted as the use of a specific new generation universal bonding systems may confer a stable dentine-bonded interface over time. Nevertheless, the use of modern ion-releasing restorative materials such as RMGIC or ACTIVA may preserve the bonding performance of those universal adhesives that are more prone to degradation after aging.

The effects of the load-cycle aging protocol on the bonding performance of the SCU system applied in ER mode and restored with the conventional RC were relevant; the bond strength of this group of specimens dropped significantly ($p < 0.05$). Moreover, only the specimens bonded using SCU applied both in ER and SE mode and then restored with the conventional composite showed a significant drop in bond strength compared to the specimens in the control (CTR, 24 h) group after prolonged aging in artificial saliva. The ultramorphology analysis performed in the specimens of the control group (24 h), created using the SCU system applied in ER mode and restored with RC showed the presence of demineralized-acid-etched dentine collagen that was not well resin-infiltrated (Figure 1A,B). While the same specimens submitted to LC aging showed no exposed collagen, but mineralized dentine with resin tags that obliterated the dentinal tubules (Figure 2). This was an interesting result, so we hypothesize that a possible explanation to the difference in bonding performance observed between these two latter situations (LC-only aging vs. CTR) may be attributed to the fact that the hybrid layer created using simplified adhesives applied in etch-and-rinse mode can represent the critical part of the resin–dentine interface, as it probably remains only partially polymerized [33–35]. Indeed, it has been advocated that during cycling loading such un-polymerized monomers within the hybrid layer, created with simplified, highly hydrophilic etch-and-rinse adhesives, may be mechanically "intruded" into the demineralized dentine causing a more compact and performant hybrid layer. However, such a morphological change within the resin–dentine interface may favor higher stress concentrations during the cycling load at the bottom of the hybrid layer, causing an accelerated mechanically-induced degradation phenomenon in this specific zone that often remains partially demineralized and poorly infiltrated by adhesive monomers [34,35]. Indeed, the absence of a proper, partially demineralized bottom of the hybrid layer may explain why the dentine bonded with SCU or FTB in SE mode showed no bond-strength drop after load-cycle aging, regardless of the restorative material and the protocol employed for aging [33,34]. Our results seem to be in accordance with those of Dorfer et al. [36] who demonstrated water diffusion within the resin–dentine interface and hybrid layer during flexure; this promoted chemical/mechanical degradation and washout of "poorly" polymerized water-soluble monomers.

Apparently, such type of degradation mentioned above was improbable in dentine etched with phosphoric acid and bonded using the same simplified adhesive (SCU), but restored with RMGIC or ACTIVA. Indeed, such restorative materials may have absorbed some of the stress generated by the load-cycle aging due to their lower modulus of elasticity, thereby reducing the risk for degradation at the bonding interface [14,15]. The fact that the two GIC-based materials with lower moduli of elasticity may have distributed stresses within their bulk structure lowering the tension concentration at the interface created with the SCU adhesive, applied both in ER and SE mode and subsequently submitted to a cycling load followed by prolonged storage in AS. This observation was supported by the absence of reduction in bonding performance compared to those specimens restored with the conventional composite; this latter group presented a significant bond strength drop ($p < 0.05$) after such a prolonged aging protocol. In addition to the significant bond strength reduction (Table 3), the results of this current study also showed the presence of funneled dentinal tubules, with no presence of collagen fibrils and no residual of restorative material on the dentine surface (Figure 1), which are all typical morphological signs that indicate collagen hydrolysis and proteolytic denaturation caused by the activity of proteases such as MMPs and cathepsins [34,35,37]. Conversely, the SCU adhesive applied in ER mode and restored with ACTIVA failed mainly in mixed mode or in cohesive/mixed mode when restored with the RMGIC. The SEM fractographic analysis highlighted in those specimens the presence of exposed dentine due to a fracture that occurred underneath the hybrid layer, which left behind a well mineralized dentine with no sign of collagen degradation. Indeed, in this latter case, mineralized peri-tubular dentine around the lumen of the dentine tubules and with no demineralized and exposed collagen fibrils was often observed; this is a typical ultramorphological aspect of failure occurring away from the hybrid layer in resin–dentine interface characterized by high bonding stability [37].

Furthermore, mineral debris were detected as a possible result of the bioactivity of ACTIVA and RMGIC (Figure 3D). Indeed, glass-ionomer materials are considered the main bioactive ion-releasing restorative materials currently available in clinics, since they may be able to induce mineral growth within the bonded-dentine interface [18]. We speculate that the results of this study may be somehow correlated to the those hypothesized by Toledano et al. [22,33], who showed that when bioactive materials are submitted to mechanical cycling load, they may promote diffusion of ions through the adhesive-bonded dentine due to the permeable nature of simplified all-in-one bonding systems [37], increasing the mineral–matrix ratio, and reduce nanoleakage and permeability at the resin–dentine interface. Moreover, it has been demonstrated that fluoride ions may inhibit both pro- and active metalloproteinases (MMP-2 and MMP-9) [38], thus reducing the enzymatic degradation at the bonding interface. It may be also possible that in the case of diffusion of calcium and phosphate ions through permeable hybrid layers, these may precipitate and crystallize in complex calcium-phosphates and inhibit MMPs through the formation of a Ca-PO/MMP complex [39].

On the other hand, a possible explanation for the differences in bonding performance attained in this study with the two simplified universal adhesives when restored with a conventional RC may be related to their different chemical compositions. Unlike FTB, the SCU system, which was the only adhesive that both when applied in ER and SE mode in combination with RC presented a significant bond strength drop after prolonged aging protocol, contains a polyalkenoic acid copolymer (PAC). It has been shown that PAC contained in adhesives tends to accumulate primarily on the outer surface of the hybrid layer and creates "isles" between dentine and the adhesive layer [39]. It is also well known that PAC has multiple pendent carboxylic acids along a linear backbone that bind water, which causes important water sorption and solubility. Moreover, the high molecular weight of PAC [40] precludes its penetration into interfibrillar spaces within the acid-etched dentine.

Several reports indicated that simplified adhesives containing relatively high amount of bisphenol A diglycidyl methacrylate (Bis-GMA) in combination with PAC and 2-hydroxyethyl methacrylate (HEMA) do not infiltrate well into acid-etched dentine, so creating HEMA-rich/Bis-GMA-poor hybrid layers. It is also believed that HEMA, mixing with water within the hybrid layer, may produce hydrogels able to absorb water, which in turn enable hydrolytic and enzymatic degradation processes that jeopardize the longevity of resin–dentine interfaces [41–43]. Furthermore, it is generally well known that water-containing and acidic, single-bottle, pre-hydrolyzed silane coupling agents have a relatively short shelf life because both water and lower pH media can cause silane to degrade over time [44]. A modern, universal adhesive such as SCU contains both free silane and silaned nanofillers. Thus, we believe that water sorption at the adhesive layer may have accelerated polymer hydrolysis and filler debonding, reducing the durability of its bonding performance [44,45].

The information obtained in this study, along with all the observations discussed above, may also be relevant to the contemporary philosophy in atraumatic restorative dentistry. This is based on the preparation of minimally invasive cavities in order to preserve as much sound dental tissue as possible. However, such an ultraconservative intervention should always be followed by restorative treatments performed using therapeutic restorative approaches that protect the resin–dentine interface from degradation processes and prevent the reoccurrence of secondary carious lesions [46,47]. It is well known that the bonding performance of adhesive systems applied to caries-affected dentine (CAD) is not as strong as that attained when such materials are used in sound dentine; the bonding performance seems correlated to the low biomechanical properties of CAD (e.g., modulus of elasticity) [47]. Therefore, such a situation leads to failure of the restoration over time, so that improvements and suitable alternative restorative procedures are necessary in order to improve the durability of the bonding between adhesives and CAD. Wang et al [48] demonstrated distinct differences in the depth of dentine demineralization and degree of adhesive infiltration in non-carious and CAD. Because of the structural alteration and porosities in CAD, deeper, demineralized layers occurred. The deeper the demineralized collagen, the poorer the resin infiltration into the deepest part of the CAD. This resulted in phase separation of resin adhesives and "weak" bond strength. However, Tekçe et al. [49] showed that in such circumstances, the use of flowable resin-based composites, RMGICs, and compomers may provide stronger dentine-bond strength and better margin sealing than conventional glass-ionomer cement and resin composites due to the ability of such materials to dissipate the occlusal stress and the therapeutic effect of ions released over time.

In conclusion, within the limitations of this study, it is possible to affirm that the choice of appropriate materials from a chemical and mechanical point of view can make a difference on the bonding performance/durability of dentine-bonded interfaces. Indeed, the application of well-formulated modern adhesive systems in combination with ion-releasing dentine-replacement materials might offer to clinicians the possibility to perform more long-lasting adhesive restorations. However, these concepts must be corroborated by future in vivo and/or clinical trial studies in order to evaluate their true suitability in a clinical scenario.

Author Contributions: Conceptualization, S.S. and V.F.-L.; Data curation, F.F., M.E., and M.M.; Formal Analysis, A.A.N. and M.G.; Investigation, P.M.P. and I.M.; Methodology, M.G. and S.S.; Resources I.M.; Project Administration, V.F.-M.; Supervision, S.S.; Validation, M.G.; Writing–Original Draft Preparation, S.S. and F.F.; Writing–Review and Editing, S.S., F.F., and M.G.

Funding: This research received no external funding.

Conflicts of Interest: The authors declare no conflict of interest.

References

1. Arhun, N.; Celik, C.; Yamanel, K. Clinical evaluation of resin-based composites in posterior restorations: Two-year results. *Oper. Dent.* **2010**, *35*, 397–404. [CrossRef] [PubMed]
2. Ferracane, J.L. Resin composite–state of the art. *Dent. Mater.* **2011**, *27*, 29–38. [CrossRef] [PubMed]
3. Kakaboura, A.; Rahiotis, C.; Watts, D.; Silikas, N.; Eliades, G. 3D-marginal adaptation versus setting shrinkage in light-cured microhybrid resin composites. *Dent. Mater.* **2007**, *23*, 272–278. [CrossRef] [PubMed]
4. Boaro, L.C.; Froes-Salgado, N.R.; Gajewski, V.E.; Bicalho, A.A.; Valdivia, A.D.; Soares, C.J.; Miranda Junior, W.G.; Braga, R.R. Correlation between polymerization stress and interfacial integrity of composites restorations assessed by different in vitro tests. *Dent. Mater.* **2014**, *30*, 984–992. [CrossRef] [PubMed]
5. Van Dijken, J.W.; Lindberg, A. A 15-year randomized controlled study of a reduced shrinkage stress resin composite. *Dent. Mater.* **2015**, *31*, 1150–1158. [CrossRef] [PubMed]
6. He, Z.; Shimada, Y.; Sadr, A.; Ikeda, M.; Tagami, J. The effects of cavity size and filling method on the bonding to Class I cavities. *J. Adhes. Dent.* **2008**, *10*, 447–453. [PubMed]
7. Sakaguchi, R.L.; Peters, M.C.; Nelson, S.R.; Douglas, W.H.; Poort, H.W. Effects of polymerization contraction in composite restorations. *J. Dent.* **1992**, *20*, 178–182. [CrossRef]
8. Davidson, C.L.; de Gee, A.J.; Feilzer, A. The competition between the composite-dentin bond strength and the polymerization contraction stress. *J. Dent. Res.* **1984**, *63*, 1396–1399. [CrossRef] [PubMed]
9. De Munck, J.; Van Landuyt, K.; Coutinho, E.; Poitevin, A.; Peumans, M.; Lambrechts, P.; Van Meerbeek, B. Micro-tensile bond strength of adhesives bonded to Class-I cavity-bottom dentin after thermo-cycling. *Dent. Mater.* **2005**, *21*, 999–1007. [CrossRef] [PubMed]
10. Fleming, G.J.; Cara, R.R.; Palin, W.M.; Burke, F.J. Cuspal movement and microleakage in premolar teeth restored with resin-based filling materials cured using a 'soft-start' polymerisation protocol. *Dent. Mater.* **2007**, *23*, 637–643. [CrossRef] [PubMed]
11. Bernardo, M.; Luis, H.; Martin, M.D.; Leroux, B.G.; Rue, T.; Leitao, J.; DeRouen, T.A. Survival and reasons for failure of amalgam versus composite posterior restorations placed in a randomized clinical trial. *J. Am. Dent. Assoc.* **2007**, *138*, 775–783. [CrossRef] [PubMed]
12. He, Z.; Shimada, Y.; Tagami, J. The effects of cavity size and incremental technique on micro-tensile bond strength of resin composite in Class I cavities. *Dent. Mater.* **2007**, *23*, 533–538. [CrossRef] [PubMed]
13. Chen, C.; Niu, L.N.; Xie, H.; Zhang, Z.Y.; Zhou, L.Q.; Jiao, K.; Chen, J.H.; Pashley, D.H.; Tay, F.R. Bonding of universal adhesives to dentine–Old wine in new bottles? *J. Dent.* **2015**, *43*, 525–536. [CrossRef] [PubMed]
14. Nikolaenko, S.A.; Lohbauer, U.; Roggendorf, M.; Petschelt, A.; Dasch, W.; Frankenberger, R. Influence of c-factor and layering technique on microtensile bond strength to dentin. *Dent. Mater.* **2004**, *20*, 579–585. [CrossRef] [PubMed]
15. Irie, M.; Suzuki, K.; Watts, D.C. Immediate performance of self-etching versus system adhesives with multiple light-activated restoratives. *Dent. Mater.* **2004**, *20*, 873–880. [CrossRef] [PubMed]
16. Irie, M.; Suzuki, K.; Watts, D.C. Marginal gap formation of light-activated restorative materials: Effects of immediate setting shrinkage and bond strength. *Dent. Mater.* **2002**, *18*, 203–210. [CrossRef]
17. Sampaio, P.C.; de Almeida Junior, A.A.; Francisconi, L.F.; Casas-Apayco, L.C.; Pereira, J.C.; Wang, L.; Atta, M.T. Effect of conventional and resin-modified glass-ionomer liner on dentin adhesive interface of Class I cavity walls after thermocycling. *Oper. Dent.* **2011**, *36*, 403–412. [CrossRef] [PubMed]
18. Sauro, S.; Faus-Matoses, V.; Makeeva, I.; Nunez Marti, J.M.; Gonzalez Martinez, R.; Garcia Bautista, J.A.; Faus-Llacer, V. Effects of Polyacrylic Acid Pre-Treatment on Bonded-Dentine Interfaces Created with a Modern Bioactive Resin-Modified Glass Ionomer Cement and Subjected to Cycling Mechanical Stress. *Materials* **2018**, *11*, 1884. [CrossRef] [PubMed]
19. Boksman, L.; Jordan, R.E.; Suzuki, M.; Charles, D.H. A visible light-cured posterior composite resin: Results of a 3-year clinical evaluation. *J. Am. Dent. Assoc.* **1986**, *112*, 627–631. [CrossRef] [PubMed]
20. Boksman, L.; Jordan, R.E.; Suzuki, M. Posterior composite restorations. *Compend. Contin. Educ. Dent.* **1984**, *367*, 372–373.
21. Jordan, R.E.; Suzuki, M.; Gwinnett, A.J. Conservative applications of acid etch-resin techniques. *Dent. Clin. N. Am.* **1981**, *25*, 307–336. [PubMed]
22. Toledano, M.; Cabello, I.; Aguilera, F.S.; Osorio, E.; Osorio, R. Effect of in vitro chewing and bruxism events on remineralization, at the resin-dentin interface. *J. Biomech.* **2015**, *48*, 14–21. [CrossRef] [PubMed]

23. Khvostenko, D.; Salehi, S.; Naleway, S.E.; Hilton, T.J.; Ferracane, J.L.; Mitchell, J.C.; Kruzic, J.J. Cyclic mechanical loading promotes bacterial penetration along composite restoration marginal gaps. *Dent. Mater.* **2015**, *31*, 702–710. [CrossRef] [PubMed]
24. Khvostenko, D.; Hilton, T.J.; Ferracane, J.L.; Mitchell, J.C.; Kruzic, J.J. Bioactive glass fillers reduce bacterial penetration into marginal gaps for composite restorations. *Dent. Mater.* **2016**, *32*, 73–81. [CrossRef] [PubMed]
25. Browning, W.D. The benefits of glass ionomer self-adhesive materials in restorative dentistry. *Compend. Contin. Educ. Dent.* **2006**, *27*, 308–314. [PubMed]
26. Forsten, L. Resin-modified glass ionomer cements: Fluoride release and uptake. *Acta Odontol. Scand.* **1995**, *53*, 222–225. [CrossRef] [PubMed]
27. Forss, H.; Jokinen, J.; Spets-Happonen, S.; Seppa, L.; Luoma, H. Fluoride and mutans streptococci in plaque grown on glass ionomer and composite. *Caries Res.* **1991**, *25*, 454–458. [CrossRef] [PubMed]
28. Cho, S.Y.; Cheng, A.C. A review of glass ionomer restorations in the primary dentition. *J. Can. Dent. Assoc.* **1999**, *65*, 491–495. [PubMed]
29. Fuss, M.; Wicht, M.J.; Attin, T.; Derman, S.H.M.; Noack, M.J. Protective Buffering Capacity of Restorative Dental Materials In Vitro. *J. Adhes. Dent.* **2017**, *19*, 177–183. [PubMed]
30. Tezvergil-Mutluay, A.; Agee, K.A.; Hoshika, T.; Tay, F.R.; Pashley, D.H. The inhibitory effect of polyvinylphosphonic acid on functional matrix metalloproteinase activities in human demineralized dentin. *Acta Biomater.* **2010**, *6*, 4136–4142. [CrossRef] [PubMed]
31. Osorio, R.; Yamauti, M.; Sauro, S.; Watson, T.F.; Toledano, M. Experimental resin cements containing bioactive fillers reduce matrix metalloproteinase-mediated dentin collagen degradation. *J. Endod.* **2012**, *38*, 1227–1232. [CrossRef] [PubMed]
32. Sauro, S.; Watson, T.; Moscardo, A.P.; Luzi, A.; Feitosa, V.P.; Banerjee, A. The effect of dentine pre-treatment using bioglass and/or polyacrylic acid on the interfacial characteristics of resin-modified glass ionomer cements. *J. Dent.* **2018**, *73*, 32–39. [CrossRef] [PubMed]
33. Toledano, M.; Aguilera, F.S.; Sauro, S.; Cabello, I.; Osorio, E.; Osorio, R. Load cycling enhances bioactivity at the resin-dentin interface. *Dent. Mater.* **2014**, *30*, e169–e188. [CrossRef] [PubMed]
34. Sauro, S.; Osorio, R.; Watson, T.F.; Toledano, M. Assessment of the quality of resin-dentin bonded interfaces: An AFM nano-indentation, muTBS and confocal ultramorphology study. *Dent. Mater.* **2012**, *28*, 622–631. [CrossRef] [PubMed]
35. Sauro, S.; Toledano, M.; Aguilera, F.S.; Mannocci, F.; Pashley, D.H.; Tay, F.R.; Watson, T.F.; Osorio, R. Resin-dentin bonds to EDTA-treated vs. acid-etched dentin using ethanol wet-bonding. Part II: Effects of mechanical cycling load on microtensile bond strengths. *Dent. Mater.* **2011**, *27*, 563–572. [CrossRef] [PubMed]
36. Dorfer, C.E.; Staehle, H.J.; Wurst, M.W.; Duschner, H.; Pioch, T. The nanoleakage phenomenon: Influence of different dentin bonding agents, thermocycling and etching time. *Eur. J. Oral Sci.* **2000**, *108*, 346–351. [CrossRef] [PubMed]
37. Sauro, S.; Pashley, D.H.; Mannocci, F.; Tay, F.R.; Pilecki, P.; Sherriff, M.; Watson, T.F. Micropermeability of current self-etching and etch-and-rinse adhesives bonded to deep dentine: A comparison study using a double-staining/confocal microscopy technique. *Eur. J. Oral Sci.* **2008**, *116*, 184–193. [CrossRef] [PubMed]
38. Tezvergil-Mutluay, A.; Seseogullari-Dirihan, R.; Feitosa, V.P.; Cama, G.; Brauer, D.S.; Sauro, S. Effects of Composites Containing Bioactive Glasses on Demineralized Dentin. *J. Dent. Res.* **2017**, *96*, 999–1005. [CrossRef] [PubMed]
39. Makowski, G.S.; Ramsby, M.L. Differential effect of calcium phosphate and calcium pyrophosphate on binding of matrix metalloproteinases to fibrin: Comparison to a fibrin-binding protease from inflammatory joint fluids. *Clin. Exp. Immunol.* **2004**, *136*, 176–187. [CrossRef] [PubMed]
40. Larraz, E.; Deb, S.; Elvira, C.; Roman, J.S. A novel amphiphilic acrylic copolymer based on Triton X-100 for a poly(alkenoate) glass-ionomer cement. *Dent. Mater.* **2006**, *22*, 506–514. [CrossRef] [PubMed]
41. Wang, Y.; Spencer, P. Quantifying adhesive penetration in adhesive/dentin interface using confocal Raman microspectroscopy. *J. Biomed. Mater. Res.* **2002**, *59*, 46–55. [CrossRef] [PubMed]
42. Wang, Y.; Spencer, P. Effect of acid etching time and technique on interfacial characteristics of the adhesive-dentin bond using differential staining. *Eur. J. Oral Sci.* **2004**, *112*, 293–299. [CrossRef] [PubMed]
43. Sattabanasuk, V.; Vachiramon, V.; Qian, F.; Armstrong, S.R. Resin-dentin bond strength as related to different surface preparation methods. *J. Dent.* **2007**, *35*, 467–475. [CrossRef] [PubMed]

44. Yoshihara, K.; Nagaoka, N.; Sonoda, A.; Maruo, Y.; Makita, Y.; Okihara, T.; Irie, M.; Yoshida, Y.; Van Meerbeek, B. Effectiveness and stability of silane coupling agent inc'orporated in universal adhesives. *Dent. Mater.* **2016**, *32*, 1218–1225. [CrossRef] [PubMed]

45. Van Landuyt, K.L.; De Munck, J.; Mine, A.; Cardoso, M.V.; Peumans, M.; Van Meerbeek, B. Filler debonding & subhybrid-layer failures in self-etch adhesives. *J. Dent. Res.* **2010**, *89*, 1045–1050. [PubMed]

46. Eick, J.D.; Robinson, S.J.; Chappell, R.P.; Cobb, C.M.; Spencer, P. The dentinal surface: Its influence on dentinal adhesion. Part III. *Quintessence Int.* **1993**, *24*, 571–582. [PubMed]

47. Erhardt, M.C.; Toledano, M.; Osorio, R.; Pimenta, L.A. Histomorphologic characterization and bond strength evaluation of caries-affected dentin/resin interfaces: Effects of long-term water exposure. *Dent. Mater.* **2008**, *24*, 786–798. [CrossRef] [PubMed]

48. Wang, Y.; Spencer, P.; Walker, M.P. Chemical profile of adhesive/caries-affected dentin interfaces using Raman microspectroscopy. *J. Biomed. Mater. Res. A* **2007**, *81*, 279–286. [CrossRef] [PubMed]

49. Tekçe, N.; Tuncer, S.; Demirci, M.; Pashaev, D. The bonding effect of adhesive systems and bulk-fill composites to sound and caries-affected dentine. *J. Adhes. Sci. Technol.* **2016**, *30*, 171–185. [CrossRef]

materials

MDPI

Article

Bone Healing in Rabbit Calvaria Defects Using a Synthetic Bone Substitute: A Histological and Micro-CT Comparative Study

Minas Leventis [1,*], **Peter Fairbairn** [2], **Chas Mangham** [3], **Antonios Galanos** [4], **Orestis Vasiliadis** [1], **Danai Papavasileiou** [1] **and Robert Horowitz** [5]

[1]   Laboratory of Experimental Surgery and Surgical Research N. S. Christeas, Medical School, University of Athens, 75 M. Assias Street, Athens 115 27, Greece; orestis@vasiliadis.net (O.V.); d.pap.mes@gmail.com (D.P.)

[2]   Department of Periodontology and Implant Dentistry, School of Dentistry, University of Detroit Mercy, 2700 Martin Luther King Jr Boulevard, Detroit, MI 48208, USA; peterdent66@aol.com

[3]   Manchester Molecular Pathology Innovation Centre, The University of Manchester, Nelson Street, Manchester M13 9NQ, UK; D.C.Mangham@manchester.ac.uk

[4]   Laboratory of Research of the Musculoskeletal System, Medical School, University of Athens, 2 Nikis Street, Athens 145 61, Greece; galanostat@yahoo.gr

[5]   Departments of Periodontics, Implant Dentistry, and Oral Surgery, New York University College of Dentistry, 345 E 24th Street, New York, NY 10010, USA; rah7@nyu.edu

*   Correspondence: mlevent@dent.uoa.gr

Received: 31 August 2018; Accepted: 15 October 2018; Published: 17 October 2018

Abstract: Bioactive alloplastic materials, like beta-tricalcium phosphate (β-TCP) and calcium sulfate (CS), have been extensively researched and are currently used in orthopedic and dental bone regenerative procedures. The purpose of this study was to compare the performance of EthOss versus a bovine xenograft and spontaneous healing. The grafting materials were implanted in standardized 8 mm circular bicortical bone defects in rabbit calvariae. A third similar defect in each animal was left empty for natural healing. Six male rabbits were used. After eight weeks of healing, the animals were euthanized and the bone tissue was analyzed using histology and micro-computed tomography (micro-CT). Defects treated with β-TCP/CS showed the greatest bone regeneration and graft resorption, although differences between groups were not statistically significant. At sites that healed spontaneously, the trabecular number was lower ($p < 0.05$) and trabecular separation was higher ($p < 0.05$), compared to sites treated with β-TCP/CS or xenograft. Trabecular thickness was higher at sites treated with the bovine xenograft ($p < 0.05$) compared to sites filled with β-TCP/CS or sites that healed spontaneously. In conclusion, the novel β-TCP/CS grafting material performed well as a bioactive and biomimetic alloplastic bone substitute when used in cranial defects in this animal model.

Keywords: bone regeneration; β-tricalcium phosphate; calcium sulfate; bone substitutes; animal study

1. Introduction

Bone grafting procedures are performed to manage osseous defects of the jaw due to pathological processes or trauma, to preserve the alveolar ridge after extraction, and to augment the bone around dental implants. For this purpose, a wide variety of bone substitutes, barrier membranes, and growth-factor preparations are routinely used, and several different surgical methods have been proposed [1,2]. Autogenous bone is still considered the gold standard among bone grafting materials as it possess osteoconductive, osteoinductive, and osteogenetic properties; it neither transmits diseases nor triggers immunologic reactions; and is gradually absorbed and replaced by newly-formed high quality osseous tissue. The disadvantages of using autogenous bone include restricted availability, the need for additional surgical site, increased morbidity, and extended operating time [3,4]. As an alternative solution, bone graft substitutes are widely used in bone reconstructive surgeries, and the science of biomaterials has become one of the fastest growing scientific fields in recent years [5]. By definition, bone substitutes are any "synthetic, inorganic or biologically organic combination which can be inserted for the treatment of a bone defect instead of autogenous or allogenous bone" [6]. This definition applies to a wide variety of materials of different origins, composition, and biological mechanisms of function regarding graft resorption and new bone formation. Thus, the selection of biomaterials in clinical practice must be based on their biocompatibility, biodegradability, bioactivity, and mechanical properties, as well as the resulting cell behavior [7–11]. Parameters like the physicochemical characteristics, hydrophilicity and hydrophobicity, and molecular weight may influence the handling and performance of bone substitutes [12,13]. In general, the ideal grafting material should act as a substrate for bone ingrowth into the defect, to be ultimately fully replaced by host bone with an appropriate degradation rate in relation to new bone development for complete regeneration up to the condition of *restitutio ad integrum* [1,14]. The grafting material should also be able to retain the volume stability of the augmented area [1].

Bioactivity is a characteristic of chemical bonding between bone grafts and host biological tissues. Calcium phosphate ceramics and calcium sulfates are considered bioactive materials as they have the ability to evoke a controlled action and reaction to the host tissue environment with a controlled chemical dissolution and resorption, to ultimately be fully replaced by regenerated bone [5,15].

Among bioactive ceramics, β-TCP and hydroxyapatite ($Ca10(PO4)6(OH)2$) are frequently utilized in dental bone regenerative procedures [13]. Their composition is similar to that of natural bone, they are biocompatible and osteoconductive materials, can osseointegrate with the defect site, and due to their non-biologic origin, their use does not involve any risk of transmitting infections or diseases [16–22]. The degradation process of these biomaterials produces and releases ions that can create an alkaline environment that seems to enhance cell activity and accelerate bone reconstruction [13]. Recent in vitro and in vivo experimental studies have shown that such alloplastic bone substitutes can stimulate stem cells to differentiate to osteogenic differentiation of stem cells, as well as ectopic bone induction [23–27]. β-TCP may promote the proliferation and differentiation of endothelial cells and improve neovascularization in the grafted site, having clear benefits for osteogenic processes [13,28].

The ability of the bacteriostatic CS to set is well documented. Adding CS to β-TCP produces a compound alloplastic biomaterial that hardens in situ and binds directly to the host bone, helping maintain the space and shape of the grafted site, and acts as a stable scaffold [29–35]. The improved mechanical stability of the graft is a crucial factor for bone healing and differentiation of mesenchymal cells to osteoblasts [36], thus contributing to enhanced regeneration of high quality hard tissue [37,38]. The in situ hardening CS element may act as a cell occlusive barrier membrane, halting soft connective tissue proliferation into the graft during the first stages of healing [39–41].

Both CS and β-TCP are fully resorbable bone substitutes, leading to the regeneration of high quality vital host bone without the long-term presence of graft remnants. The CS element resorbs over a three- to six-week period, depending on patient physiology, creating a vascular porosity in the β-TCP scaffold for improved vascular ingrowth and angiogenesis. The β-TCP element resorbs by hydrolysis and enzymatic and phagocytic processes, usually over a period of 9–16 months. Although evaluating these resorptive mechanisms is difficult, it seems that cell-based degradation might be more important than dissolution, and macrophages and osteoclasts may be involved in phagocytosis, again largely dependent on host physiology [22,41–43].

As recent studies in bone reconstruction are gradually shifting their focus to biodegradable and bioactive materials, resorbable alloplastic bone substitutes might be a potential alternative to autogenous bone or bovine xenografts in dental bone reconstructive procedures. However, limited information is available in the recent literature. Therefore, the aim of this study was to compare the performance of a novel alloplastic bone substitute composed of β-TCP and CS, versus a bovine xenograft and spontaneous healing, in cranial bone defects in rabbits.

2. Experimental Section

2.1. Animals

Six adult male New Zealand White rabbits, each weighing 3 kg (±250 g), were used in this study with the approval of the Institutional Animal Care and Use Committee of the Veterinary Department, Greek Ministry of Rural Development and Veterinary, Attica Prefecture, Greece (project identification code: 5176/10-10-2017). Animals were provided with an appropriate balanced dry diet and water ad libitum, and caged individually in a standard manner at the N. S. Christeas animal research facility, Medical School, University of Athens, Greece. All animals were allowed seven days from their arrival to the facility in order to acclimatize to their new environment.

2.2. Surgical Procedures

Surgical procedures are shown in Figure 1. Under general anesthesia by orotracheal intubation, a longitudinal midline linear incision was made in the skin over the top of the cranial vault to expose the skull. The overlying periosteum was then excised, and three separate and identical 8-mm-diameter bicortical cranial round defects were created in the calvaria of each animal using a trephine drill with an internal diameter of 8 mm (Komet Inc., Lemgo, Germany) on a slow-speed electric handpiece by applying 0.9% physiologic saline irrigation. During the osteotomy, care was taken not to injure the dura mater under the bone. Then, using a thin periotome, the circular bicortical bone segment was mobilized and luxated.

Following a randomization technique using cards, the three resultant bone defects in each animal were randomly assigned treatment: (1) one defect was filled with 150 mg of the test alloplastic biomaterial (group 1), (2) one defect was filled with 150 mg of bovine xenograft (group 2), and (3) one sham defect remained unfilled (group 3).

The test bone graft substitute used in group 1 (EthOss, Ethoss Regeneration Ltd., Silsden, UK) is a self-hardening biomaterial consisting of β-TCP (65%) and CS (35%), preloaded in a sterile plastic syringe. In accordance with the manufacturer's instructions, prior to applying the alloplastic graft into the bone defect, the particles of the biomaterial were mixed in the syringe with sterile saline. After application, a bone plunger was used to gently condense the moldable graft particles in order to occupy the entire volume of the site up to the level of the surrounding host bone. A saline-wet gauze was used to further compact the graft particles and accelerate the in situ hardening of the CS element of the graft. As a result, after a few minutes, the alloplastic bone substitute formed a stable, porous scaffold for host osseous regeneration.

As a xenograft, a bovine deproteinized cancellous bone graft with a particle size of 0.25–1 mm (Bio-Oss, Geistleich Pharmaceutical, Wollhausen, Switzerland) was used in group 2. Bio-Oss consists of loose particles. According to the manufacturer's instructions, before application, the material was mixed with sterile saline and then placed into the bone defect, avoiding excessive compression.

Interrupted resorbable 4-0 sutures (Vicryl, Ethicon, Johnson & Johnson, Somerville, NJ, USA) were used to close the overlying soft tissues in layers.

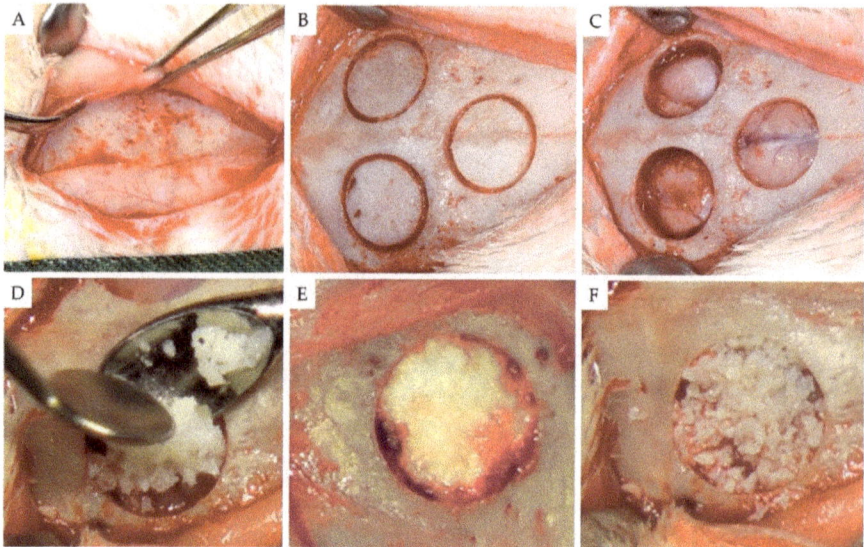

Figure 1. The surgical process. (**A**) Surgical exposure of the rabbit calvaria; (**B**) using a trephine burr, three identical circular osteotomies were performed; and (**C**) after removing the bicortical bone segments. The circular three-defect model was utilized with a frontal bone defect affecting the inter-frontal suture plus two bilateral defects affecting the parietal bones. (**D**) Two sites were treated with bone substitutes and the third left unfilled. (**E**) EthOss and (**F**) Bio-Oss.

Each experimental animal received antibiotics (30 mg/kg of Zinadol, GlaxoWellcome, Athens, Greece) every 24 h and analgesics (15 mg/kg of Depon; Bristol-Myers Squibb, Athens, Greece) for 2 days postoperatively. An intravenous injection of sodium thiopental (100 mg/kg of Pentothal; Abbott Hellas, Athens, Greece) was used to euthanize all animals after an 8-week healing period. The calvaria bones containing the healed sites were surgically harvested and immediately fixed in neutral buffered formalin (10%) for 24 h.

2.3. Micro-CT Evaluation

Each calvaria was scanned using a micro-CT scanner (Skyscan 1076, Bruker, Belgium) at 50 kV, 200 µA, and a 0.5 mm aluminum filter. The pixel size was 18.26 µm. Two images were captured every $0.7°$ through $180°$ rotation of the sample; the exposure time per image was 420 ms. The X-ray images were reconstructed using the NRecon software (Skyscan, Bruker, Belgium) and analyzed using Skyscan CT analysis software. Specific thresholds were set on segmenting the micro-CT images in order to distinguish the newly-formed bone from the connective tissue and the grafting materials. A lower threshold (level 60) was used for all groups to segment the bone tissue, whereas higher threshold levels were used to segment the Bio-Oss and the EthOss particles (level 90 and level 120, respectively). Analysis was performed using an 8-mm-diameter circular region that was placed in the center of the initial defect area. Trabecular bone analysis was performed, and based on the micro-CT results several parameters regarding new bone formation, residual graft, and the microarchitecture of the regenerated bone were calculated (Table 1).

Table 1. Parameters assessed by analysis of the micro-computed tomography (CT) data.

Parameter	Abbreviation	Description	Standard Unit
Bone volume fraction	BV/TV	Ratio of the segmented newly-formed bone volume to the total volume of the region of interest	%
Residual material volume fraction	RMVF	Ratio of the residual grafting material volume to the total volume of the region of interest	%
Trabecular number	Tb.N	Measure of the average number of trabeculae per unit length	1/mm
Trabecular thickness	Tb.Th	Mean thickness of trabeculae, assessed using direct 3D methods	mm
Trabecular separation	Tb.Sp	Mean distance between trabeculae, assessed using direct 3D methods	mm

2.4. Histology

After micro-CT analysis, bone specimens were decalcified in bone decalcification solution (Diapath S.p.a., Martinengo, Italy) for 14 days. After routine processing, slices were obtained from the central part of each healed bone defect using a saw (Exakt saw 312, Exakt Apparatebau GmbH, Norderstedt, Germany), embedded in paraffin, sectioned longitudinally into multiple 3-μm-thick sections and stained with Hematoxylin and Eosin stain. For qualitative analysis of the bone regenerative process, the stained preparations were examined under a light microscope (Nikon Eclipse 80, Nikon, Tokyo, Japan) and the entire section was evaluated. Images of each section were acquired with a digital camera microscope unit (Nikon DS-2MW, Nikon, Tokyo, Japan).

2.5. Statistics

Statistical analysis was performed using SPSS software (v. 17, SPSS Inc., Chicago, IL, USA). Data are expressed as mean ± standard deviation (SD). The Shapiro-Wilk test was utilized for normality analysis of the parameters. The comparison of variables among the 3 groups was performed using the one-way ANOVA model. Pairwise comparisons were performed using the Bonferroni test. All tests were two-sided, and statistical significance was set at $p < 0.05$.

3. Results

3.1. Overall

All animals survived for the duration of the study without complications. At eight weeks, there were no clinical signs of infection, hematoma, or necrosis at the defect sites. The dura mater and brain tissues under all bone defect sites exhibited no clinical evidence of inflammation, scar formation, or an adverse tissue reaction to the bone grafting materials (Figure 2A). Closure of the cortical window and filling of the defects with new bone were macroscopically observed at all defect sites. At bone defect sites grafted with bovine xenograft (Bio-Oss), graft particles embedded in newly formed hard tissue were clinically observed, whereas the newly-formed hard tissue occupying the sites treated with the alloplastic biomaterial (EthOss) was macroscopically homogeneous, without clear clinical distinction of residual graft particles. Clinical observation of sites left empty revealed that the spontaneously healed circular bone defects were bridged by a thin layer of newly-formed hard tissue (Figure 2B, Video S1).

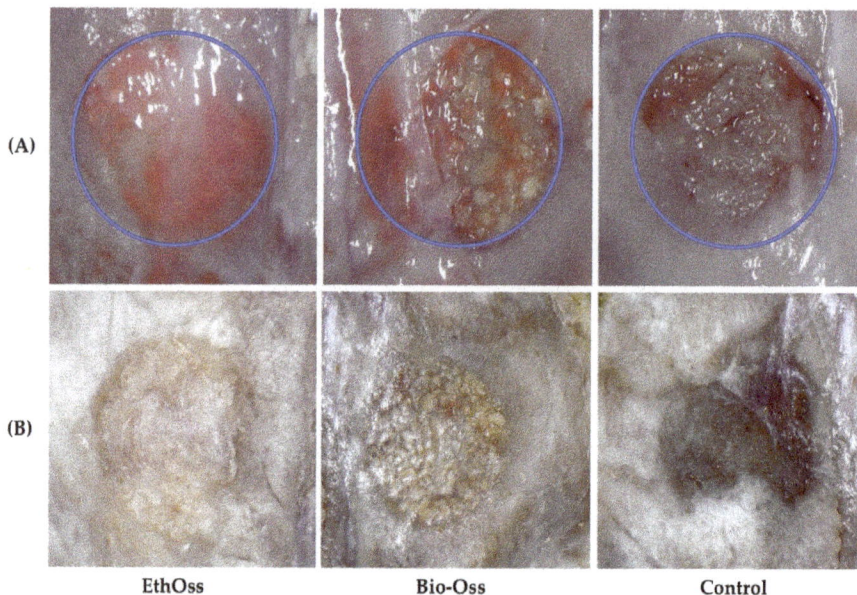

Figure 2. Gross observations of the 8-mm-diameter calvaria bone defect sites after eight weeks of healing: (**A**) fresh harvested rabbit calvaria and (**B**) after removing the dura mater and fixed in neutral buffered formalin (10%) for 24 h. Clinical observation revealed a different pattern of healing of the osseous defect between groups.

3.2. Micro-CT Evaluation

The radiological imaging results acquired from the micro-CT after eight weeks of healing are shown in Figure 3.

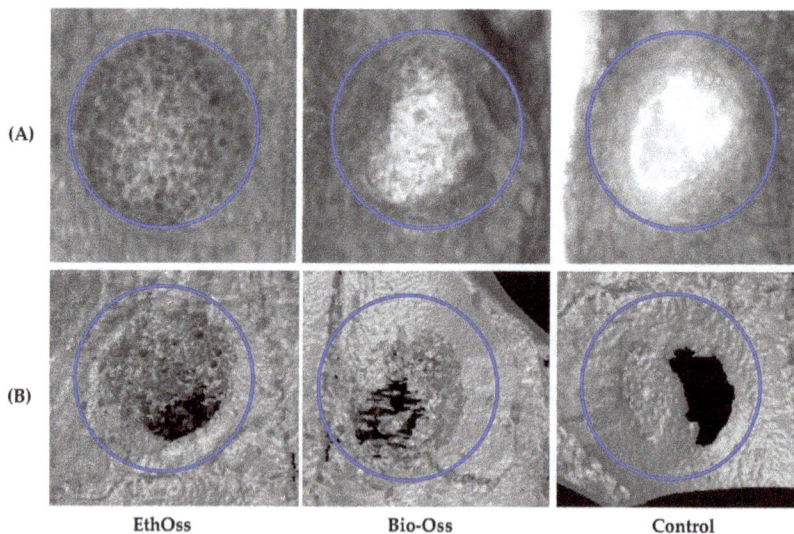

Figure 3. (**A**) Axial sections and (**B**) reconstructed three-dimensional (3D) micro-computed tomography (CT) images of the 8-mm-diameter defect sites after eight weeks of healing.

There were no significant differences between parameters that expressed new bone regeneration (BV/TV), between the three groups. At this time point, there were no statistically significant differences regarding the percentage of the residual graft material (RMVF) between groups 1 and 2, where EthOss and Bio-Oss were used as bone substitutes, respectively (Figure 4 and Table 2).

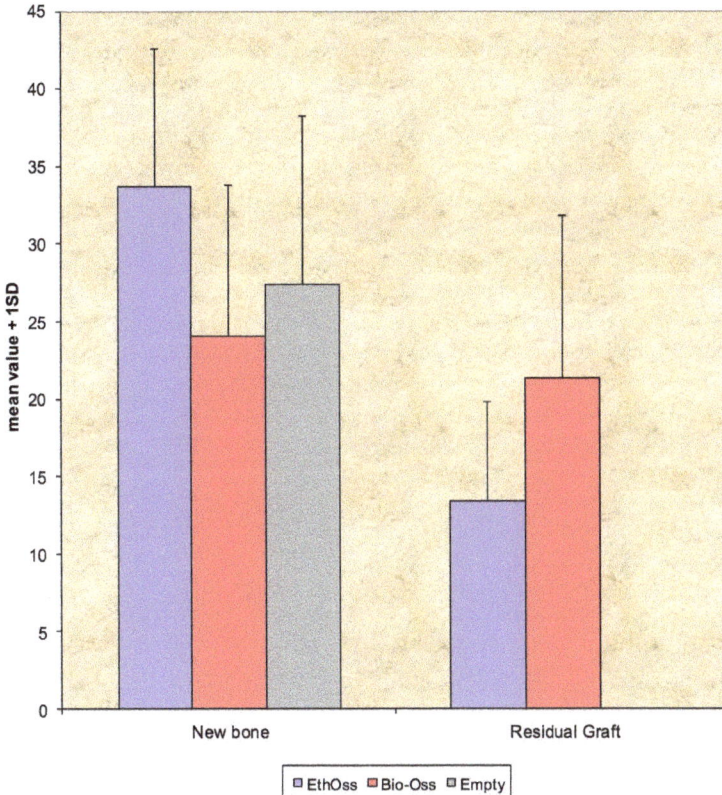

Figure 4. The percentage of new bone (BV/TV) between the three groups, and the percentage of residual graft (RMVF) in sites treated with EthOss and Bio-Oss, after eight weeks of healing. Data are presented as means. The differences between groups were not statistically significant ($p > 0.05$).

Table 2. Comparison of parameters associate with the newly-formed bone (BV/TV) and the residual grafting material (RMVF). The differences between groups were not statistically significant ($p > 0.05$).

Parameter	Site	N	Mean	SD	*p*-Value
BV/TV	EthOss	6	33.70	8.94	
	Bio-Oss	6	24.07	9.69	0.525
	Control	6	27.36	10.95	
RMVF	EthOss	6	13.41	6.43	
	Bio-Oss	6	21.36	10.05	0.070
	Control	6	-	-	-

At eight weeks, there were statistically significant differences between the three groups in the parameters associated with the microarchitecture of the newly-formed hard tissue (Table 3). Regarding the parameter Tb.N, pairwise comparisons indicated statistically significant difference between group 3 (Empty) and group 1 (EthOss) ($p < 0.001$), and group 2 (Bio-Oss) ($p < 0.001$), whereas there was no difference between group 1 (EthOss) and group 2 (Bio-Oss) ($p = 0.126$). Regarding the parameter Tb.Th, pairwise comparisons indicated statistically significant difference between group 2 (Bio-Oss) and group1 (EthOss) ($p = 0.001$), and Empty ($p < 0.001$), whereas there was no difference between group 1 (EthOss) and group 3 (Empty) ($p = 1.000$). For parameter Tb.S, pairwise comparisons indicated statistically significant difference between group 3 (Empty) and group 1 (EthOss) ($p = 0.001$), and group 2 (Bio-Oss) ($p < 0.001$), whereas there was no difference between group 1 (EthOss) and group 2 (Bio-Oss) ($p = 0.662$).

Table 3. Comparison of parameters associated with the microarchitecture of the newly-formed hard tissue in the three groups (a: $p < 0.05$ vs. control; b: $p < 0.05$ vs. Bio-Oss).

Parameter	Site	N	Mean	SD	*p*-Value
	EthOss	6	1.511 [a]	0.255 [a]	
Tb.N	Bio-Oss	6	1.213 [a]	0.198 [a]	<0.001
	Control	6	0.541	0.239	
	EthOss	6	0.219 [b]	0.018 [b]	
Tb.Th	Bio-Oss	6	0.291	0.029	<0.001
	Control	6	0.210 [b]	0.028 [b]	
	EthOss	6	0.486 [a]	0.136 [a]	
Tb.S	Bio-Oss	6	0.713 [a]	0.238 [a]	<0.001
	Control	6	1.686	0.455	

3.3. Histology

The histological slides for groups 1 (EthOss) and 2 (Bio-Oss) after eight weeks of healing are shown in Figure 5. Histologically, the analyzed biopsy contained newly-formed bone, residual grafting material, and vascularized uninflamed connective tissue. In all specimens, no significant inflammatory response, no necrosis, or foreign body reactions were observed. The graft particles were surrounded by or in contact with lamellar bone, demonstrating good osteoconductivity and biocompatibility.

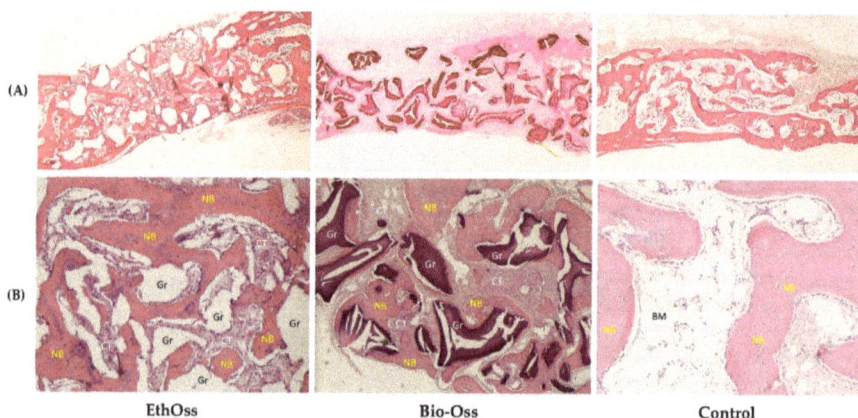

Figure 5. Histological specimens at eight weeks of healing (Hematoxylin and Eosin staining). (**A**) Cross-sections of the grafted and nongrafted sites (original magnification 5×); (**B**) EthOss and Bio-Oss particles (Gr) are embedded in well-perfused connective tissue (CT) and newly-formed bone (NB). Control group showing newly-formed bone trabeculae, bone marrow, and connective tissue (original magnification 50×).

4. Discussion

The aim of this animal study was to evaluate the host response after implantation of a novel bioactive and fully resorbable alloplastic grafting material in comparison to the effect of a bovine xenograft or spontaneous natural healing, in surgically-created calvaria bone defects in rabbits.

The β-TCP/CS material was compared to a bovine xenograft. Anorganic bovine bone substitutes have been extensively studied and used in oral surgery and implantology. Numerous pre-clinical studies and clinical trials in dentistry have shown and described in detail their osteoconductive properties and their ability to maintain the volume of the augmented site in the long-term [44–50]. However, controversy still remains as to whether this graft source is truly resorbable [51,52]. Mordenfeld et al. [51] performed histological and histomorphometrical analyses of human biopsies that were harvested 11 years after sinus floor augmentation with deproteinized bovine and autogenous bone. They found that the xenograft particles were not resorbed but were well-integrated in lamellar bone with no significant changes in particle size. Another important issue is that there are still significant concerns that bone grafts of bovine origin may carry a possible risk of transmitting prions to patients [53]. According to Kim et al. [54], screening prions within the animal genome is difficult. Moreover, there is a long latency period to manifestation of bovine spongiform encephalopathy (from one year to over 50 years) in infected patients. These factors provide a framework for the discussion of possible long-term risks of the xenografts that are used so extensively in dentistry. Thus, the authors suggested abolishing the use of bovine bone. They also highlighted that patient counseling should always include a clear description of the bone grafting material origin in bone reconstructive procedures.

In our study, no fibrosis developed between the particles of the biomaterial and the regenerated bone, nor was an inflammatory response observed, confirming the biocompatibility of EthOss. Our results indicate that β-TCP/CS can support new bone formation in parallel with biomaterial dissolution. The test alloplastic graft (β-TCP/CS = 65/35) presented in this study showed pronounced new bone formation (BV/TV = 33.70%) at eight weeks after implantation in circular calvaria bone defects in rabbits. Previous experimental animal studies using similar β-TCP/CS materials reported new bone fractions varying from 26% to 49% after a healing period of three weeks to four months [32,33]. Yang et al. [55], using micro-CT analysis to study the performance of a β-TCP/CS bone substitute in a sheep vertebral bone defect model, reported a 59% hard tissue volume at 36 weeks.

The degradation of β-TCP/CS biomaterials has been demonstrated in other pre-clinical studies. Using histomorphometry, Leventis et al. [33] found a statistically significant decrease in the percentage of residual material between three and six weeks (4.54% and 1.67%, respectively) in grafted rabbit calvaria defects. Podaropoulos et al. [32] reported 21.62% of residual β-TCP/CS four months after implantation of the material in surgically created bone defects on the iliac crest of Beagle dogs, whereas complete biodegradation of the β-TCP/CS graft after 36 weeks of implantation was documented in a previous animal study [55]. In accordance with the above findings, the present study demonstrated the degradation of the β-TCP/CS test biomaterial, showing a mean graft fraction area of 13.41% at eight weeks post-implantation.

In a clinical report, Fairbairn et al. [35] used β-TCP/CS for alveolar ridge preservation. Twelve weeks after socket grafting, a trephine biopsy was performed before the placement of the implant, and the authors histologically and histomorphometrically analyzed the sample of the regenerated bone, revealing 50.28% newly-formed bone and 12.27% remnant biomaterial.

In this study, we used micro-CT to three-dimensionally observe the structure of the newly-formed bone and to analyze important parameters of bone architecture. Micro-CT is a non-invasive, non-destructive analytical method that allows a significantly larger region of interest in the sample to be directly analyzed in three dimensions, compared to traditional histological methods. In combination with the histological findings, a comprehensive image of the regenerated bone can be surveyed to provide a representative and complete description of the healing outcome in the defect site [56].

The analysis of the micro-CT data in the present study revealed statistically significant differences regarding parameters associated with the microarchitecture of the healed sites at eight weeks post-operation. At sites that healed spontaneously, the trabecular number was lower and trabecular separation was higher. These findings in the control group indicate that the grafting materials used for filling the bone defects in the other groups acted as an osteoconductive scaffold, which facilitated the development of a larger number of trabeculae in a denser three-dimensional arrangement. In parallel, defects treated with β-TCP/CS showed the greatest bone regeneration and graft resorption, although differences between groups were not statistically significant.

The use of grafting materials to treat bone defects might have an important effect on the amount of regenerated bone tissue, and the presence of the graft particles may alter the microarchitecture of the newly-formed tissue. The resorption rate and the ability of a given grafting material to assist bone reconstruction seem to affect the bone healing mechanism and the geometry of the newly-formed tissue. Such differences might affect the overall quality of the newly-formed bone [7]. The capacity of the regenerated bone to remodel and adapt to the transmitted occlusal forces, thus minimizing the risk of failure under load, depends on the amount of bone, as well as its shape and microarchitecture (spatial distribution of the bone mass) and the intrinsic properties of the materials that comprise this hard tissue. Although a moderate to strong correlation between trabecular bone volume/architecture and biomechanical properties of trabecular bone has been shown [57], it is still unclear how the long-term presence of remnant non-resorbable or slowly resorbable particles of the graft, and the associated differences in structural parameters of trabecular bone and bone microarchitecture, might interfere with the remodeling and the strength of the new tissue when regenerating bone around dental implants. In a systematic review of the alterations in bone quality after alveolar ridge preservation with different bone graft substitutes, Chan et al. [7] reported significant variations in vital bone formation utilizing different grafting materials and discussed the concern that the presence of residual biomaterial might interfere with normal bone healing and remodeling, reducing the bone-to-implant contacts, and possibly negative affecting the overall quality and architecture of the bone that surrounds the implants. However, whether changes in bone quality influence implant success and peri-implant tissue stability remains unknown.

The outcomes of the present study revealed the highest vital bone content for defects grafted with the test alloplastic material (33.70%), followed by sockets with no graft material (27.36%), and the bovine xenograft (24.07%), whereas the amount of residual graft was higher (21.36%) for the bovine xenograft compared to the alloplast (13.41%). Our findings, although not statistically significant, are in accordance with results from human clinical studies on flapless socket grafting. In a recent systematic review, Jambhekar et al. [10] analyzed the outcomes of randomized controlled trials that reported that, after a minimum healing period of 12 weeks, sockets filled with alloplastic biomaterials had the highest amount of newly-formed bone (45.53%) compared to sites subjected to spontaneous healing with no graft material (41.07%) and xenografts (35.72%). The amount of remnant biomaterial was highest for sites treated with xenografts (19.3%) compared to alloplastic materials (13.67%).

5. Conclusions

The present histological and micro-CT investigation of rabbit cranial bone defects treated with the test resorbable alloplastic β-TCP/CS graft demonstrated excellent biocompatibility of the biomaterial and pronounced new bone formation after a healing period of eight weeks.

Supplementary Materials: The following are available online at http://www.mdpi.com/1996-1944/11/10/2004/s1, Video S1: Video of representative harvested samples fixed in neutral buffered formalin revealing macroscopically the different pattern of bone healing between groups.

Author Contributions: All of the named authors were involved in the work leading to the publication of this paper and have read the paper before this submission. M.L., P.F., O.V. and D.P. performed the whole in vivo experiment; C.M. made the histological slides and performed the histological analysis; A.G. performed the statistical analysis; and M.L., P.F. and R.H. designed the experiment and have given final approval of the version to be published with full management of this manuscript.

Acknowledgments: The authors wish to thank Les A. Coulton, MD for performing the micro-CT examination and analysis, and the animal care team at the Laboratory of Experimental Surgery and Surgical Research, N.S. Christeas, Athens, Greece, for assistance during surgery.

Conflicts of Interest: M.L. is the Research Manager at Ethoss Regeneration Ltd., and P.F. is the Clinical Director at Ethoss Regeneration Ltd., having a potential conflict of interest regarding the presented alloplastic bone grafting material (EthOss). M.L. and P.F. claim to have no financial interest in the rest of the companies and products mentioned. The rest of the authors declare no conflict of interest.

References

1. Yip, I.; Ma, L.; Mattheos, N.; Dard, M.; Lang, N.P. Defect healing with various bone substitutes. *Clin. Oral Implants Res.* **2015**, *26*, 606–614. [CrossRef] [PubMed]
2. Wang, R.E.; Lang, N.P. Ridge preservation after tooth extraction. *Clin. Oral Implants Res.* **2012**, *23* (Suppl. 6), 147–156. [CrossRef]
3. Misch, C.M. Autogenous bone: Is it still the gold standard? *Implant Dent.* **2010**, *19*, 361. [CrossRef] [PubMed]
4. Le, B.Q.; Nurcombe, V.; Cool, S.M.; van Blitterswijk, C.A.; de Boer, J.; La Pointe, V.L.S. The Components of Bone and What They Can Teach Us about Regeneration. *Materials* **2017**, *11*, E14. [CrossRef] [PubMed]
5. Henkel, J.; Woodruff, M.A.; Epari, D.R.; Steck, R.; Glatt, V.; Dickinson, I.C.; Choong, P.F.; Schuetz, M.A.; Hutmacher, D.W. Bone regeneration based on tissue engineering conceptions—A 21st century perspective. *Bone Res.* **2013**, *1*, 216–248. [CrossRef] [PubMed]
6. Schlickewei, W.; Schlickewei, C. The Use of Bone Substitutes in the Treatment of Bone Defects-the Clinical View and History. *Macromol. Symp.* **2007**, *253*, 10–23. [CrossRef]
7. Chan, H.L.; Lin, G.H.; Fu, J.H.; Wang, H.L. Alterations in bone quality after socket preservation with grafting materials: A systematic review. *Int. J. Oral Maxillofac. Implants* **2013**, *28*, 710–720. [CrossRef] [PubMed]
8. Iocca, O.; Farcomeni, A.; Pardiñas Lopez, S.; Talib, H.S. Alveolar ridge preservation after tooth extraction: A Bayesian Network meta-analysis of grafting materials efficacy on prevention of bone height and width reduction. *J. Clin. Periodontol.* **2017**, *44*, 104–114. [CrossRef] [PubMed]
9. Danesh-Sani, S.A.; Engebretson, S.P.; Janal, M.N. Histomorphometric results of different grafting materials and effect of healing time on bone maturation after sinus floor augmentation: A systematic review and meta-analysis. *J. Periodontal Res.* **2017**, *52*, 301–312. [CrossRef] [PubMed]
10. Jambhekar, S.; Kernen, F.; Bidra, A.S. Clinical and histologic outcomes of socket grafting after flapless tooth extraction: A systematic review of randomized controlled clinical trials. *J. Prosthet. Dent.* **2015**, *113*, 371–382. [CrossRef] [PubMed]
11. Horowitz, R.A.; Leventis, M.D.; Rohrer, M.D.; Prasad, H.S. Bone grafting: History, rationale, and selection of materials and techniques. *Compend. Contin. Educ. Dent.* **2014**, *35* (Suppl. 4), 1–6.
12. Pryor, L.S.; Gage, E.; Langevin, C.J.; Herrera, F.; Breithaupt, A.D.; Gordon, C.R.; Afifi, A.M.; Zins, J.E.; Meltzer, H.; Gosman, A.; et al. Review of bone substitutes. *Craniomaxillofac. Trauma Reconstr.* **2009**, *2*, 151–160. [CrossRef] [PubMed]
13. Gao, C.; Peng, S.; Feng, P.; Shuai, C. Bone biomaterials and interactions with stem cells. *Bone Res.* **2017**, *5*, 17059. [CrossRef] [PubMed]
14. Glenske, K.; Donkiewicz, P.; Köwitsch, A.; Milosevic-Oljaca, N.; Rider, P.; Rofall, S.; Franke, J.; Jung, O.; Smeets, R.; Schnettler, R.; et al. Applications of Metals for Bone Regeneration. *Int. J. Mol. Sci.* **2018**, *19*, E826. [CrossRef] [PubMed]
15. Dorozhkin, S.V. Calcium orthophosphates in nature, biology and medicine. *Materials* **2009**, *2*, 399–498. [CrossRef]
16. Araújo, M.G.; Liljenberg, B.; Lindhe, J. β-tricalcium phosphate in the early phase of socket healing: An experimental study in the dog. *Clin. Oral Implants Res.* **2010**, *21*, 445–454. [CrossRef] [PubMed]
17. Trisi, P.; Rao, W.; Rebaudi, A.; Fiore, P. Histologic effect of pure-phase beta-tricalcium phosphate on bone regeneration in human artificial jawbone defects. *Int. J. Periodontics Restorative Dent.* **2003**, *23*, 69–78. [PubMed]
18. Roh, J.; Kim, J.Y.; Choi, Y.M.; Ha, S.M.; Kim, K.N.; Kim, K.M. Bone Regeneration Using a Mixture of Silicon-Substituted Coral HA and β-TCP in a Rat Calvarial Bone Defect Model. *Materials* **2016**, *9*, E97. [CrossRef] [PubMed]

19. Gregori, G.; Kleebe, H.J.; Mayr, H.; Ziegler, G. EELS characterisation of β-tricalcium phosphate and hydroxyapatite. *J. Eur. Ceram. Soc.* **2006**, *26*, 1473–1479. [CrossRef]
20. Daculsi, G.; Laboux, O.; Malard, O.; Weiss, P. Current state of the art of biphasic calcium phosphate bioceramics. *J. Mater. Sci. Mater. Med.* **2003**, *14*, 195–200. [CrossRef] [PubMed]
21. Harel, N.; Moses, O.; Palti, A.; Ormianer, Z. Long-term results of implants immediately placed into extraction sockets grafted with β-tricalcium phosphate: A retrospective study. *J. Oral Maxillofac. Surg.* **2013**, *71*, E63–E68. [CrossRef] [PubMed]
22. Kucera, T.; Sponer, P.; Urban, K.; Kohout, A. Histological assessment of tissue from large human bone defects repaired with β-tricalcium phosphate. *Eur. J. Orthop. Surg. Traumatol.* **2014**, *24*, 1357–1365. [CrossRef] [PubMed]
23. Yuan, H.; Fernandes, H.; Habibovic, P.; de Boer, J.; Barradas, A.M.; de Ruiter, A.; Walsh, W.R.; van Blitterswijk, C.A.; de Bruijn, J.D. Osteoinductive ceramics as a synthetic alternative to autologous bone grafting. *Proc. Natl. Acad. Sci. USA* **2010**, *107*, 13614–13619. [CrossRef] [PubMed]
24. Tang, Z.; Li, X.; Tan, Y.; Fan, H.; Zhang, X. The material and biological characteristics of osteoinductive calcium phosphate ceramics. *Regen. Biomater.* **2018**, *5*, 43–59. [CrossRef] [PubMed]
25. Miron, R.J.; Zhang, Q.; Sculean, A.; Buser, D.; Pippenger, B.E.; Dard, M.; Shirakata, Y.; Chandad, F.; Zhang, Y. Osteoinductive potential of 4 commonly employed bone grafts. *Clin. Oral Investig.* **2016**, *20*, 2259–2265. [CrossRef] [PubMed]
26. Barradas, A.M.; Yuan, H.; van Blitterswijk, C.; Habibovic, P. Osteoinductive biomaterials: Current knowledge of properties, experimental models and biological mechanisms. *Eur. Cell. Mater.* **2010**, *21*, 407–429. [CrossRef]
27. Cheng, L.; Shi, Y.; Ye, F.; Bu, H. Osteoinduction of calcium phosphate biomaterials in small animals. *Mater. Sci. Eng. C Mater. Biol. Appl.* **2013**, *33*, 1254–1260. [CrossRef] [PubMed]
28. Malhotra, A.; Habibovic, P. Calcium phosphates and angiogenesis: Implications and advances for bone regeneration. *Trends Biotechnol.* **2016**, *34*, 983–992. [CrossRef] [PubMed]
29. Ruga, E.; Gallesio, C.; Chiusa, L.; Boffano, P. Clinical and histologic outcomes of calcium sulfate in the treatment of postextraction sockets. *J. Craniofac. Surg.* **2011**, *22*, 494–498. [CrossRef] [PubMed]
30. Anson, D. Using calcium sulfate in guided tissue regeneration: A recipe for success. *Compend. Contin. Educ. Dent.* **2000**, *21*, 365–370. [PubMed]
31. Eleftheriadis, E.; Leventis, M.D.; Tosios, K.I.; Faratzis, G.; Titsinidis, S.; Eleftheriadi, I.; Dontas, I. Osteogenic activity of β-tricalcium phosphate in a hydroxyl sulphate matrix and demineralized bone matrix: A histological study in rabbit mandible. *J. Oral Sci.* **2010**, *52*, 377–384. [CrossRef] [PubMed]
32. Podaropoulos, L.; Veis, A.A.; Papadimitriou, S.; Alexandridis, C.; Kalyvas, D. Bone regeneration using b-tricalcium phosphate in a calcium sulfate matrix. *J. Oral Implantol.* **2009**, *35*, 28–36. [CrossRef] [PubMed]
33. Leventis, M.D.; Fairbairn, P.; Dontas, I.; Faratzis, G.; Valavanis, K.D.; Khaldi, L.; Kostakis, G.; Eleftheriadis, E. Biological response to β-tricalcium phosphate/calcium sulfate synthetic graft material: An experimental study. *Implant Dent.* **2014**, *23*, 37–43. [CrossRef] [PubMed]
34. Fairbairn, P.; Leventis, M. Protocol for Bone Augmentation with Simultaneous Early Implant Placement: A Retrospective Multicenter Clinical Study. *Int. J. Dent.* **2015**, *2015*, 589135. [CrossRef] [PubMed]
35. Fairbairn, P.; Leventis, M.; Mangham, C.; Horowitz, R. Alveolar Ridge Preservation Using a Novel Synthetic Grafting Material: A Case with Two-Year Follow-Up. *Case Rep. Dent.* **2018**, *2018*, 6412806. [CrossRef] [PubMed]
36. Giannoudis, P.V.; Einhorn, T.A.; Marsh, D. Fracture healing: The diamond concept. *Injury* **2007**, *38*, S3–S6. [CrossRef]
37. Troedhan, A.; Schlichting, I.; Kurrek, A.; Wainwright, M. Primary implant stability in augmented sinuslift-sites after completed bone regeneration: A randomized controlled clinical study comparing four subantrally inserted biomaterials. *Sci. Rep.* **2014**, *4*, 5877. [CrossRef] [PubMed]
38. Dimitriou, R.; Mataliotakis, G.I.; Calori, G.M.; Giannoudis, P.V. The role of barrier membranes for guided bone regeneration and restoration of large bone defects: Current experimental and clinical evidence. *BMC Med.* **2012**, *10*, 81. [CrossRef] [PubMed]
39. Pecora, G.; Andreana, S.; Margarone, J.E.; Covani, U.; Sottosanti, J.S. Bone regeneration with a calcium sulfate barrier. *Oral Surg. Oral Med. Oral Pathol. Oral Radiol. Endod.* **1997**, *84*, 424–429. [CrossRef]

40. Mazor, Z.; Mamidwar, S.; Ricci, J.L.; Tovar, N.M. Bone repair in periodontal defect using a composite of allograft and calcium sulfate (DentoGen) and a calcium sulfate barrier. *J. Oral Implantol.* **2011**, *37*, 287–292. [CrossRef] [PubMed]

41. Strocchi, R.; Orsini, G.; Iezzi, G.; Scarano, A.; Rubini, C.; Pecora, G.; Piattelli, A. Bone regeneration with calcium sulfate: Evidence for increased angiogenesis in rabbits. *J. Oral Implantol.* **2002**, *28*, 273–278. [CrossRef]

42. Artzi, Z.; Weinreb, M.; Givol, N.; Rohrer, M.D.; Nemcovsky, C.E.; Prasad, H.S.; Tal, H. Biomaterial Resorption Rate and Healing Site Morphology of Inorganic Bovine Bone and β-Tricalcium Phosphate in the Canine: A 24-month Longitudinal Histologic Study and Morphometric Analysis. *Int. J. Oral Maxillofac. Implants* **2004**, *19*, 357–368. [PubMed]

43. Palti, A.; Hoch, T. A concept for the treatment of various dental bone defects. *Implant Dent.* **2002**, *11*, 73–78. [CrossRef] [PubMed]

44. Artzi, Z.; Tal, H.; Dayan, D. Porous bovine bone mineral in healing of human extraction sockets. Part 1: Histomorphometric evaluations at 9 months. *J. Periodontol.* **2000**, *71*, 1015–1023. [CrossRef] [PubMed]

45. Artzi, Z.; Tal, H.; Dayan, D. Porous bovine bone mineral in healing of human extraction sockets: 2. Histochemical observations at 9 months. *J. Periodontol.* **2001**, *72*, 152–159. [CrossRef] [PubMed]

46. Pang, C.; Ding, Y.; Zhou, H.; Qin, R.; Hou, R.; Zhang, G.; Hu, K. Alveolar ridge preservation with deproteinized bovine bone graft and collagen membrane and delayed implants. *J. Craniofac. Surg.* **2014**, *25*, 1698–1702. [CrossRef] [PubMed]

47. Aludden, H.C.; Mordenfeld, A.; Hallman, M.; Dahlin, C.; Jensen, T. Lateral ridge augmentation with Bio-Oss alone or Bio-Oss mixed with particulate autogenous bone graft: A systematic review. *Int. J. Oral Maxillofac. Surg.* **2017**, *46*, 1030–1038. [CrossRef] [PubMed]

48. Jensen, T.; Schou, S.; Stavropoulos, A.; Terheyden, H.; Holmstrup, P. Maxillary sinus floor augmentation with Bio-Oss or Bio-Oss mixed with autogenous bone as graft: A systematic review. *Clin. Oral Implants Res.* **2012**, *23*, 263–273. [CrossRef] [PubMed]

49. Scheyer, E.T.; Velasquez-Plata, D.; Brunsvold, M.A.; Lasho, D.J.; Mellonig, J.T. A clinical comparison of a bovine-derived xenograft used alone and in combination with enamel matrix derivative for the treatment of periodontal osseous defects in humans. *J. Periodontol.* **2002**, *73*, 423–432. [CrossRef] [PubMed]

50. Chappuis, V.; Rahman, L.; Buser, R.; Janner, S.F.M.; Belser, U.C.; Buser, D. Effectiveness of contour augmentation with guided bone regeneration: 10-year results. *J. Dent. Res.* **2018**, *97*, 266–274. [CrossRef] [PubMed]

51. Mordenfeld, A.; Hallman, M.; Johansson, C.B.; Albrektsson, T. Histological and histomorphometrical analyses of biopsies harvested 11 years after maxillary sinus floor augmentation with deproteinized bovine and autogenous bone. *Clin. Oral Implants Res.* **2010**, *21*, 961–970. [CrossRef] [PubMed]

52. Heberer, S.; Al-Chawaf, B.; Hildebrand, D.; Nelson, J.J.; Nelson, K. Histomorphometric analysis of extraction sockets augmented with Bio-Oss Collagen after a 6-week healing period: A prospective study. *Clin. Oral Implants Res.* **2008**, *19*, 1219–1225. [CrossRef] [PubMed]

53. Kim, Y.; Nowzari, H.; Rich, S.K. Risk of Prion Disease Transmission through Bovine-Derived Bone Substitutes: A Systematic Review. *Clin. Implant Dent. Relat. Res.* **2013**, *15*, 645–653. [CrossRef] [PubMed]

54. Kim, Y.; Rodriguez, A.E.; Nowzari, H. The risk of prion infection through bovine grafting materials. *Clin. Implant Dent. Relat. Res.* **2016**, *18*, 1095–1102. [CrossRef] [PubMed]

55. Yang, H.L.; Zhu, X.S.; Chen, L.; Chen, C.M.; Mangham, D.C.; Coulton, L.A.; Aiken, S.S. Bone healing response to a synthetic calcium sulfate/β-tricalcium phosphate graft material in a sheep vertebral body defect model. *J. Biomed. Mater. Res. B Appl. Biomater.* **2012**, *100*, 1911–1921. [CrossRef] [PubMed]

56. Bouxsein, M.L.; Boyd, S.K.; Christiansen, B.A.; Guldberg, R.E.; Jepsen, K.J.; Müller, R. Guidelines for assessment of bone microstructure in rodents using micro–computed tomography. *J. Bone Miner. Res.* **2010**, *25*, 1468–1486. [CrossRef] [PubMed]

57. Bouxsein, M.L. Determinants of skeletal fragility. *Best Pract. Res. Clin. Rheumatol.* **2005**, *19*, 897–911. [CrossRef] [PubMed]

materials

MDPI

Article

Effects of Polyacrylic Acid Pre-Treatment on Bonded-Dentine Interfaces Created with a Modern Bioactive Resin-Modified Glass Ionomer Cement and Subjected to Cycling Mechanical Stress

Salvatore Sauro [1,2,](*), Vicente Faus-Matoses [3], Irina Makeeva [2], Juan Manuel Nuñez Martí [1], Raquel Gonzalez Martínez [1], José Antonio García Bautista [1] and Vicente Faus-Llácer [3]

[1]   Departamento de Odontologia, Facultad de Sciencia de la Salud, Universidad CEU Cardenal Herrera, 46115 Valencia, Spain; juan.nunez@uchceu.es (J.M.N.M.); raquel.gonzalez@uchceu.es (R.G.M.); joseanto@uchceu.es (J.A.G.B.)
[2]   Department of Therapeutic Dentistry, Sechenov University Russia, 119435 Moscow, Russia; irina_markovina@mail.ru
[3]   Departamento de Estomatología, Facultad de Medicina y Odontología, Universitat de Valencia, 46010 Valencia, Spain; fausvj@uv.es (V.F.M.); Vicente.J.Faus@uv.es (V.F.L.)
*   Correspondence: salvatore.sauro@uchceu.es; Tel.: +34-96-136-9000

Received: 31 August 2018; Accepted: 30 September 2018; Published: 2 October 2018

Abstract: Objectives: Resin-modified glass ionomer cements (RMGIC) are considered excellent restorative materials with unique therapeutic and anti-cariogenic activity. However, concerns exist regarding the use of polyacrylic acid as a dentine conditioner as it may influence the bonding performance of RMGIC. The aim of this study was to evaluate the effect of different protocols for cycling mechanical stress on the bond durability and interfacial ultramorphology of a modern RMGIC applied to dentine pre-treated with/without polyacrylic acid conditioner (PAA). **Methods:** The RMGIC was applied onto human dentine specimens prepared with silicon-carbide (SiC) abrasive paper with or without the use of a PAA conditioner. The specimens were immersed in deionised water for 24 h then divided in 3 groups. The first group was cut into matchsticks (cross-sectional area of 0.9 mm^2) and tested immediately for microtensile bond strength (MTBS). The second was first subjected to load cycling (250,000 cycles; 3 Hz; 70 N) and then cut into matchsticks and tested for MTBS. The third group was subjected to load cycling (250,000 cycles; 3 Hz; 70 N), cut into matchsticks, and then immersed for 8 months storage in artificial saliva (AS); these were finally tested for MTBS. The results were analysed statistically using two-way ANOVA and the Student–Newman–Keuls test ($\alpha = 0.05$). Fractographic analysis was performed using FE-SEM, while further RMCGIC-bonded dentine specimens were aged as previously described and used for interfacial ultramorphology characterisation (dye nanoleakage) using confocal microscopy. **Results:** The RMGIC applied onto dentine that received no pre-treatment (10% PAA gel) showed no significant reduction in MTBS after load cycling followed by 8 months of storage in AS ($p > 0.05$). The RMGIC–dentine interface created in PAA-conditioned SiC-abraded dentine specimens showed no sign of degradation, but with porosities within the bonding interface both after load cycling and after 8 months of storage in AS. Conversely, the RMGIC–dentine interface of the specimens with no PAA pre-treatment showed no sign of porosity within the interface after any of the aging protocols, although some bonded-dentine interfaces presented cohesive cracks within the cement after prolonged AS storage. However, the specimens of this group showed no significant reduction in bond strength ($p < 0.05$) after 8 months of storage in AS or load cycling ($p > 0.05$). After prolonged AS storage, the bond strength value attained in RMGIC–dentine specimens created in PAA pre-treated dentine were significantly higher than those observed in the specimens created with no PAA pre-treatment in dentine. **Conclusions:** PAA conditioning of dentine prior to application of RMGIC induces no substantial effect on the bond strength after short-term storage, but its use may increase the risk of collagen degradation at the bonding interface after prolonged aging. Modern RMGIC applied without PAA dentine pre-treatment may have greater therapeutic synergy with saliva during cycle occlusal load, thereby enhancing the remineralisation and protection of the bonding interface.

Keywords: adhesion; bioactive; cycling mechanical stress; dentine; longevity; resin-modified glass ionomer cements; polyacrylic acid treatment

---

1. Introduction

Glass ionomer cements (GICs) were introduced for the first time in dentistry by Wilson and Kent in 1969 [1] as an innovative class of dental material able to set via an acid-base reaction after mixing fluoro-aluminosilicate glass particles (FAS) with a polyacrylic acid solution (PAA) [2]. Low viscosity polyacids, such as maleic and itaconic acids were incorporated within the PAA solution to improve the handling and setting of GICs [3–5]. Tartaric acid was also incorporated into the PAA solution to enhance the handling properties and increase the working time [6,7].

It is well known that the initial setting occurs due to a gelation reaction between the fluoro-aluminosilicate glass particles and polyalkenoate acids [8], followed by a proper hardening phase characterised by the cross-linking of the carboxylic groups present in the polymeric chains with calcium and aluminium ions present in the FAS. However, the final chemical reaction for complete setting occurs during the following 48 h [6,7], although the final "maturation" of the cement may take several months due to the slow release of aluminium ions from the glass particles. It is also important to highlight that sodium and fluoride ions are not usually involved in the setting reaction, but rather, these ions remain unreacted within the matrix, and are released gradually into the surrounding environment (e.g., bioactivity) [9,10]. Indeed, for this reason, GICs present unique therapeutic anti-cariogenic activity, which is mainly attributed to the release of fluoride (F^-) ions and to their buffering properties [11–14].

Glass ionomer cements are being used for a wide range of applications in dentistry [2]. These include the restoration of deciduous teeth [15], anterior class III and V restorations [16,17], cementation (luting) of crowns, bridges and orthodontic appliances [18–20], restorations of non-carious teeth with minimal preparation [21,22], and sandwich technique restorations [23,24]. Furthermore, they comprise the main material for atraumatic restorative therapy (ART) [25]. Indeed, subsequent to selective removal of the caries-infected tissues, GICs are applied as therapeutic ion-releasing materials to remineralise the caries-affected tissue left behind inside the dental cavity [26,27]. GICs may also exhibit a number of drawbacks, such as brittleness [28], poor wear resistance, inadequate surface properties [29,30], and sensitivity to high moisture in the oral cavity when newly placed [31].

In order to overcome such drawbacks, several modifications have been introduced to conventional GICs [32–34]. A key modification was the reinforcement of GICs through the incorporation of urethane monomers to produce resin-modified glass ionomer cements (RMGICs) [35,36]. Unlike conventional GICs, RMGICs can be self-activated (self-polymerisation) or light-cured (photo-polymerisation reaction). These "hybrid" materials have been generated to combine the mechanical properties of resin monomers with the anti-carious potential of GICs [37]. Indeed, it has been observed that RMGICs not only release fluoride, but they may also have greater flexural strength and lower solubility compared to conventional GICs [36,37]. RMGICs have a decreased fluoride release and higher creep relative to conventional powder-based ionomers [7]. Although first generation of RMGICs presented slight expansion due to water sorption (from 3.4% up to 11.3%) 24 h after placement [38], modern formulations have overcome this problem [36]. Conventional RMGICs are also characterised by lower mechanical (e.g., Young's modulus and flexural strength) and "inferior" aesthetic properties compared to resin composites [7,38].

A key advance of glass ionomer-based materials is their self-adhesive properties to bind chemically to calcium ions (Ca^{2+}) in the apatite of enamel and dentine through chelation of carboxyl group of acidic polymeric chains [39–41]. However, the self-adhesive mechanism of GIC-based materials to dentine is also due to the micromechanical interlocking achieved by shallow hybridization of the micro-porous collagen network. A 10% solution of PAA is mostly used as an enamel/dentine conditioner to remove the smear layer prior to the application of GIC-based restorative materials. Nevertheless, concerns exist regarding its use, application time, and concentration, as these factors may interfere with the overall bonding performance. Indeed, a high number of adhesive failures between a RMGIC and resin composite have been reported when a polyalkenoic conditioner was used on smear-layer covered dentine [40,41]. Cycling occlusal stress occurring during mastication, swallowing, as well as in cases of parafunctional habits, can affect the integrity of the bonding interface, making it more susceptible to short and long term degradation in the oral environment [42].

The aim of this study was to evaluate the microtensile bond strength (MTBS), after short-term load-cycle aging or after load cycle followed by prolonged aging (8 months) in artificial saliva (AS), of a modern bioactive RMGIC applied to dentine with or without surface pre-conditioning using 10% polyacrylic acid (PAA). Fractographic analysis and interfacial dye-assisted nanoleakage assessment of the bonded interfaces were evaluated using field-emission scanning electron microscopy (FE-SEM) and confocal laser-scanning microscopy (CLSM), respectively.

The tested null hypotheses were that the durability of RMGIC applied with or without the use of a PAA conditioner would be affected by: (i) short-term load-cycle aging; (ii) or load cycle followed by prolonged aging (8 months) in AS.

2. Materials and Methods

2.1. Preparation of Dentine Specimens

Sound human molars were extracted for periodontal or orthodontic reasons and stored in distilled water at 5 °C for no longer than 3 months. The roots were removed 1 mm beneath the cemento–enamel junction using a diamond-embedded blade (high concentration XL 12205; Benetec, London, UK) mounted on a low speed microtome (Remet evolution, REMET, Bologna, Italy). A second parallel cut was made to remove the occlusal enamel and expose mid-coronal dentine.

Two main groups (n = 30 specimens/group) were created based on dentine pre-treatment. Group 1: Specimens were abraded using 320-grit SiC abrasive paper (1 min) under continuous irrigation, followed by a water rinse (20 s), and air-drying (3 s); they were then restored with a light-cured RMGIC (no PAA conditioning). Group 2: Specimens were abraded with 320-grit SiC abrasive paper (1 min), conditioned with 10% PAA gel for 20 s rinsed with water (20 s), dried (3 s), and restored with a light-cured RMGIC (PAA conditioning).

The restorative procedure was performed using the content of two mono-dose capsules of a commercial RMGIC (RIVA light cure HV, Bayswater, VIC, Australia), mixed for 10 s in a trituration unit, and applied in bulk on to the dentine surface and light-cured for 30 s with a light-curing unit (Radii plus, SDI Ltd, Bayswater VIC, Australia) with a LED light source (>1000 mW/cm^2).

The experimental design required that each main group be subsequently subdivided into three sub-groups (n = 10 specimens) based on the aging protocol: (1) CRT: no aging (control, 24 h in deionised water); (2) LC: Load cycling (250,000 cycles in artificial saliva); (3) LC-AS: Load cycling (250,000 cycles in artificial saliva), followed by prolonged water storage (8 months in artificial saliva).

The composition of the artificial saliva was (AS: 0.103 g·L^{-1} of CaCl$_2$, 0.019 g·L^{-1} of MgCl$_2$·6H$_2$O, 0.544 g·L^{-1} of KH$_2$PO$_4$, 30 g·L^{-1} of KCl and 4.77 g·L^{-1} HEPES (acid) buffer, pH 7.4). The specimens in the subgroup LC and LC-AS were mounted in plastic rings with acrylic resin for load cycle testing (250,000 cycles; 3 Hz; 70 N). A compressive load was applied to the flat surface of the RMGIC using a 5-mm diameter spherical stainless steel plunger attached to a cyclic loading machine (model S-MMT-250NB; Shimadzu, Tokyo, Japan) while immersed in AS [43].

2.2. Micro-Tensile Bond Strength (MTBS) and Fracture Analysis (FE-SEM)

The specimens were sectioned using a hard-tissue microtome (Remet evolution, REMET, Bologna, Italy) in both the X and Y planes across the dentine-RMGIC interface, obtaining approx. 20 matchstick-shaped specimens from each tooth with cross-sectional areas of 0.9 mm^2. All the specimens were stored at 100% humidity, and were then (i) immediately cut into matchsticks, or (ii) load cycled and then cut into matchsticks, or (iii) load cycled, cut into matchsticks, and then stored for 8 months in AS; specimens were finally subjected to an MTBS test. The latter was performed using a microtensile bond strength device with a stroke length of 50 mm, peak force of 500 N, and a displacement resolution of 0.5 mm. Modes of failure were classified as a percentage of adhesive (A), mixed (M) or cohesive (C) failures when the failed interfaces were examined at 30X magnification by stereoscopic microscopy. Five representative fractured specimens from each sub-group were critical-point dried and mounted on aluminium stubs with carbon cement. The specimens were gold-sputter-coated and imaged using field-emission scanning electron microscopy (FE-SEM S-4100; Hitachi, Wokingham, UK) at 10 kV and a working distance of 15 mm.

Bond strength values in MPa were initially assessed for normality distribution and variances homogeneity using Kolmogorov-Smirnov and Levene's tests, respectively. To analyse if the substrate pre-treatment approaches had an influence on the bond strength, two-way analysis of variance (pre-treatment and substrate condition) was performed. Chi-square analysis was performed to compare the results of failure mode between groups. The significance level was set at $p \leq 0.05$. SPSS V16 for Windows (SPSS Inc., Chicago, IL, USA) was used.

2.3. Ultramorphology of the Bonded-Dentine Interfaces: Confocal Microscopy Evaluation

One dentine-bonded matchstick sample (Ø 0.9 mm^2) was selected from the centre of each tooth in every experimental sub-group. These were coated with a fast-setting nail varnish, applied 1 mm from the bonded interface. They were immersed in a Rhodamine B (Merck KGaA, Darmstadt, Germany) water solution (0.1 wt.%) for 24 h. Subsequently, the specimens were ultrasonicated in distilled water for 5 min and then polished for 30 s each side with a 2400-grit SiC paper. The specimens were finally ultrasonicated again in distilled water for 5 min and submitted for confocal microscopy analysis. Using a confocal scanning microscope (Olympus FV1000, Olympus Corp., Tokyo, Japan) equipped with a 63X/1.4 NA oil-immersion lens and a 543 nm LED illumination, reflection and fluorescence images were obtained with a 1-μm z-step to optically section the specimens to a depth of up to 20 μm below the surface. The z-axis scan of the interface surface was pseudo-coloured arbitrarily for improved exposure and compiled into both single and topographic projections using the CLSM image-processing software (Fluoview Viewer, Olympus Corp., Tokyo, Japan). The configuration of the system was standardised and used at constant settings for the entire investigation [43]. Each dentine interface was investigated completely, and then five optical images were randomly captured. Micrographs representing the most common morphological features observed along the bonded interfaces were captured and recorded.

3. Results

3.1. Micro-Tensile Bond Strength (MTBS) and Failure Mode Analysis

Microtensile bond strength means and standard deviation are expressed in MPa in Table 1. Dentine surface treatments and aging in AS (8 months) had no significant influence on the MTBS results ($p > 0.01$). Interactions between factors were not significant ($p > 0.05$). The MTBS performed at 24 h with the non-load-cycled specimens showed that the use of PAA dentine conditioning induced no significant increase in bond strength ($p > 0.05$), compared to the specimens created by applying the RMGIC on smear layer-covered dentine (no PAA-treatment).

Table 1. The results show the mean ± SD of the MTBS (MPa) to dentine when resin-modified glass ionomer cement was applied after different dentine pre-treatments.

Main Groups Dentine Etching (10% PAA gel)	24 h AS (CTR)	Load Cycling in AS (LC)	Load Cycling and 8-Month in AS (LC-AS)
No PAA (95/5)	16.3 ± 5.9 (A1) (5/25/70)	16.4 ± 4.1 (A1) (2/10/88)	13.1 ± 4.6 (A1) (10 */55 */35 *)
Yes PAA (100/0)	21.5 ± 4.8 (A1) (0/15/85)	21.1 ± 5.5 (A1) (3/17/80)	14.2 ± 5.2 (A2) (13 */65 */22 *)

Percentage (%) of total number of beams (intact sticks/pre-failed sticks) in the dentine treatment groups and percentage of failure modes (adhesive/mix/cohesive). The same letter indicates no differences in columns with different dentine treatments maintained in the same aging conditions. The same number indicates differences in rows for the same dentine treatment but different aging conditions ($p > 0.05$). The symbol (*) indicates significant differences in the mode of failure in the same treatment group after different aging conditions.

Likewise, after load cycling, the specimens created in dentine pre-treated using PAA showed a bond strength comparable ($p > 0.05$) to that obtained with the specimens created with the RMGIC applied onto dentine surfaces that received no PAA conditioning.

The specimens created in dentine pre-treated with PAA and those without PAA conditioning showed no significant difference ($p > 0.05$) after LC-AS aging compared to the specimens in the control group (24 h) or those in the group where the specimens where subjected to load cycling only. The only significant difference ($p < 0.05$) in terms of bond strength was observed after LC-AS aging; this occurred between the specimens created in dentine pre-treated with PAA and those without PAA conditioning.

3.2. Failure and Fractographic FE-SEM Analysis

Most of the specimens from all groups failed predominantly in cohesive mode within RMGIC (range: 70–88%) and in mixed mode (10–25%) after 24 h and load cycling aging (Table 1), while most of the specimens tested after LC-AS aging failed prevalently in mixed mode (range: 55–65%) compared to those tested after 24 h or load cycling only ($p < 0.05$). The percentage of adhesive failures after LC-AS aging was significantly higher ($p < 0.05$) (range: 10–13%) in the specimens in both PAA and no- PAA groups compared to those tested after 24 h or load cycling (range 2–5%).

The SEM fractographic results at 24 h and after load cycling aging are shown in Figure 1. In short, the specimens created without PAA pre-treatment that failed during a microtensile bong strength test mainly in cohesive mode after 24 h of storage in water (Figure 1A) showed a surface covered by residual RMGIC (Figure 2B,C). The specimens created with the use of PAA re-treatment applied on dentine that failed in mixed mode after 24 h storage showed some areas with the presence of unprotected dentinal tubules, which were totally exposed and characterised by the presence of partially demineralised collagen fibrils (Figure 1F,1-F1). Also, the specimens created with the use of no PAA and then subjected to load cycling prevalently showed a surface covered by RMGIC with no exposure of the dentinal tubules (Figure 1G). The specimens created with the use of PAA applied on dentine showed after load cycling only the presence of totally exposed unprotected dentinal tubules (Figure 1F); at higher magnification, it was possible to observe the presence of partially-demineralised collagen fibrils (Figure 1H-1).

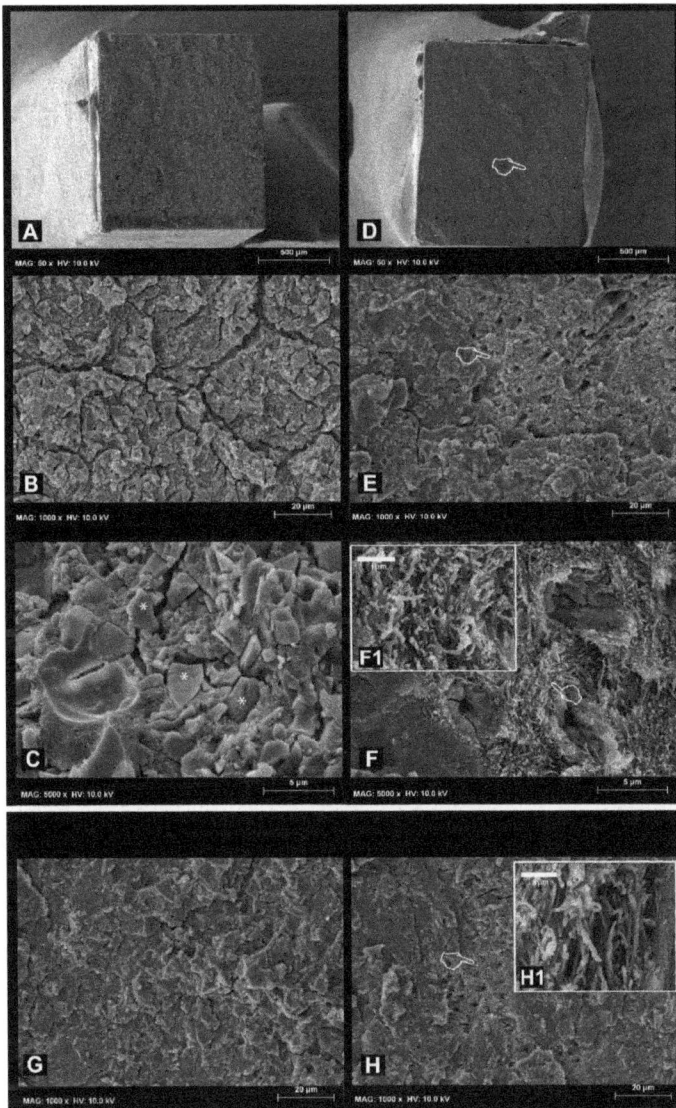

Figure 1. (**A**) Representative SEM micrograph of a specimen created with the use of no PAA applied on dentine that failed during the microtensile bong strength test in cohesive mode after 24 h of storage in water. At higher magnification, it is possible to note a surface covered by residual RMGIC (**B**) characterized by the presence of particles (*) of fluoroaluminosilicate glass (**C**). (**D**) Representative SEM fractographic analysis of specimens created with the use of PAA applied on dentine that failed in mixed mode after 24 h storage. At higher magnification, it is possible to see the presence of totally exposed unprotected dentinal tubules (pointer), and the presence of some partially demineralised residual collagen fibrils (pointer) (**E,F,F**-1). (**G**) Representative SEM micrograph of specimen created with the use of no PAA applied on dentine that failed during microtensile bond strength test in cohesive mode after load cycling aging. Also, in this case it is possible to note a surface covered by RMGIC with no dentine exposure. (**H**): Representative SEM fractographic analysis of specimens created with the use of PAA applied on dentine that failed in mixed mode after load cycling aging. It is possible to see the presence of totally exposed unprotected dentinal tubules (pointer), and, at a higher magnification, it is possible to note the presence of partially demineralised collagen fibrils (**H**-1).

The SEM fractographic results after LC-AS aging are depicted in Figure 2. A residual presence of RMGIC (Figure 2B) and a dentine surface devoid of exposed tubules was observed in the specimens created with the use of no PAA mode (Figure 2A); the dentine surface devoid of exposed tubules was still covered by smear layer (Figure 2C). On the other hand, the specimens created with the use of PAA applied on dentine that failed in mixed mode after load cycling and 8 months of storage in AS were characterised by the presence of residual RMGIC and some exposed dentine (Figure 2D). At higher magnification, tubules which were still occluded by residual RMGI were detected, but with no presence of exposed collagen fibrils was observed; these probably degraded over time during the prolonged aging in AS.

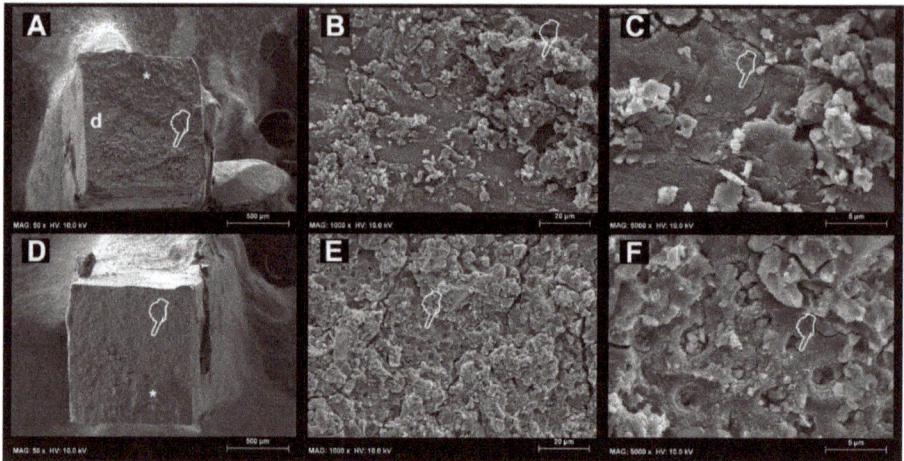

Figure 2. SEM micrographs obtained after load cycle followed by prolonged AS storage. (**A**) Representative SEM micrograph of specimen created with the use of no PAA applied on dentine that failed during microtensile bong strength test in mixed mode, where it is possible to see residual RMGIC (pointer), compact residual RMGIC (*) and some exposed dentine (d). (**B**) At higher magnification, it is possible to note a surface covered by residual RMGIC [pointer] and a dentine surface (pointer) with no exposed tubules but still covered by smear layer (**C**), (**D**); Representative SEM fractographic analysis of specimens created with the use of PAA applied on dentine that failed in mixed mode after 8 months of storage in AS which is characterised by the presence of residual RMGIC (*) and dentine (pointer). (**E**) At higher magnification, it is possible to observe the presence of exposed dentinal tubules (pointer) surrounded by residual RMGIC particles. (**F**) At even higher magnification it is possible to note the tubules are still occluded, but with no presence of exposed collagen fibrils, which probably degraded over time during prolonged aging in AS.

3.3. Ultramorphology of the Bonded-Dentine Interfaces: Confocal Microscopy Evaluation

The results of the ultramorphology and nanoleakage analysis of the RMGIC-dentine interfaces performed through dye-assisted confocal microscopy at 24 h and after load cycling only are shown in Figure 3. In short, at 24 h, the RMGIC applied onto dentine without PAA pre-treatment presented a gap-free interface characterised by a thin interdiffusion layer, which absorbed the fluorescent solution (Rhodamine B) through the dentinal tubules (Figure 3A). Conversely, the RMGIC applied onto the dentine pre-treated with PAA presented a thicker and more porous interdiffusion layer (Figure 3B). The RMGIC-dentine specimens created by applying the RMGIC onto the dentine pre-treated with no PAA and subjected to short-term load-cycle, showed an interdiffusion layer which was slightly thinner compared to that of the control specimens (24 h), (Figure 3C). This morphological features were also observed in the specimens bonded using RMGIC onto dentine pre-treated with 10% PAA and then subjected to load cycling. Indeed, such an aging protocol had no effect on the overall morphology of the interface, but the interdiffusion layer clearly appeared thinner than that observed in the specimens at 24 h (Figure 3C).

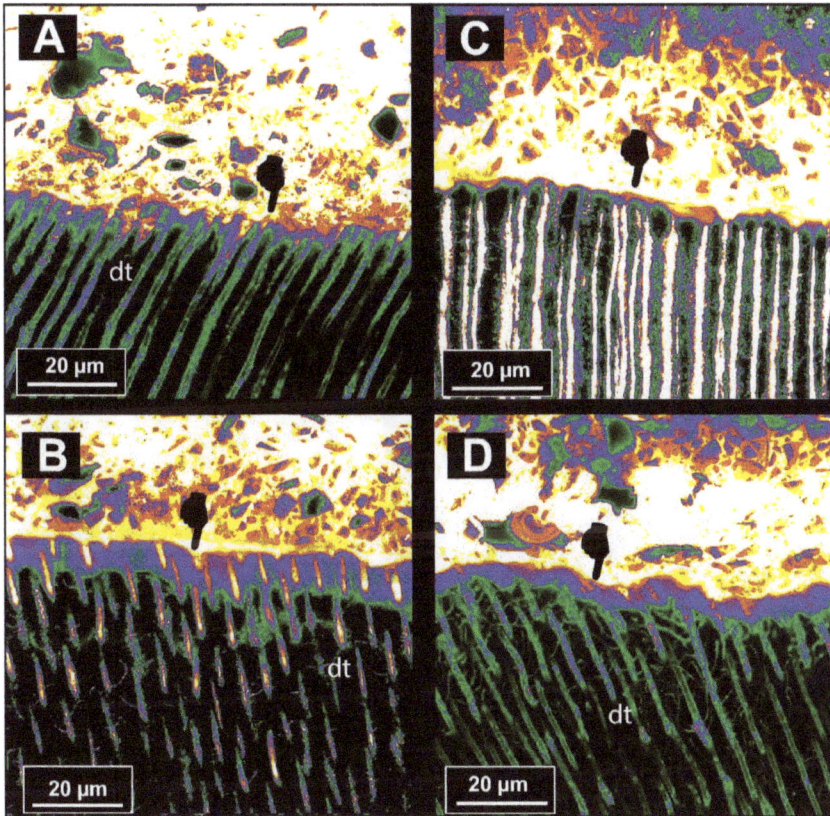

Figure 3. Confocal images of interfaces at 24 h or after short-term cycle load aging. (**A**) CLSM projection image exemplifies the interfacial characteristics at 24 h of the bond–dentine interface created by application of the resin-modified glass ionomer cement (RMGIC) onto dentine without PAA pre-treatment. It is possible to see a permeable gap-free interface that absorbed the fluorescein solution through the dentinal tubules (dt). In particular, this highlighted the existence of a thin interdiffusion layer (pointer). (**B**) CLSM projection at 24 of a representative bond-dentine interfaces created by RMGIC applied onto dentine pre-treated with PAA. In this case, it is possible to appreciate a thicker interdiffusion layer that absorbed the fluorescein solution through the dentinal tubules (dt) (**C**) A representative CLSM projection of a RMGIC-dentine interface created by applying the RMGIC onto a dentine pre-treated with no PAA and subjected to load cycling. It is possible to observe that such an aging protocol had no effect on the overall morphology of the interface, although the interdiffusion layer appears slightly thinner (pointer) than that observed in picture (**A**). (**D**) A representative CLSM projection of a RMGIC-dentine interface created by applying the RMGIC onto a dentine pre-treated with 10% PAA and subjected to load cycling. Also, in this case, it is possible to observe that such an aging protocol had no effect on the overall morphology of the interface, but the interdiffusion layer appears clearly thinner (pointer) than that observed in picture (**B**), which represents the same specimens subjected to no load-cycle aging.

The results of the ultramorphology and nanoleakage analysis of the RMGIC-dentine interfaces after LC-AS aging are shown in Figure 4. In this case, it was noted that the RMGIC-dentine interface of the RMGIC applied onto the dentine surface pre-treated with no PAA often presented cohesive fractures within the RMGIC layer; these were probably created during specimen preparation due to the brittle characteristics of such a material (Figure 4A). Conversely, the RMGIC-dentine interface created by applying the RMGIC onto a dentine pre-treated with PAA and then subjected to prolonged AS storage showed a remaining thin permeable interdiffusion layer, which may indicate the presence of porosities causing collagen degradation during subsequent prolonged water storage (Figure 4B).

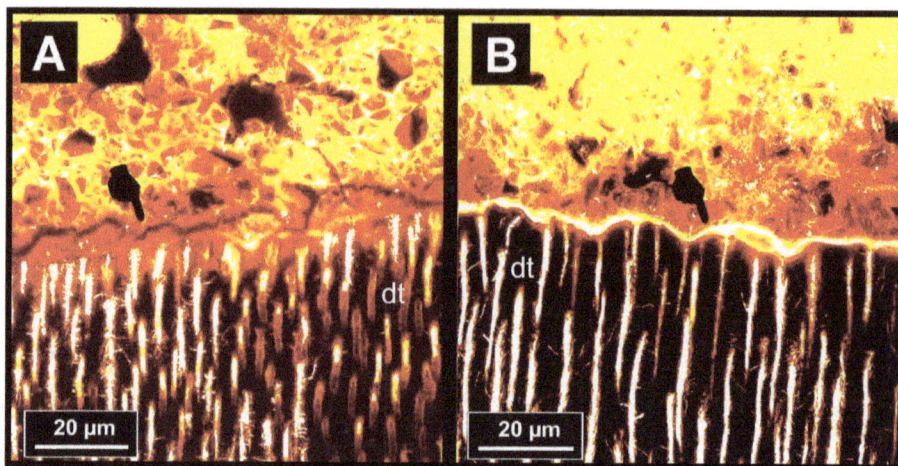

Figure 4. Confocal images of interfaces after cycle load aging and prolonged AS aging. (**A**) A representative CLSM projection of a RMGIC-dentine interface created by applying the RMGIC onto a dentine pre-treated with no PAA and subjected to prolonged AS storage. It is possible to observe the presence of a cohesive fracture within the RMGIC layer (pointer), probably created during specimen preparation (polishing) due to the brittle characteristics of such a material. This observation is supported by the absence of a permeable interfusion layer at RMGIC-dentine interface due to the maturation of the latter after prolonged storage in AS. Conversely, the RMGIC-dentine interface created by applying the RMGIC onto a dentine pre-treated with PAA and then subjected to prolonged AS storage (**B**) shows a remaining thin permeable interdiffusion layer (pointer), which indicates the presence of porosities created subsequent to collagen degradation during prolonged water storage.

4. Discussion

Therapeutic minimally invasive dentistry encompasses the philosophy of preservation of reparable dental tissues, along with the use of remineralising approaches to re-establish as much as possible of the mechanical properties of such tissues [44]. Glass-ionomer materials can be considered the main self-adhesives [45,46] and ion-releasing restorative materials available in clinics nowadays which are able to achieve such a target. However, it is believed that the overall bonding performance of such materials may be maximised if the dental substrates are pre-treated with a diluted polyacrylic acid conditioner (PAA 10%) [47,48]. Indeed, PAA conditioners remove the smear layer from dentine and enamel surfaces, consequently making the HAp directly accessible to interact with glass ionomer cements. Moreover, a slight dentine demineralisation occurs subsequent to PAA application, and a submicron interdiffusion layer is formed, which provides micromechanical retention [47–49]; the residual HAp within the demineralised collagen fibrils may also serve as receptors for additional chemical interaction [44–48]. The use of PAA as a dentine conditioner is still a theme of debate with modern resin-modified GIC [40,41], especially when considering its effect on the durability and remineralisation of dentine-bonded interfaces [43]. Furthermore, it has been recently demonstrated [50] that with conventional RMGICs such as Vitrebond Plus (3M ESPE), due to a great level energy accumulation at the dentine bonding interface during cycle load, there was evident fluorescent permeability associated with a lack of hermetic sealing.

The results of our study are in accordance those of Toledano at al. [50], as all the specimens applied in dentine pre-treated with or without PAA showed remining permeability at the bonding interface. Indeed, our current results demonstrated that the RMGIC tested in this study after the application on dentine without PAA pre-treatment presented a thin gap-free interface characterised by a thin layer of Rhodamine B absorbed through dentinal tubules (Figure 3A). In contrast, the interface of the RMGIC applied onto the dentine pre-treated with PAA was characterised by more Rhodamine B accumulation, which was due to a lack of sealing of dentinal tubules (Figure 3B), as well as a thicker layer of PAA-demineralised dentine, which remained non-infiltrated by the RMGIC [43,50].

However, after short-term load-cycle aging, the RMGIC-dentine specimens created using the RMGIC in dentine without PAA pre-treatment showed only a very thin interdiffusion layer characterised by slight fluorescence signal at the interface. As described by Toledano at al. [50], such a reduction in porosities at the bonding interface may have been due to apatite-like precipitation and remineralisation induced at the interface during mechanical cycling stress. Conversely, the specimens created with the RMGIC applied after PAA dentine pre-treatment also showed that the thickness fluorescent signal at the interface was reduced compared to the same specimens at 24 h (Figure 3C), although such a porous layer was thicker than that observed when using no PAA dentine pre-treatment. In this case, it is possible that the level of mineral precipitation was not so suitable for remineralising all the porosities within the interdiffusion layer, especially at its bottom. Indeed, Kim at al. [51] recently shown that GIC-based materials fail to completely remineralise apatite-depleted dentine due to a lack of nucleation of new apatite, even when biomimetic remineralising analogues were employed during the aging period [52].

It is important to highlight that such short-term load cycle aging was not able to induce any significant change in the microtensile bond strength, and no difference in the mode of failure in both groups of specimens (PAA vs. PAA dentine pre-treatment) was observed, compared to the control specimens at 24 h. Conversely, the results of this study showed that the specimens created with the representative RMGIC applied in dentine pre-treated with PAA showed a significant reduction in bond strength after LC-AS aging compared to specimens tested after 24 h or after short-term load cycle aging. Moreover, the LC-AS aging protocol induce also a significant change in the mode of failure; RMGIC applied in dentine pre-treated with or without PAA presented more adhesive qualities compared to all the other groups. Therefore, while the first null hypothesis is totally rejected, the second one that the durability of RMGIC applied with or without the use of PAA conditioner would be affected by load cycle followed by prolonged aging (8 months) in AS tested must be partially rejected.

Some of the current results are in accordance with those reported by Inoue et al. [45], who showed that the use of PAA conditioner before application of GIC-based materials offered no significant increase of the MTBS to dentine. Similar results were also recently reported by Sauro et al. [43], who showed that RMGIC applied onto dentine pre-treated with PAA showed significant µTBS reduction after 6 months of AS storage alone or in combination with load cycling ($p > 0.05$). Moreover, they also showed that for the RMGIC-dentine interface, specimens were affected by degradation/nanoleakage after aging, unlike the interfaces created without the use of PAA conditioning, which showed signs of remineralisation/maturation of the bonding interface.

The fractographic SEM analysis performed in this study showed that pre-treating the dentine with PAA could, in some cases, cause clear exposure of the collagen fibrils both before and after prolonged AS storage. Moreover, such fibrils were less abundant after prolonged AS storage compared to those observed in the specimens aged in AS for 24 h (Figure 2F). These specimens also showed a significant increase in the number of adhesive failures. In accordance with previously-published results [40,41,43], current fractographic SEM results showed that the specimens created with the use of no dentine PAA conditioner that failed in adhesive mode presented with dentine still covered by residual RMGIC, as the fracture occurred just above the dentine surface (Figure 2C).

It is hypothesised that such a result could be attributed to hydrolytic degradation processes that occurs over time within the collagen. Indeed, the use of PAA to pre-treat the dentine tissue may have demineralised the dentine collagen and activated endogenous matrix collagenolytic (MMP 1, MMP 8 and MMP 13) and gelatinolytic (MMP 2 and MMP 9) metalloproteinases [53]. It was also demonstrated [54] that high concentration of carboxylic groups in PAA acid conditioner may cause the formation a PAA-based polymeric gel layer within the bonding interface, which induces more water sorption at the interface. The RMGIC itself may also have degraded and become more porous over time in AS, thereby facilitating diffusion of water towards the glass-ionomer–dentine interface and causing an acceleration of the degradation processes [55].

The results of the ultramorphology and nanoleakage analysis of the RMGIC-dentine interfaces after prolonged AS storage showed that when RMGIC were applied onto the dentine surface, pre-treated with no PAA, and then subjected to prolonged AS storage, a thin permeable interdiffusion layer remained. This outcome may support the hypothesis that a bonding interface is usually characterised by the presence of porosities created subsequent to collagen degradation during prolonged water storage (Figure 4B). Conversely, the specimens created with the use of no PAA dentine conditioning showed that the absorbing layer at the interface seen at 24 h examination (Figure 3A) disappeared after prolonged AS storage (Figure 4A) due to the maturation of such areas [55], and possible remineralisation [43,56]. Indeed, the therapeutic properties (e.g., ion releasing) of RMGIC may have induced the growth of mineral crystals and remineralisation within the bonded-dentine interface, which interfered with the proteolytic action of endogenous dentine metalloproteinases [56,57]. This latter hypothesis is in accordance with previous studies that demonstrated the fluoride might inhibit both pro- and active forms of MMP 2 and MMP 9 [58]. Moreover, Makowski & Ramsby [59] reported that mineral precipitation, as well as apatite formation, may inhibit MMP activity through the formation of [Ca-PO/MMP] complexes. Sauro et al. [43] have recently reported that such a remineralising potential of RMGIC may increase if they are applied onto dentine pre-treated with bioactive glass in air-abrasion systems, and then conditioned with or without the use of a PAA conditioning gel.

In conclusion, within the limitations of this in vitro study, it is possible to affirm that the clinical decision of using a PAA conditioner should be based upon the histological features of the dentine retained after cavity preparation (e.g., sound or caries-affected dentine). However, modern RMGICs may be used for dentine restorations with or without the use of PAA pre-treatment, but it is important to consider that such a type of acid etching procedure might increase the risk of degradation at the bonding interface after prolonged service in oral cavity under mechanical cycling stress and prolonged saliva immersion. Conversely, in cases of no PAA dentine pre-treatment, it might be possible to have a synergic combination between GIC-based materials, saliva, and cycle occlusal load, which may enhance the therapeutic properties of RMGIC to induce mineralisation and protection of the bonding interface, thereby achieving more long-lasting restorations.

Author Contributions: For research articles with several authors, a short paragraph specifying their individual contributions must be provided. The following statements should be used "Conceptualization, S.S. and I.M..; Methodology, V.F.M., J.M.N.M.; Validation, S.S., V.F.L.; Formal Analysis, R.G.M., J.A.G.B.; Investigation, S.S., I.M., V.F.M., X.X.; Resources, V.F.L., I.M.; Data Curation, X.X.; Writing-Original Draft Preparation, S.S.; Writing-Review & Editing, S.S..; Supervision, V.F.L, S.S., I.M.; Project Administration, S.S., V.F.L., I.M.; Funding Acquisition, V.F.L, I.M.", please turn to the CRediT taxonomy for the term explanation. Authorship must be limited to those who have contributed substantially to the work reported.

Funding: This research received no external funding.

Acknowledgments: We declare that we have no proprietary, financial, professional or other personal interest of any nature or kind in any product, service, and/or company that could be construed as influencing the position presented in, or the review of this manuscript. The authors of this article would like to thank SDI (Australia) for donating us the RMGIC (RIVA LC) and the PAA conditioner used in this study.

Conflicts of Interest: The authors declare no conflict of interest.

References

1. Wilson, A.D.; Kent, B.E. The glass-ionomer cement, a new translucent dental filling material. *J. Appl. Chem. Biotechnol.* **1971**, *21*, 313. [CrossRef]
2. Wilson, A.D.; Nicholson, J.W. *Acid-Base Cements: Their Biomedical and Industrial Applications*; Cambridge University Press: Cambridge, UK, July 2005; Volume 3.
3. Prosser, H.J.; Powis, D.R.; Wilson, A.D. Glass-ionomer cements of improved flexural strength. *J. Dent. Res.* **1986**, *65*, 146–148. [CrossRef] [PubMed]
4. Alhalawani, A.M.; Curran, D.J.; Boyd, D.; Towler, M.R. The role of poly(acrylic acid) in conventional glass polyalkenoate cements. *J. Polym. Eng.* **2016**, *36*, 221–237. [CrossRef]
5. Guggenberger, R.; May, R.; Stefan, K.P. New trends in glass-ionomer chemistry. *Biomaterials* **1998**, *19*, 479–483. [CrossRef]
6. McCabe, J.F.; Walls, A.W. (Eds.) *Applied Dental Materials*; John Wiley & Sons: Hoboken, NJ, USA, May 2013.
7. Anusavice, K.J.; Shen, C.; Rawls, H.R. *Phillips' Science of Dental Materials*; Elsevier Health Sciences: Amsterdam, The Netherlands, 2013.
8. Wilson, A.D. The chemistry of dental cements. *Chem. Soc. Rev.* **1978**, *7*, 265–296. [CrossRef]
9. Nicholson, J.W. Chemistry of glass-ionomer cements: A review. *Biomaterials* **1998**, *19*, 485–494. [CrossRef]
10. Zainuddin, N.; Karpukhina, N.; Hill, R.G.; Law, R.V. A long-term study on the setting reaction of glass ionomer cements by 27 Al MAS-NMR spectroscopy. *Dent. Mater.* **2009**, *25*, 290–295. [CrossRef] [PubMed]
11. Preston, S.M.; Higham, S.M.; Agalamanyi, E.A.; Mair, L.H. Fluoride recharge of aesthetic dental materials. *J. Oral Rehabil.* **1999**, *26*, 936–940. [CrossRef] [PubMed]
12. Forsten, L. Resin-modified glass ionomer cements: Fluoride release and uptake. *Acta Odontol. Scand.* **1995**, *53*, 222–225. [CrossRef] [PubMed]
13. Forss, H.; Jokinen, J.; Spets-Happonen, S.; Seppä, L.; Luoma, H. Fluoride and mutans streptococci in plaque grown on glass ionomer and composite. *Caries Res.* **1991**, *25*, 454–458. [CrossRef] [PubMed]
14. Preston, A.J.; Mair, L.H.; Agalamanyi, E.A.; Higham, S.M. Fluoride release from aesthetic dental materials. *J. Oral Rehabil.* **1999**, *26*, 123–129. [CrossRef] [PubMed]
15. De Amorim, R.G.; Leal, S.C.; Frencken, J.E. Survival of atraumatic restorative treatment (ART) sealants and restorations: A meta-analysis. *Clin. Oral Investig.* **2012**, *16*, 429–441. [CrossRef] [PubMed]
16. Van Dijken, J.W. 3-Year clinical evaluation of a compomer, a resin-modified glass ionomer and a resin composite in class III restorations. *Am. J. Dent.* **1996**, *9*, 195–198. [PubMed]
17. Abdalla, A.I.; Alhadainy, H.A.; Garcia-Godoy, F. Clinical evaluation of glass ionomers and compomers in class V carious lesions. *Am. J. Dent.* **1997**, *10*, 18–20. [PubMed]
18. Leevailoj, C.; Platt, J.A.; Cochran, M.A.; Moore, B.K. In vitro study of fracture incidence and compressive fracture load of all-ceramic crowns cemented with resin-modified glass ionomer and other luting agents. *J. Prosthet. Dent.* **1998**, *80*, 699–707. [CrossRef]
19. Pascotto, R.C.; de Lima Navarro, M.F.; Capelozza Filho, L.; Cury, J.A. In vivo effect of a resin-modified glass ionomer cement on enamel demineralization around orthodontic brackets. *Am. J. Orthod. Dentofac.* **2004**, *125*, 36–41. [CrossRef]
20. Jokstad, A.; Mjör, I.A. Ten years' clinical evaluation of three luting cements. *J. Dent.* **1996**, *24*, 309–315. [CrossRef]
21. Murdoch-Kinch, C.A.; McLean, M.E. Minimally invasive dentistry. *J. Am. Dent. Assoc.* **2003**, *134*, 87–95. [CrossRef] [PubMed]
22. Peumans, M.; de Munck, J.; Mine, A.; van Meerbeek, B. Clinical effectiveness of contemporary adhesives for the restoration of non-carious cervical lesions. A systematic review. *Dent. Mater.* **2014**, *30*, 1089–1103. [CrossRef] [PubMed]
23. Terata, R.; Nakashima, K.; Kubota, M. Effect of temporary materials on bond strength of resin-modified glass-ionomer luting cements to teeth. *Am. J. Dent.* **2000**, *13*, 209–211. [PubMed]
24. McLean, J.W.; Powis, D.R.; Prosser, H.J.; Wilson, A.D. The use of glass-ionomer cements in bonding composite resins to dentine. *Br. Dent. J.* **1985**, *158*, 410–414. [CrossRef] [PubMed]
25. Andersson-Wenckert, I.E.; Kieri, C. Durability of extensive class II open-sandwich restorations with a resin-modified glass ionomer cement after 6 years. *Am. J. Dent.* **2004**, *17*, 43–50. [PubMed]

26. Yamaga, R.; Nishino, M.; Yoshida, S.; Yokomizo, I. Diammine silver fluoride and its clinical application. *J. Osaka Univ. Dent. Sch.* **1972**, *12*, 1–20. [PubMed]

27. Sauro, S.; Osorio, R.; Watson, T.F.; Toledano, M. Influence of phosphoproteins' biomimetic analogs on remineralization of mineral-depleted resin–dentin interfaces created with ion-releasing resin-based systems. *Dent. Mater.* **2015**, *31*, 759–777. [CrossRef] [PubMed]

28. Xie, D.; Brantley, W.A.; Culbertson, B.M.; Wang, G. Mechanical properties and microstructures of glass-ionomer cements. *Dent. Mater.* **2000**, *16*, 129–138. [CrossRef]

29. Peutzfeldt, A.; Garcia-Godoy, F.; Asmussen, E. Surface hardness and wear of glass ionomers and compomers. *Am. J. Dent.* **1997**, *10*, 15–17. [PubMed]

30. Ahmed, N.; Zafar, M.S. Effects of wear on hardness and stiffness of restorative dental materials. *Life Sci. J.* **2014**, *11*, 11–18.

31. Um, C.M.; Øilo, G. The effect of early water contact on glass-ionomer cements. *Quintessence Int.* **1992**, *1*, 23.

32. Moshaverinia, A.; Roohpour, N.; Chee, W.W.L.; Schricker, S.R. A review of powder modifications in conventional glass-ionomer dental cements. *J. Mater. Chem.* **2011**, *21*, 1319–1328. [CrossRef]

33. Moshaverinia, A.; Ansari, S.; Moshaverinia, M.; Roohpour, N.; Darr, J.A.; Rehman, I. Effects of incorporation of hydroxyapatite and fluoroapatite nanobioceramics into conventional glass ionomer cements (GIC). *Acta Biomater.* **2008**, *4*, 432–440. [CrossRef] [PubMed]

34. Moshaverinia, A.; Ansari, S.; Movasaghi, Z.; Billington, R.W.; Darr, J.A.; Rehman, I.U. Modification of conventional glass-ionomer cements with N-vinylpyrrolidone containing polyacids, nano-hydroxy and fluorapatite to improve mechanical properties. *Dent. Mater.* **2008**, *24*, 1381–1390. [CrossRef] [PubMed]

35. Wilson, A.D. Resin-modified glass-ionomer cements. *Int. J. Prosthodont.* **1989**, *3*, 425–429.

36. Soncini, J.A.; Maserejian, N.N.; Trachtenberg, F.; Tavares, M.; Hayes, C. The Longevity of amalgam versus compomer/composite restorations in posterior primary and permanent teeth: Findings from the new england children's amalgam trial. *J. Am. Dent. Assoc.* **2007**, *138*, 763–772. [CrossRef] [PubMed]

37. McCabe, J.F. Resin-modified glass-ionomers. *Biomaterials* **1998**, *19*, 521–527. [CrossRef]

38. Sauro, S.; Pashley, D. Strategies to stabilise dentine-bonded interfaces through remineralising operative approaches - State of The Art. *Int. J. Adhes. Adhes.* **2016**, *69*, 39–57. [CrossRef]

39. Tyas, M.J.; Burrow, M.F. Adhesive restorative materials: A review. *Aust. Dent. J.* **2004**, *49*, 112–121. [CrossRef] [PubMed]

40. Inoue, S.; Abe, Y.; Yoshida, Y.; De Munck, J.; Sano, H.; Suzuki, K.; Lambrechts, P.; Van Meerbeek, B. Effect of conditioner on bond strength of glass-ionomer adhesive to dentin/enamel with and without smear layer interposition. *Oper. Dent.* **2004**, *29*, 685–692. [PubMed]

41. De Munck, J.; Van Meerbeek, B.; Yoshida, Y.; Inoue, S.; Suzuki, K.; Lambrechts, P. Four-year water degradation of a resin-modified glass-ionomer adhesive bonded to dentin. *Eur. J. Oral Sci.* **2004**, *112*, 73–83. [CrossRef] [PubMed]

42. Toledano, M.; Cabello, I.; Aguilera, F.S.; Osorio, E.; Osorio, R. Effect of in vitrochewing and bruxism events on remineralization, at the resin-dentin interface. *J. Biomech.* **2015**, *48*, 14–21. [CrossRef] [PubMed]

43. Sauro, S.; Watson, T.; Moscardó, A.P.; Luzi, A.; Feitosa, V.P.; Banerjee, A. The effect of dentine pre-treatment using bioglass and/or polyacrylic acid on the interfacial characteristics of resin-modified glass ionomer cements. *J. Dent.* **2018**, *73*, 32–39. [CrossRef] [PubMed]

44. Watson, T.F.; Atmeh, A.R.; Sajini, S.; Cook, R.J.; Festy, F. Present and future of glass ionomersand calcium-silicate cements as bioactive materials in dentistry: biophotonics-based interfacial analyses in health and disease. *Dent. Mater.* **2014**, *30*, 50–61. [CrossRef] [PubMed]

45. Inoue, S.; Van Meerbeek, B.; Abe, Y.; Yoshida, Y.; Lambrechts, P.; Vanherle, G.; Sano, H. Effect of remaining dentin thickness and the use of conditioner on micro-tensile bond strength of a glass-ionomer adhesive. *Dent. Mater.* **2001**, *17*, 445–455. [CrossRef]

46. Yoshida, Y.; Van Meerbeek, B.; Nakayama, Y.; Snauwaert, J.; Hellemans, L.; Lambrechts, P.; Vanherle, G.; Wakasa, K. Evidence of chemical bonding at biomaterial–hard tissue interfaces. *J. Dent. Res.* **2000**, *79*, 709–714. [CrossRef] [PubMed]

47. Van Meerbeek, B.; Vargas, M.; Inoue, S.; Yoshida, Y.; Peumans, M.; Lambrechts, P.; Vanherle, G. Adhesives and cements topromote preservation dentistry. *Oper. Dent.* **2001**, *26*, 119–144.

48. Van Meerbeek, B.; Inoue, S.; Perdigao, J.; Lambrechts, P.; Vanherle, G. *Enamel and Dentin Adhesion*; Summitt, J.B., Robbins, J.W., Schwartz, R.S, Eds.; Quintessence Publishing Co.: Chicago, IL, USA, 2001; pp. 178–235.

49. Lin, A.; McIntyre, N.; Davidson, R. Studies on the adhesion of glass ionomer cements to dentin. *J. Dent. Res.* **1992**, *71*, 1836–1841. [CrossRef] [PubMed]

50. Toledano, M.; Osorio, R.; Osorio, E.; Cabello, I.; Toledano-Osorio, M.; Aguilera, F.S. In vitro mechanical stimulation facilitates stress dissipation and sealing ability at the conventional glass ionomer cement-dentin interface. *J. Dent.* **2018**, *73*, 61–69. [CrossRef] [PubMed]

51. Kim, Y.K.; Yiu, C.K.; Kim, J.R.; Gu, L.; Kim, S.K.; Weller, R.N.; Pashley, D.H.; Tay, F.R. Failure of a glass ionomer to remineralise apatite-depleted dentin. *J. Dent. Res.* **2010**, *89*, 230–235. [CrossRef] [PubMed]

52. Gu, L.S.; Kim, J.; Kim, Y.K.; Liu, Y.; Dickens, S.H.; Pashley, D.H.; Ling, J.Q.; Tay, F.R. A chemical phosphorylation-inspired design for Type I collagen biomimetic remineralization. *Dent. Mater.* **2010**, *26*, 1077–1089. [CrossRef] [PubMed]

53. Nishitani, Y.; Yoshiyama, M.; Wadgaonkar, B.; Breschi, L.; Mannello, F.; Mazzoni, A.; Carvalho, R.M.; Tjäderhane, L.; Tay, F.R.; Pashley, D.H. Activation of gelatinolytic/collagenolytic activity in dentin by self-etching adhesives. *Eur. J. Oral Sci.* **2006**, *114*, 160–166. [CrossRef] [PubMed]

54. Es-Souni, M.; Fischer-Brandies, H.; Zaporojshenko, V.; Es-Souni, M. On the interaction of polyacrylic acid as a conditionning agent with bovine enamel. *Biomaterials* **2002**, *23*, 2871–2878. [CrossRef]

55. Sidhu, S.K.; Watson, T.F. Interfacial characteristics of resin-modified glass-ionomer materials: A study on fluid permeability using confocal fluorescence microscopy. *J. Dent. Res.* **1998**, *77*, 1749–1759. [CrossRef] [PubMed]

56. Sauro, S.; Watson, T.F.; Thompson, I.; Toledano, M.; Nucci, C.; Banerjee, A. Influence of air-abrasion executed with PAA-bioglass 45S5 on the bonding performance of a resin-modified glass ionomer cement. *Eur. J. Oral Sci.* **2012**, *120*, 168–177. [CrossRef] [PubMed]

57. Brackett, M.G.; Agee, K.A.; Brackett, W.W.; Key, W.O.; Sabatini, C.; Kato, M.T.; Buzalaf, M.A.; Tjäderhane, L.; Pashley, D.H. Effect of sodium fluoride on the endogenous MMP activity of dentin matrices. *J. Nat. Sci.* **2015**, *1*, e118. [PubMed]

58. Tezvergil-Mutluay, A.; Seseogullari-Dirihan, R.; Feitosa, V.P.; Cama, G.; Brauer, D.S.; Sauro, S. Effects of Composites Containing Bioactive Glasses on Demineralized Dentin. *J. Dent. Res.* **2017**, *96*, 999–1005. [CrossRef] [PubMed]

59. Makowski, G.S.; Ramsby, M.L. Differential effect of calcium phosphate and calcium pyrophosphate on binding of matrix metalloproteinases to fibrin: comparison to a fibrin-binding protease from inflammatory joint fluids. *Clin. Exp. Immunol.* **2004**, *136*, 176–187. [CrossRef] [PubMed]

materials

MDPI

Article

Biomimetic Mineralizing Agents Recover the Micro Tensile Bond Strength of Demineralized Dentin

Luiz Filipe Barbosa-Martins [1], Jossaria Pereira de Sousa [1], Lívia Araújo Alves [2], Robert Philip Wynn Davies [3] and Regina Maria Puppin-Rontanti [4,*]

[1]   Department of Pediatric Dentistry, Piracicaba Dental School, State University of Campinas, Piracicaba 13414-903; Brazil; flpmarttins@gmail.com (L.F.B.-M.); jossariasousa@gmail.com (J.P.d.S.)
[2]   Department of Oral Diagnosis, Piracicaba Dental School, State University of Campinas, Piracicaba 13414-903, Brazil; liviaaalves@hotmail.com
[3]   Division of Oral Biology, School of Dentistry, Faculty of Medicine & Health, University of Leeds, Leeds LS9 7TF, UK; R.P.W.Davies@leeds.ac.uk
[4]   Departments of Pediatric Dentistry and Restorative Dentistry, Piracicaba Dental School, University of Campinas, Piracicaba 13414-903, Brazil
*   Correspondence: rmpuppin@unicamp.br; Tel.: +55-19-2106-5286

Received: 10 July 2018; Accepted: 11 September 2018; Published: 14 September 2018

Abstract: Biomimetic remineralization is an approach that mimics natural biomineralization, and improves adhesive procedures. The aim of this paper was to investigate the influence of Dentin Caries-like Lesions (DCLL)-Producing Model on microtensile bond strength (μTBS) of etch and rinse adhesive systems and investigate the effect of remineralizing agents such as Sodium Fluoride (NaF), MI Paste™ (MP) and Curodont™ Repair (CR) on caries-affected dentin (n = 6). Nine groups were established: (1) Sound dentin; (2) Demineralized dentin/Chemical DCLL: (3) Demineralized dentin/Biological DCLL; (4) Chemical/DCLL + NaF; (5) Chemical/DCLL + MP; (6) Chemical/DCLL + CR; (7) Biological/DCLL + NaF; (8) Biological/DCLL + MP; (9) Biological/DCLL + CR. Then all dentin blocks were subjected to a bonding procedure with Adper™ Single Bond 2 adhesive system/Filtek Z350XT 4 mm high block, following this they were immersed in deionized water/24 h and then sectioned with $\cong$1 mm^2 beams. The μTBS test was conducted at 1 mm/min/500 N loading. Failure sites were evaluated by SEM (scanning electron microscopy (150$\times$). μTBS data were submitted to factorial ANOVA and Tukey's test ($p < 0.05$). The highest values were found when demineralized dentin was treated with MP and CR, regardless caries lesion depth ($p < 0.05$). There was a predominance of adhesive/mixed in the present study. It was concluded that the use of the artificial dentin caries production models produces differences in the μTBS. Additionally MP and CR remineralizing agents could enhance adhesive procedures even at different models of caries lesion.

Keywords: dentin; desmineralization; microtensile bond strength

1. Introduction

During the execution of routine dental restorations, the hybrid structure formed by the dental bonding procedure occurs through the interaction and subsequent polymerization of monomers around the demineralized collagen matrix [1]. The oral cavity is a severe environment for the resin-dental bond to survive for a reasonable length of time, with thermomechanical changes, chemical attacks by acids and enzymes and other factors posing routine daily challenges. Therefore, to achieve effective and stable bonding, the preservation of dentin collagen is critical, since collagen represents the major organic component of the dentin matrix [2].

Caries is among the most common diseases worldwide [3], and the immediate bond strengths to caries-affected dentin are commonly 20–50% lower than to sound dentin [4–6]. The restoration of the normal conditions of the mineral content of the caries-affected dentin, prevents the action of enzymes in addition to providing an increased bond durability [7].

Biomimetic remineralization mimics the process of natural biomineralization by replacing demineralized collagen matrix water with apatite crystallites [7]. Caries-affected dentin is comprised of about 14–53% of water compared with sound dentin, which exhibits a much lower value [8]. Therefore, by replacing water with minerals at the dentin–resin interface, this would increase the mechanical properties and inhibit water-related hydrolysis [9].

It has recently been demonstrated that the use of remineralizing agents in dentin could recover the mechanical properties of the substrate [10]. In addition to sodium fluoride (NaF) and sodium phosphate (Na_3PO_4), casein phosphopeptide amorphous calcium phosphate (CPP-ACP), which is derived from milk protein, can release calcium phosphate assisting in enamel and dentin remineralization [11]. It acts mainly by inhibiting demineralization and enzymatic degradation [12]. Furthermore, recent studies have shown that the use CPP-ACP has no negative effect on bond strength [13,14]. The peptidic biomimetic matrix 'P$_{11}$-4', which has been incorporated into a clinical product (Curodont™ Repair) has shown encouraging results in early clinical trials. It has been shown to improve the visual appearance of carious lesions and increases the opacity on X-rays after treatment of proximal caries [15,16]. Additionally, following the application of P$_{11}$-4 and subsequent bonding procedures an increased resin-dentin bond strength has been observed [17].

The restructured demineralized collagen matrix found in caries-affected dentin process by interventions such as Sodium Fluoride (NaF), CPP-ACP contained in MI Paste™ (GC International) and P$_{11}$-4 peptide contained in Curodont™ Repair (Credentis AG), prior to adhesive procedures by an etch-and-rinse adhesive system (Adper™ Single Bond 2 (3M ESPE) in the demineralized dentin could be a promising proposal for adhesive clinical procedures.

Clinical binding procedures simulated by mechanical methods (i.e., TMBS, μTBS) often use artificial demineralized dentin. However, the lack of standardization of caries lesions creates technical difficulties for evaluation [18]. In vitro models have been used to produce demineralized dentin under controlled conditions [19–22]. Chemical methods provide superficial dentin demineralization, resulting in a substrate with similar hardness compared to natural caries-affected dentin [19]. Conversely, the microbiological method promotes an excessive softening of dentin, but with a more comparable morphological pattern of collagen degradation to natural caries lesions [19–22]. Pacheco et al., 2013 [23], evaluating molecular and structural lesions related to dental caries, produced by the chemical (GC), biological (GB), in situ (GIS) and natural (CNG) approaches, showed similar and lower surface hardness between CNG and GB of which GC and GIS, lower mineral content (Ca^{2+} and $PO_4{}^{3-}$) for GB and GNC than GC and GIS. Therefore, the structure and mechanical properties are different with respect to the caries model production and the remineralizing agents may act differently in the adhesion procedures depending on the model of caries lesions used.

To ascertain the potential efficacy of the bonding procedure carried out on remineralized dentin we aimed to evaluate different remineralization treatments and the method used in producing the simulated dentin-like caries lesions on the micro tensile bond strength of remineralized dentin. The hypothesis was that the Dentin Caries-like Lesions Producing Model and the remineralizing agents (Sodium Fluoride-(NaF), MI Paste™-(MP) and Curodont™ Repair-(CR)) affect microtensile bond strength-μTBS of etch-and-rinse adhesive system on caries-affected dentin.

2. Materials and Methods

Sixty-three sound human third molars were collected with patients' informed consent, as approved by the Ethics Committee of Piracicaba Dental School, University of Campinas (Protocol number 37634814.5.0000.5418). The teeth were stored in 0.1% thymol solution at 4 °C for no longer than 2 months after extraction. A 4.0 mm coronal dentin slice from each tooth was obtained by sectioning 2.0 mm below cement-enamel junction (CEJ), and 2.0 mm above CEJ, using a slow-speed water-cooled diamond saw (Isomet 1000, Buehler Ltd., Lake Bluff, IL, USA). Six dentin slices were used for sound dentin (control group-CG), and the others were randomly assigned into 2 groups (n = 24), according to the caries method production. The dentin surface of each specimen was wet polished with a 600-grit SiC paper (Arotec, São Paulo, Brazil) for 30 s to create a standardized smear layer. The dentin surfaces were carefully examined under a stereomicroscope at ×50 magnification to confirm the absence of enamel islets. The specimens were immediately subjected to production of caries in vitro. The group distribution is illustrated in Figure 1. To check the caries dentin depth, three teeth were chosen from each group and probed using a polarized light microscope (Figure 2).

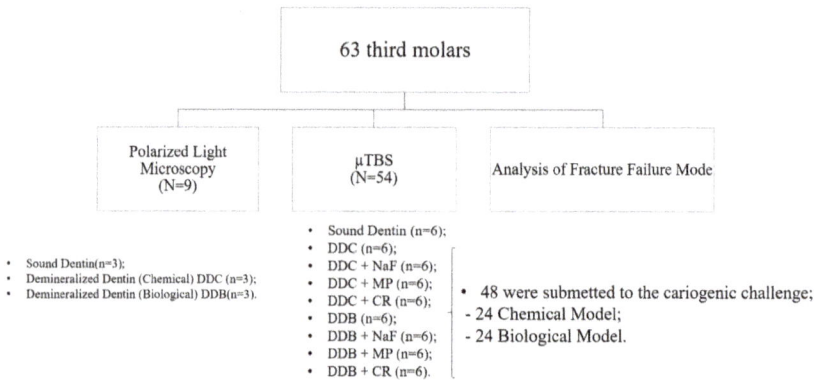

Figure 1. Experimental design. DDC—demineralized dentin provided by chemical model; DDC/NaF-DDC + 2% NaF; DDC/MP-DDC + MI Paste™; DDC/CR-DDC + Curodont™ Repair; DDB—demineralized dentin provided by biological model; DDB/NaF-DDB + 2% NaF; DDB/MP-DDB + MI Paste™; DDB/CR-DDB + Curodont™ Repair.

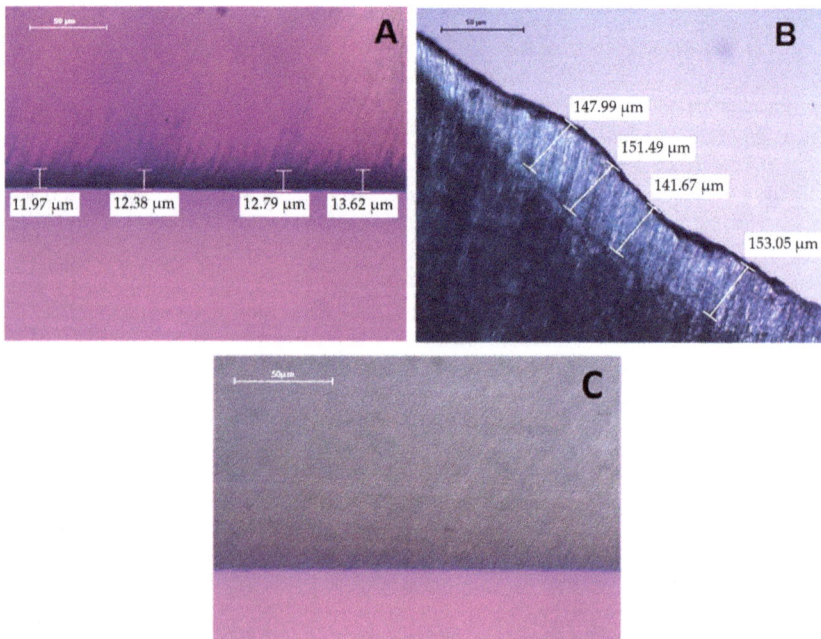

Figure 2. (**A**) Artificial caries lesions provided by Chemical Model; (**B**) Artificial caries lesions provided by Biological Model after removing the softened tissue; (**C**) Sound Dentin.

2.1. Artificial Dentin Caries-Like Lesions (DCLL) Production Protocols

Sixty dentin slices (54 teeth for μTBS and 6 for polarized light microscopy) were randomly assigned into 2 groups according to dentin-like caries lesions producing models: chemical (carboxymethylcellulose acid gel) and biological (Streptococcus mutans—UA159 biofilm).

2.1.1. Chemical Model

The specimens were submerged in vials containing 5 mL of 6% carboxymethylcellulose acid gel (0.1 M lactic acid titrated to pH 5.0 in a KOH solution) at pH 5.0 and 37 °C. The specimens remained in the gel for 48 h without renewal [23]. This model has been reported to supposedly provide a demineralized dentin similar to that of caries affected dentin.

2.1.2. Biological Model

The specimens were fixed with orthodontic wire on the lids of glass vials containing 250 mL of sterile deionized water and were sterilized with gamma radiation (14.5 kGy dose) for 60 h (Pacheco et al., 2013). Then, they were transferred to another glass vial containing 250 mL of sterile brain-heart infusion (BHI) broth (LabCenter, São Paulo, Brazil) supplemented with 0.5% yeast extract (LabCenter, São Paulo, Brazil), 0.5% glucose (LabCenter, São Paulo, Brazil), 1% sucrose (LabCenter, São Paulo, Brazil) and 2% *S. mutans* (UA159, ATCC, Oklahoma, OK, USA) incubated at 37 °C and supplemented with 10% CO_2, pH of around 4.0. Starter culture was transferred into 250 mL of fresh BHI and grown for 4 h at 37 °C under aerobic conditions. Optical density at 550 nm (A550) of all bacterial suspensions was adjusted to 0.05 prior to inoculation. Inoculation occurred only in the first day of the experiment, but the broth was renewed every 48 h over a 7-day period. The broth was Gram stained daily to monitor contamination. The resulting biofilm formed over the teeth was removed with gauze and the softer dentin layer was removed using #6 carbide drills; the removal ceased when dentin appearance was like that of caries-affected dentin [23].

2.2. Polarized Light Microscopy (PLM)

After providing dentin caries-like lesions, three specimens of each dentin caries-like lesions-producing models and control group were sectioned perpendicular to the occlusal surface to obtain slices. The two more central slices of each tooth were selected and polished with #800, #1200, #2400 and #4000-grit silicon carbide (SiC) paper (Buehler, Lake Buff, IL, USA), obtaining a 0.15 μm dentin thickness. Dentin slices were analyzed for the depth of demineralization on PLM (Leica DMLP, Leica microsystems, Wetzlar, Germany). The dentin slices were kept in 100% humidity throughout the investigation. Depth caries lesions were measured in a PLM (Leica DMLP, Leica microsystems) using 20×/0.4 (corr.) objective. Standard settings for contrast, brightness and light were used for all images. Four measurements were made in different parts of the same lesion from the lesion border to the deepest part of the lesion, for each dentin caries-like lesion model. An average depth for each specimen was calculated from the individual values based on depth difference between dentin caries-like lesions and sound dentin on the same specimen. Dentin caries-like lesions observed in the specimens subjected to Chemical Model presented x = 12.69 μm average depth and Biological Model x = 148.55 μm, measured using PLM.

2.3. Dentin Surface Treatment

Thirty-six teeth were assigned to 6 groups according to the remineralization treatment: demineralized dentin by chemical model (DDC) + treated with 0.2% NaF Solution (1 min)—DDC/NaF; DDC + treated with MI Paste™ (1 min)—DDC/MP; and DDC + treated with Curodont™ Repair—DDC/CR applied (5 min) plus Ca^{2+} and PO_4^{3-} Solution (1 min); demineralized dentin by biological model (DDB) + 0.2% NaF Solution—DDB/NaF; DDB + treated with MI Paste™ applied

(1 min)—DDB/MP; and DDB + treated with Curodont™ Repair—DDB/CR applied (5 min) plus Ca^{2+} and PO_4^{3-} Solution (1 min).

The specimens of groups DDC/NaF and DDB/NaF were remineralized using 0.1 mL of the 0.2% NaF. 0.2% NaF solution was applied on demineralized dentin surface and left dry for 1 min at room temperature for the groups DDC/NaF and DDB/NaF. 0.1 mL of the MI Paste™ was applied onto the surface of the specimens from DDC/MP and DDB/MP groups with microbrush for 1 min at room temperature, the excess paste was removed by washing with deionized water. For DDC/CR and DDB/CR groups, 50 μL of the Curodont™ Repair was applied and left for 5 min, then, a Ca^{2+} and PO_4^{3-} solution was applied and left onto surface for 1 min. For all treatments, the solution excess was removed with absorbent paper.

2.4. Bonding Procedures

A single operator applied the adhesive according to the manufacturer's instruction (Table 1). An LED light-curing unit (Bluephase, Ivoclar Vivadent; Schaan, Liechtenstein) was set to the low power mode with a light intensity of 650 mW/cm^2. A nanohybrid resin composite (Filtek Z350 XT, A2 (3M ESPE, St. Paul, MN, USA)) was used to create resin composite buildups in four layers of 1 mm each [17]. Each layer was light cured for 20 s, followed by a final polymerization of 60 s. The specimens were then stored at 100% humidity at 37 °C for 24 h.

Table 1. Materials, manufactures, components, batch numbers and application mode of tested materials.

Materials (Manufactures)	Main Components	Batch Number	Application Mode	
0.2% NaF Solution	0.2 g of NaF in 100 mL deionized water	Made in the Lab *	1.	Apply 1.0 mL of 0.2% NaF solution
Ca^{2+} and PO_4^{3-} Solution	Saturated solution of Ca^{2+} and PO_4^{3-} (1.5 mmol/L calcium, 0.9 mmol/L phosphate, and 150 mmol/L KCl in 20 mmol/L cacodylic buffer, pH 7.0) [24].	Made in the Lab	1.	Apply 0.1 mL of Ca^{2+} and PO_4^{3-} solution
MI™ Paste—GC Internacional, Itabashi-ku, Tóquio, Japão	Glycerol, CPP-ACP, D-Sorbitol, Propylene glycol, Titanium dioxide and silicon	N2347319	1.	Apply 0.1 mL of MI™ Paste
Curodont™ Repair—Credentis AG, Dorfstrasse, Windisch, Switzerland	P_{11}-4 peptide—amino acid sequence— (Ace-Gln-Gln-Arg-Phe-Glu-Trp-Glu-Phe-Glu-Gln-Gln-NH$_2$)	N342x	1. 2.	Apply 50 μL of Curodont™ Repair for 5 min Apply 0.1 mL of Ca^{2+} and PO_4^{3-} solution
Scotchbond™ Universal Etchant—3M ESPE; St Paul, MN, USA	32% phosphoric acid	N345	1. 2.	Apply etchant for 15 s Rinse for 10 s
Adper Single Bond 2.0—3M ESPE; St Paul, MN, USA	HEMA, water, ethanol, Bis-GMA, dimethacrylates, amines, metacrylate functional copolymer of polyacrylic and polyitaconic acids, 10% by weight of 5 nanometer-diameter spherical sílica particles	N42912	3. 4. 5.	Blot water excess Apply 2 consecutive coats of adhesive for 15 s with gentle agitation Gently air dry for 5 s 6. Light-cure for 10 s
Filtek™ Z350 XT—3M ESPE; St Paul, MN, USA	BIS-GMA, Bis-EMA, UDMA, TEG-DMA, camphorquinone, non-agglomerated silica nanoparticles	N98354	1. 2.	Incremental insertion 2 mm Light-cure for 20 s

* Pediatric Dentistry Laboratory.

2.5. Microtensile Bond Strength Test (μTBS)

After storage, the specimens were sectioned perpendicularly to the resin/dentin interface to produce dentin–resin beams with 1 mm^2 at cross-sectional area, using a low speed diamond saw (ISOMET 1000, Buehler Ltd., Lake Buff, IL, USA). From six to eight beams were obtained per tooth, each beam was measured with a digital caliper (Mitutoyo; Kawasaki, Japan) to determine the cross-sectional area. All beams were kept in deionized water for 24 h.

For μTBS measurements, each beam was fixed to a microtensile device with cyanocrylate glue (Super Bonder (#1883519), Loctite, Henkel Corp., Rocky Hill, CT, USA), and tested in a universal testing machine (DL 2000, EMIC, Equipment and Systems Ltda., São José dos Pinhais, Brazil). The test was carried out with a 500 N load at 1.0 mm/min cross speed until failure. The μTBS values were expressed in MPa.

2.6. Analysis of Failure Mode

All the fractured specimens from the microtensile bond strength analysis were assessed to determine the failure mode using SEM at ×50 and ×150 magnifications. The fractured surfaces of the beams were paired, air dried, mounted on aluminum stubs, gold coated, and examined by SEM (JSM-5600LV, JEOL; Tokyo, Japan), operated at 15 kV. The failure patterns were classified according to the following categories: adhesive, mixed (involving resin composite, adhesive and/ or dentin), cohesive failure in the resin composite, and cohesive failure in dentin [17,25,26].

2.7. Statistical Analysis

Bond strength values for each group were analyzed by Shapiro–Wilk test (R Software version 3.4.3, The R Foundation for Statistical Computing, Vienna, Austria) in order to assess the normality of the data distribution. Factorial ANOVA and post hoc Tukey test (R Software version 3.4.3, The R Foundation for Statistical Computing, Vienna, Austria) were used to determine statistically significant differences between factors: the dentin-like caries lesions model (two levels—chemical and biological models) and dentin remineralization treatment (four levels—SD, DD, NaF + DD, CPP-ACP + DD and P$_{11}$-4 + DD) on dentin/resin bond strength, and additional Dunnett test to determine statistically significant differences between the experimental groups and the control group (sound dentin). The Kruskal Wallis test was used to evaluate the failure mode. The R Software version 3.4.3 (The R Foundation for Statistical Computing, Vienna, Austria), was used to perform the tests. Statistical difference was set at $\alpha = 5\%$.

3. Results

The images of the lesions observed in Chemical Model x = 12.69 (average depth) (A) and Biological Model x = 148.55 (B) using polarized light microscopy are displayed in Figure 2.

Factorial ANOVA revealed a significant interaction between studied factors: dentin caries-like lesion model and remineralization treatment ($p < 0.001$). In addition, there was a statistically significant difference concerning artificial dentin caries-like lesion model ($p < 0.001$), and also between treatments ($p < 0.001$).

As shown in Table 2, a Tukey test (R Software version 3.4.3, The R Foundation for Statistical Computing, Vienna, Austria) revealed that μTBS values of NaF to demineralized dentin for both, chemical and biological dentin caries-like lesion models, were significantly lower than other remineralizing agents ($p < 0.001$). Biological DCLL model significantly reduced the microtensile bond strength when the dentin was treated by NaF and Curodont™ Repair ($p < 0.001$). Chemical DCLL model provided higher μTBS than the biological one, when demineralized dentin was treated by NaF and Curodont Repair ($p < 0.05$). In addition, dentin demineralized by the chemical DCLL treated with Curodont™ Repair provided the highest bond strength, and there was no significant difference from MI Paste™ for the same dentin condition, and they were significantly higher than sound dentin

($p < 0.05$). However, when DCLL was treated with MI Paste™, there was no influence regardless of the DCLL model ($p > 0.05$) and they were significantly higher than sound dentin ($p < 0.05$). However, only when NaF was used there was an observed lower µTBS than sound dentin ($p < 0.05$). For both DCLL models, demineralized dentin treated with NaF, MI Paste™ and Curodont™ Repair showed significant higher µTBS than demineralized dentin ($p < 0.01$). Demineralized dentin group showed the lowest bond strength for all groups ($p < 0.01$) (Table 2).

Table 2. Average and standard deviation of µTBS of demineralized dentin considering the Artificial Caries Development Models.

Experimental Groups	Artificial Caries Development Models	
	Chemical Model	Biological Model
Sound Dentin	43.32 ± 4.35	
Demineralized Dentin	21.96 ± 5.92 Ca *	22.89 ± 2.68 Da *
Demineralized Dentin + NaF	33.43 ± 10.42 Ba *	26.94 ± 6.70 Cb *
Demineralized Dentin + MI Paste™ (CPP-ACP)	45.25 ± 8.83 Aa *	47.95 ± 6.69 Aa *
Demineralized Dentin + Curodont™ Repair (P₁₁-4)	46.42 ± 12.03 Aa *	42.07 ± 7.83 Bb

Uppercase letters represent statistically significant difference in the column ($p < 0.001$). Lowercase letters represent no statistically significant difference in the row ($p > 0.05$). * indicates statistically significant difference with the control group (sound dentin) ($p < 0.05$) by additional Dunnett's test.

The failure modes of specimens are shown in Figure 3. The failure modes of DDC (76%) and DDB (85%) specimens were predominantly adhesive failure; Mixed failure were found for DDC-NaF (84%), DDB-NaF (61%), DDC-MI Paste™ (54%), DDB-MI Paste™ (68%), DDC-Curodont™ Repair (71%) and DDB-Curodont™ Repair (70%); cohesive failure in composite resin was observed in the DDB-NaF (4%); cohesive failure in dentin was observed in the DDC-MI Paste™ (6%) DDB-MI Paste™ (2%) and DDB-Curodont™ Repair (5%) groups. The failure patterns were often adhesive and mixed for all groups. There was no statistically significant difference between the fracture type by Kruskal-Wallis's test, concerning DCLL model ($p = 0.9967$).

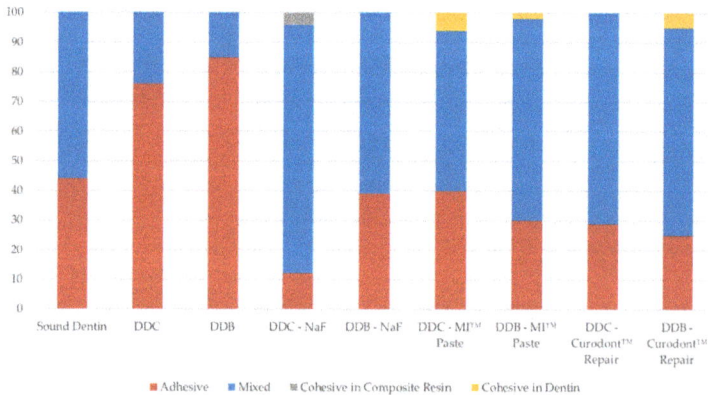

Figure 3. Distribution of failure modes. DDC—Demineralized Dentin by chemical model; DDB—Demineralized Dentin by biological model; NaF—Sodium Floride; MI Paste™-CPP-ACP—Casein phospopeptide-amorphous calcium phosphate; Curodont™ Repair—P₁₁-4—Peptide self-assembly.

Figure 4 shows representative SEMs for the fracture patterns observed for the different groups. Figure 3A—Cohesive failure on composite; 3B—Adhesive failure; 3C—Mixed failure; and 3D—cohesive failure on dentin.

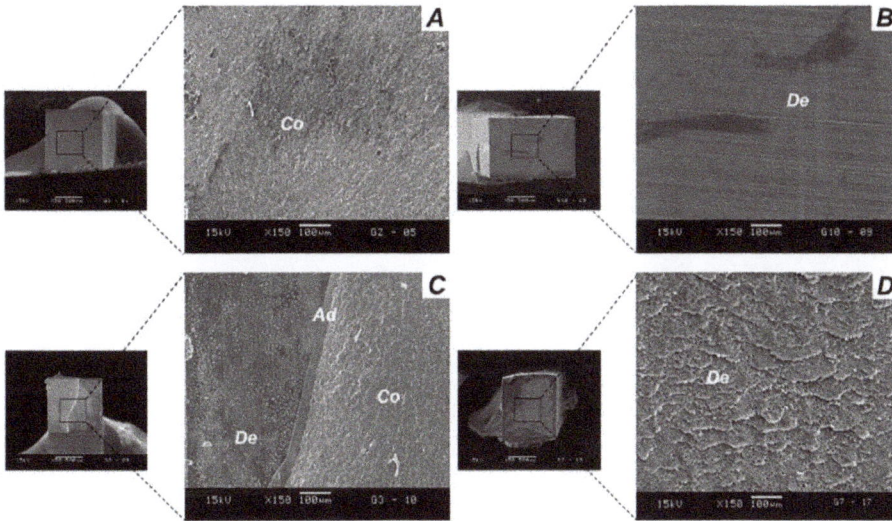

Figure 4. SEM image of failure modes. (**A**) Cohesive failure in Resin Composite; (**B**) Adhesive failure; (**C**) Mixed failure and (**D**) Cohesive failure in Dentin. Abbreviations shows areas of **Co**. Composite; **Ad**. Adhesive; **De**. Dentin.

4. Discussion

Considering the studies of bond strength on resin/dentin interfaces, the quality of the dentin substrate can play a key role in the longevity of the bonding [24,27–29]. In the present study, we have evaluated the influence of artificial caries development models (chemical and biological) and substrate conditions on resin/dentin bond strength. In this study, the first null hypothesis that there is a difference between DCLL models producing was proved, since there was no significant influence of the DCLL model on µTBS of demineralized dentin. However, the second hypothesis that there is influence of DCLL model and demineralized dentin treatment on µTBS of an etch and rinse adhesive system was rejected, since the µTBS values were dependent of DCLL model and mineralizing agent type. The highest µTBS were found when demineralized dentin was treated with for Curodont™ Repair and NaF ($p < 0.001$), although there was no significant difference on µTBS when dentin was treated with MI Paste ($p > 0.05$).

This study corroborates previous investigations demonstrating that the bonding procedures on demineralized dentin [30–32] present lower µTBS when compared to a sound one, regardless of the artificial caries development model (Table 2). Morphological changes in the substrate provided by caries production process can induce a decreased µTBS [20,33]. This reduction can be associated with changes in physical and chemical properties of the demineralized substrate when compared to sound dentin [34]. Demineralized dentin provides a high porosity in the inter-tubular dentin, exposure of collagen fibers along with decrease in mineral content [35] and partial penetration of resin monomers and a non-homogeneous hybrid layer [36]. In a porous hybrid layer, over time, mineral and organic matrix would be degraded giving rise to gaps which may be visible using a SEM which show a higher rate of degradation [36]. In addition, the demineralization of dentin surface results in a more hydrophobic surface, avoiding the wettability of the adhesive [17].

The caries lesion provided by a biological model, which uses *S. mutans* biofilm, seems to be quite similar to the natural ones, based on molecular and structural evaluations [23]. Another model used for providing dentin caries-like lesions is the chemical one, and it can be used to simulate caries-affected dentin [5].

It is desirable that bonding between mineralized tooth tissues, such as dentin, and the restorative materials must be sufficiently effective to resist varied challenges, such as biofilm attack, hydrolytic and enzymatic degradation, thermal and mechanical stress from repeated loading over many months or years [37,38]. The reinforcement of demineralized collagen matrix can be achieved using remineralizing agents [10,39]. The current biomimetic remineralization approach provides a proof-of-concept that utilizes nanotechnological principles to mimic natural biomineralization, extending the longevity of resin–dentin bonds [10]. The mineral reinforcement of collagen matrix found in demineralized dentin appears to be a strategy that restores conditions found on sound dentin, as seen in the present study. In the present study the biomimetic remineralization strategy provided a higher (CPP-ACP + Chemical and Biological DCCL models; P_{11}-4 + Biological DCCL) or similar (P_{11}-4) µTBS to demineralized dentin than sound one, while the NaF remineralization provided higher µTBS values than demineralized dentin, but lower than sound one. It can be observed that only for CCP-ACP, regardless DCLL model, the mineralizing agent provided a higher µTBS than sound dentin.

Despite it, concerning affected dentin, Bahari et al. (2014) [40] showed that 5 consecutive days of CPP-ACP application for 15 min did not have any significant effect on µTBS of SB to demineralized dentin. However, it has been considered that the methodology used in that study was quite different from that used in the present one. Firstly, according to the methodology description (Bahari et al, 2014) even sound dentin was submitted to the CPP-ACP action. Casein phosphopeptides (CPP) have been described to bind amorphous calcium phosphate, forming nano-complexes of casein phosphopeptide–amorphous calcium phosphate (CPP-ACP), thereby stabilizing in calcium phosphates [41,42]. Calcium and phosphate ions can easily diffuse into the porous lesion and deposits in the partially demineralized crystals and rebuild hydroxy-apatite crystals [43]. This further substantiates the theory that CPP-ACP is considered a biomaterial [44]. The presence of bioavailable calcium and phosphate present the MI Paste™ can maintain a supersaturated state in dental substrate [11]. Studies have demonstrated that CPP-ACP could reduce demineralization and increase remineralization of dentin [44,45]. It is possible that in that study the CPP-ACP could have been impregnated onto caries-affected dentin and sound dentin [13,46] in the same way.

It is well-established that the collagen matrix serves as a scaffold for crystal deposition but does not provide a mechanism for orderly nucleation of hydroxyapatite [39]. The results of the present study can be attributed to the ability of CPP-ACP to increase deposition of crystals on the dentin surface [47]. Furthermore, the CPP also has the capacity to stabilize nano-ACP [48]. Therefore, the deposition and stabilization may result in a restructuring of the characteristics found in sound dentin, showing the highest µTBS when demineralized dentin was treated with MI Paste. This reinforcement approach of demineralized collagen matrix structure may favor the bonding procedure. Further studies should be carried out to verify the stability of the bonding strength when CPP-ACP is used.

It is generally believed that extracellular matrix proteins, which play an important role in controlling apatite nucleation and growth in the dentin remineralization process [49], mediate a biomineralization process. Biomimetic remineralization represents a different approach to this by attempting to backfill the demineralized dentin collagen with liquid-like ACP nanoprecursor particles that are stabilized by biomimetic analogs of noncollagenous proteins [10,50]. In this way, maybe this particular nucleation would provide a regular and feasible restructuration of the demineralized dentin and also would provide a favorable substrate for bonding, due to the more hydrophyllic nature of the substrate [23].

Another interesting finding of the present study was the fact that the µTBS means of MI Paste™ group was higher than those found in sound dentin, and did not show a significant difference between either artificial caries development models. The remineralization process and the artificial caries development model, studied in this article, showed that the artificial caries development model affects only the µTBS of the demineralized dentin treated with NaF and P_{11}-4.

Similar to CCP-ACP, the P_{11}-4 approach using the DCLL chemical model provided significantly higher µTBS than sound dentin. The chemical model of DCLL provided a lower content of type I collagen and higher content of calcium and phosphate ions, than the Biological one [23]. The collagen is the precursor for mineralization, acting as a scaffold for mineral aggregation. However, when NaF is used, only a deposition of ions and CaF formation occurs on dentin surface. Possibly the high content of mineral ions decreased the surface energy of the demineralized dentin [23]. The opposite can be observed when biomimetic remineralization happens, using CPP-ACP or P_{11}-4. In this case, the mineralization occurs by organized crystal formation guided by the scaffold. This kind of surface can experience a high surface energy and a high level of wettability by resin monomers providing the highest µTBS [23].

The results of the study indicated that treatment with P_{11}-4, with a single application showed significant improvement on µTBS. Despite the fact that there was significant difference between the DCLL models, when P_{11}-4 is used, the µTBS values still show high values. For the biological model when the peptide P_{11}-4 was used no significant difference from sound dentin was observed. It is suggested that the use of the P_{11}-4 is able to nucleate hydroxyapatite and to promote repair of caries-like lesions in vitro. We have no knowledge in the literature of any report in which treatment with P_{11}-4 has been conducted in demineralized dentin, associating its effects with bonding procedures. However, research groups using other peptides, observed that this strategy mimics the functions of non-collagenous proteins (NCPs) [51,52]. This suggests that the action of the P_{11}-4 peptide reflects the reinforcing of demineralized collagen matrix.

The potential for enamel lesion repair of P_{11}-4 may mimic the functions of NPCs [15,53,54]. Several studies indicate that P_{11}-4 forms three dimensional fibrillar hierarchical structures resulting in gels in response to specific environmental triggers [53,54]. Assembled P_{11}-4 forms scaffold-like structures with negative charge domains, mirroring biological macromolecules in mineralized tissue extracellular matrices (ECM) [54].

Fluoride is well-known for its proved anti-cariogenic and antimicrobial capacity. Its ability to prevent demineralization and promote remineralization by calcium phosphate precipitation on dental surface by reducing the dissolution of hydroxyapatite [55]. Thus, the effect of demineralized dentin treated with NaF on the µTBS, improved the µTBS compared to demineralized dentin, but did not reach the sound dentin µTBS. It has to be considered that the dentin etching with 35% phosphoric acid increases the bonding efficacy of dental adhesives and removes the smear layer and the superficial part of the dentin, opening dentin tubules, demineralizing the dentin surface and increasing the microporosity of the intertubular dentin [56]. Despite the phosphoric acid used in the etching procedure removing part of the mineral deposits, possibly the reinforcement provided by NaF on demineralized dentin structure would improve the µTBS compared to demineralized dentin [57,58]. However, the differences found with µTBS of Chemical and Biological models can be explained by the deeper demineralization provided by biological model than the chemical one. Therefore, it is known that the mineral deposition of fluorides occurs on surface and generally causes hyper mineralization of the dentin and in dentin tubules [10,59–61]. The disorganized precipitation and deposit of mineral on dentin may mechanically obliterate the tubules reducing the performance of the restorative material [61], providing less efficiency of NaF treatment.

Another important aspect is that the NaF and Curodont Repair treatments showed a statistically significant difference between the DCLL. With regard to the treatment with NaF, a previous study [62] showed that this treatment is able to reduce the subsurface dentin demineralization compared with the control from 30 to 50 µm depth. At the other depths (60–220 µm) NaF showed no positive effect. Such results may be attributed to a possible reaction between NaF and demineralized dentin, producing soluble fluoride. In the present study, the demineralized dentin associated with NaF treatment exhibit mixed failure type. This result may be associated with a surface reinforcement, which was partially removed by the etching with the phosphoric acid during bonding procedures.

With the exception of the treatment with NaF, demineralized dentin treated with Curodont Repair has shown no negative influence on bond strength, as the biological DCLL did not differ from sound dentin. This behavior can be attributed to different interactions with the demineralized substrate, since the caries depth produced by the biological model DCLL was higher than chemical one. It has been reported that the P_{11}-4 scaffold can act as nucleator for hydroxyapatite, infiltrating into the porous lesions and increase the mineral diffusion within the lesion, restructuring the affected tissue [15].

Moreover, regardless of the artificial caries development model, the results presented in this study suggest that the use of remineralizing agents can reinforce the mechanical properties of demineralized dentin and would favor the durability of resin-dentin bonds, since the treated demineralized substrate provides an organized mineral surface. However, the degree of improvement in bonding strength is dependent on the artificial caries development model and dentin treatment. Further studies have to be carried out in order to observe the long-term efficacy of the remineralized dentin bonded to adhesive systems.

5. Conclusions

Based on the results of this study it can be concluded that:

MI Paste™ and Curodont™ Repair recovered higher µTBS values when compared to sound dentin. Each agent shows a different interaction with each artificial caries development model used. The remineralizing treatment of demineralized dentin is a potential approach for increasing bond strength of etch and rinse adhesive systems.

Author Contributions: Conceptualization, L.F.B.-M. and R.M.P.-R.; Data curation, L.F.B.-M.; Formal analysis, L.F.B.-M., J.P.d.S. and R.M.P.-R.; Funding acquisition, R.M.P.-R.; Investigation, L.F.B.-M. and R.M.P.-R.; Methodology, L.F.B.-M., J.P.d.S., L.A.A. and R.M.P.-R.; Project administration, L.F.B.-M., J.P.d.S. and R.M.P.-R.; Resources, R.M.P.-R.; Supervision, R.P.W.D. and R.M.P.-R.; Validation, L.F.B.-M. and J.P.d.S.; Visualization, L.F.B.-M., J.P.d.S., L.A.A. and R.M.P.-R.; Writing: original draft, L.F.B.-M., J.P.d.S., L.A.A., R.P.W.D. and Regina M.P.-R.; Writing: review and editing, L.F.B.-M., J.P.d.S., L.A.A., R.P.W.D. and R.M.P.-R.

Funding: This work was supported by grants of the Cordination for the Improvement of Higher Education Personnel (CAPES) and São Paulo Research Foundation (FAPESP) (2011/16634-3, 2015/12660-0) by the research financial support.

Acknowledgments: The authors are grateful to Credentis (AG, Dorfstrasse, Windisch, Switzerland) for providing one of the materials tested.

Conflicts of Interest: The authors declare no conflict of interest.

References

1. Nakabayashi, N.; Kojima, K.; Masuhara, E. The promotion of adhesion by the infiltration of monomers into tooth substrates. *J. Biomed. Mater. Res.* **1982**, *16*, 265–273. [CrossRef] [PubMed]

2. Van, M.B.; Perdigão, J.; Lambrechts, P.; Vanherle, G. The clinical performance of adhesives. *J. Dent.* **1998**, *26*, 1–20.

3. Kassebaum, N.J.; Bernabé, E.; Dahiya, M.; Bhandari, B.; Murray, C.J.; Marcenes, W. Global burden of untreated caries: A systematic review and metaregression. *J. Dent. Res.* **2015**, *94*, 650–658. [CrossRef] [PubMed]

4. Spencer, P.; Wang, Y. Adhesive phase separation at the dentin interface under wet bonding conditions. *J. Biomed. Mater. Res.* **2002**, *62*, 447–456. [CrossRef] [PubMed]

5. de Carvalho, F.G.; de Fucio, S.B.; Sinhoreti, M.A.; Correr-Sobrinho, L.; Puppin-Rontani, R.M. Confocal laser scanning microscopic analysis of the depth of dentin caries-like lesions in primary and permanent teeth. *Braz. Dent. J.* **2008**, *19*, 139–144. [CrossRef] [PubMed]

6. Perdigão, J.; Reis, A.; Loguercio, A.D. Dentin adhesion and MMPs: A comprehensive review. *J. Esthet. Restor. Dent.* **2013**, *25*, 219–241. [CrossRef] [PubMed]

7. Tjäderhane, L. Dentin bonding: Can we make it last? *Oper. Dent.* **2015**, *40*, 4–18. [CrossRef] [PubMed]

8. Nakajima, M.; Kunawarote, S.; Prasansuttiporn, T.; Tagami, J. Bonding to caries-affected dentin. *Jpn. Dent. Sci. Rev.* **2011**, *47*, 102–114. [CrossRef]

9. Niu, L.N.; Zhang, W.; Pashley, D.H.; Breschi, L.; Mao, J.; Chen, J.H.; Tay, F.R. Biomimetic remineralization of dentin. *Dent. Mater.* **2014**, *30*, 77–96. [CrossRef] [PubMed]

10. Tay, F.R.; Pashley, D.H. Biomimetic remineralization of resin-bonded acid-etched dentin. *J. Dent. Res.* **2009**, *88*, 719–724. [CrossRef] [PubMed]

11. Reynolds, E.C. Calcium phosphate-based remineralization systems: Scientific evidence? *Aust. Dent. J.* **2008**, *53*, 268–273. [CrossRef] [PubMed]

12. Brackett, M.G.; Agee, K.A.; Brackett, W.W.; Key, W.O.; Sabatini, C.; Kato, M.T.; Buzalaf, M.A.; Tjäderhane, L.; Pashley, D.H. Effect of Sodium Fluoride on the endogenous MMP Activity of Dentin Matrices. *J. Nat. Sci.* **2015**, *1*, 1–11.

13. Borges, B.C.; Souza-Junior, E.J.; da Costa Gde, F.; Pinheiro, I.V.; Sinhoreti, M.A.; Braz, R.; Montes, M.A. Effect of dentin pre-treatment with a casein phosphopeptide-amorphous calcium phosphate (CPP-ACP) paste on dentin bond strength in tridimensional cavities. *Acta Odontol. Scand.* **2013**, *71*, 271–277. [CrossRef] [PubMed]

14. Shafiei, F.; Derafshi, R.; Memarpour, M. Bond strength of self-adhering materials: Effect of dentin-desensitizing treatment with a CPP-ACP paste. *Int. J. Periodontics Restor. Dent.* **2017**, *37*, 337–343. [CrossRef] [PubMed]

15. Brunton, P.A.; Davies, R.P.; Burke, J.L.; Smith, A.; Aggeli, A.; Brookes, S.J.; Kirkham, J. Treatment of early caries lesions using biomimetic self-assembling peptides—A clinical safety trial. *Br. Dent. J.* **2013**, *215*, 1–6. [CrossRef] [PubMed]

16. Schlee, M.; Schad, T.; Koch, J.H.; Cattin, P.C.; Rathe, F. Clinical performance of self-assembling peptide P_{11}-4 in the treatment of initial proximal carious lesions: A practice-based case series. *J. Investig. Clin. Dent.* **2018**, *9*, e12286. [CrossRef] [PubMed]

17. Barbosa-Martins, L.F.; de Sousa, J.P.; de Castilho, A.R.F.; Puppin-Rontani, J.; Davies, R.P.W.; Puppin-Rontani, R.M. Enhancing bond strength on demineralized dentin by pre-treatment with selective remineralising agents. *J. Mech. Behav. Biomed. Mater.* **2018**, *81*, 214–221. [CrossRef] [PubMed]

18. Erhardt, M.C.; Rodrigues, J.A.; Valentino, T.A.; Ritter, A.V.; Pimenta, L.A. In vitro µTBS of one-bottle adhesive systems: Sound versus artificially-created caries-affected dentin. *J. Biomed. Mater. Res. B Appl. Biomater.* **2008**, *86*, 181–187. [CrossRef] [PubMed]

19. Marquezan, M.; Corrêa, F.N.; Sanabe, M.E.; Rodrigues Filho, L.E.; Hebling, J.; Guedes-Pinto, A.C.; Mendes, F.M. Artificial methods of dentine caries induction: A hardness and morphological comparative study. *Arch. Oral Biol.* **2009**, *54*, 1111–1117. [CrossRef] [PubMed]

20. Zanchi, C.H.; Lund, R.G.; Perrone, L.R.; Ribeiro, G.A.; Del Pino, F.A.; Pinto, M.B.; Demarco, F.F. Microtensile bond strength of two-step etch-and-rinse adhesive systems on sound and artificial caries-affected dentin. *Am. J. Dent.* **2010**, *23*, 152–156. [PubMed]

21. Sanabe, M.E.; de Souza Costa, C.A.; Hebling, J. Exposed collagen in aged resin–dentin bonds produced on sound and caries-affected dentin in the presence of chlorhexidine. *J. Adhes. Dent.* **2011**, *13*, 117–124. [PubMed]

22. Joves, G.J.; Inoue, G.; Nakashima, S.; Sadr, A.; Nikaido, T.; Tagami, J. Mineral density, morphology and bond strength of natural versus artificial caries-affected dentin. *Dent. Mater. J.* **2013**, *32*, 138–143. [CrossRef] [PubMed]

23. Pacheco, L.F.; Banzi, É.; Rodrigues, E.; Soares, L.E.; Pascon, F.M.; Correr-Sobrinho, L.; Puppin-Rontani, R.M. Molecular and structural evaluation of dentin caries-like lesions produced by different artificial models. *Braz. Dent. J.* **2013**, *24*, 610–618. [CrossRef] [PubMed]

24. Zancopé, B.R.; Rodrigues, L.P.; Parisotto, T.M.; Steiner-Oliveira, C.; Rodrigues, L.K.A.; Nobre-dos-Santos, M. CO_2 laser irradiation enhances CaF_2 formation and inhibits lesion progression on demineralized dental enamel—In vitro study. *Lasers Med. Sci.* **2016**, *31*, 539–547. [CrossRef] [PubMed]

25. Armstrong, S.; Breschi, L.; Özcan, M.; Pfefferkorn, F.; Ferrari, M.; Van Meerbeek, B. Academy of Dental Materials guidance on in vitro testing of dental composite bonding effectiveness to dentin/enamel using micro-tensile bond strength (µTBS) approach. *Dent. Mater.* **2017**, *33*, 133–143. [CrossRef] [PubMed]

26. Bacchi, A.; Abuna, G.; Babbar, A.; Sinhoreti, M.A.C.; Feitosa, V.P. Influence of 3-month simulated pulpal pressure on the microtensile bond strength of simplified resin luting systems. *J. Adhes. Dent.* **2015**, *17*, 265–271. [PubMed]

27. Peixoto, A.C.; Bicalho, A.A.; Isolan, C.P.; Maske, T.T.; Moraes, R.R.; Cenci, M.S.; Soares, C.J.; Faria-e-Silva, A.L. Bonding of adhesive luting agents to caries-affected dentin induced by a microcosm biofilm model. *Oper. Dent.* **2015**, *40*, 102–111. [CrossRef] [PubMed]

28. Qi, Y.P.; Li, N.; Niu, L.N.; Primus, C.M.; Ling, J.Q.; Pashley, D.H.; Tay, F.R. Remineralization of artificial dentinal caries lesions by biomimetically modified mineral trioxide aggregate. *Acta Biomater.* **2012**, *8*, 836–842. [CrossRef] [PubMed]

29. Robinson, C.; Brookes, S.J.; Kirkham, J.; Wood, S.R.; Shore, R.C. In vitro studies of the penetration of adhesive resins into artificial caries-like lesions. *Caries Res.* **2001**, *35*, 136–141. [CrossRef] [PubMed]

30. Lenzi, T.L.; Calvo, A.F.; Tedesco, T.K.; Ricci, H.A.; Hebling, J.; Raggio, D.P. Effect of method of caries induction on aged resin–dentin bond of primary teeth. *BMC Oral Health* **2015**, *15*, 79. [CrossRef] [PubMed]

31. Yoshiyama, M.; Doi, J.; Nishitani, Y.; Itota, T.; Tay, F.R.; Carvalho, R.M.; Pashley, D.H. Bonding ability of adhesive resins to caries-affected and caries-infected dentin. *J. Appl. Oral Sci.* **2004**, *12*, 171–176. [CrossRef] [PubMed]

32. Doozandeh, M.; Firouzmandi, M.; Mirmohammadi, M. The simultaneous effect of extended etching time and casein phosphopeptide-amorphous calcium phosphate containing paste application on shear bond strength of etch-and-rinse adhesive to caries-affected dentin. *J. Contemp. Dent. Pract.* **2015**, *16*, 794–799. [CrossRef] [PubMed]

33. Ceballos, L.; Camejo, D.G.; Victoria Fuentes, M.; Osorio, R.; Toledano, M.; Carvalho, R.M.; Pashley, D.H. Microtensile bond strength of total-etch and self-etching adhesives to caries-affected dentine. *J. Dent.* **2003**, *31*, 469–477. [CrossRef]

34. Fusayama, T.; Kurosaki, N. Structure and removal of carious dentin. *Int. Dent. J.* **1972**, *22*, 401–411. [PubMed]

35. Wang, S.; Huang, C.; Zheng, T.L.; Zhang, Z.X.; Wang, Y.N.; Cheng, X.R. Microtensile bond strength and morphological evaluations of total-etch and self-etch adhesives to caries-affected dentin. *Zhonghua Kou Qiang Yi Xue Za Zhi* **2006**, *41*, 323–326. [PubMed]

36. Sano, H.; Yoshiyama, M.; Ebisu, S.; Burrow, M.F.; Takatsu, T.; Ciucchi, B.; Carvalho, R.; Pashley, D.H. Comparative SEM and TEM observations of nanoleakage within the hybrid layer. *Oper. Dent.* **1995**, *20*, 160–167. [PubMed]

37. Stanislawczuk, R.; Pereira, F.; Muñoz, M.A.; Luque, I.; Farago, P.V.; Reis, A.; Loguercio, A.D. Effects of chlorhexidine-containing adhesives on the durability of resin–dentine interfaces. *J. Dent.* **2014**, *42*, 39–47. [CrossRef] [PubMed]

38. Nassar, M.; Hiraishi, N.; Shimokawa, H.; Tamura, Y.; Otsuki, M.; Kasugai, S.; Ohya, K.; Tagami, J. The inhibition effect of non-protein thiols on dentinal matrix metalloproteinase activity and HEMA cytotoxicity. *J. Dent.* **2014**, *42*, 312–318. [CrossRef] [PubMed]

39. Liu, Y.; Kim, Y.K.; Dai, L.; Li, N.; Khan, S.O.; Pashley, D.H.; Tay, F.R. Hierarchical and non-hierarchical mineralisation of collagen. *Biomaterials* **2011**, *32*, 1291–1300. [CrossRef] [PubMed]

40. Bahari, M.; Savadi Oskoee, S.; Kimyai, S.; Pouralibaba, F.; Farhadi, F.; Norouzi, M. Effect of casein phosphopeptide-amorphous calcium phosphate treatment on microtensile bond strength to carious affected dentin using two adhesive strategies. *J. Dent. Res. Dent. Clin. Dent. Prospects* **2014**, *8*, 141–147. [PubMed]

41. Reynolds, E.C. Anticariogenic complexes of amorphous calcium phosphate stabilized by casein phosphopeptides: A review. *Spec. Care Dent.* **1998**, *18*, 8–16. [CrossRef]

42. Cross, K.J.; Huq, N.L.; Reynolds, E.C. Casein phosphopeptides in oral health–chemistry and clinical applications. *Curr. Pharm. Des.* **2007**, *13*, 793–800. [CrossRef] [PubMed]

43. Kumar, V.L.; Itthagarun, A.; King, N.M. The effect of casein phosphopeptide-amorphous calcium phosphate on remineralization of artificial caries-like lesions: An in vitro study. *Aust. Dent. J.* **2008**, *53*, 34–40. [CrossRef] [PubMed]

44. Cao, Y.; Mei, M.L.; Xu, J.; Lo, E.C.; Li, Q.; Chu, C.H. Biomimetic mineralisation of phosphorylated dentine by CPP-ACP. *J. Dent.* **2013**, *41*, 818–825. [CrossRef] [PubMed]

45. Rahiotis, C.; Vougiouklakis, G. Effect of a CPP-ACP agent on the demineralization and remineralization of dentine in vitro. *J. Dent.* **2007**, *35*, 695–698. [CrossRef] [PubMed]

46. Sattabanasuk, V.; Burrow, M.F.; Shimada, Y.; Tagami, J. Resin bonding to dentine after casein phosphopeptide-amorphous calcium phosphate (CPP-ACP) treatments. *J. Adhes. Sci. Technol.* **2009**, *23*, 1149–1161. [CrossRef]

47. Adebayo, O.A.; Burrow, M.F.; Tyas, M.J. Resin-dentine interfacial morphology following CPP-ACP treatment. *J. Dent.* **2010**, *38*, 96–105. [CrossRef] [PubMed]

48. Rahiotis, C.; Vougiouklakis, G.; Eliades, G. Characterization of oral films formed in the presence of a CPP-ACP agent: An in situ study. *J. Dent.* **2008**, *36*, 272–280. [CrossRef] [PubMed]

49. Jee, S.S.; Thula, T.T.; Gower, L.B. Development of bone-like composites via the polymer-induced liquid-precursor (PILP) process. Part 1: Influence of polymer molecular weight. *Acta Biomater.* **2010**, *6*, 3676–3686. [CrossRef] [PubMed]

50. Kim, J.; Gu, L.; Breschi, L.; Tjäderhane, L.; Choi, K.K.; Pashley, D.H.; Tay, F.R. Implication of ethanol wet-bonding in hybrid layer remineralization. *J. Dent. Res.* **2010**, *89*, 575–580. [CrossRef] [PubMed]

51. Wang, D.Y.; Zhang, L.; Chen, J.H. The role of dentinal matrix metalloproteinases in collagenous degeneration of tooth tissue. *Zhonghua Kou Qiang Yi Xue Za Zhi* **2011**, *46*, 379–381. [PubMed]

52. Cao, Y.; Liu, W.; Ning, T.; Mei, M.L.; Li, Q.L.; Lo, E.C.; Chu, C.H. A novel oligopeptide simulating dentine matrix protein 1 for biomimetic mineralization of dentine. *Clin. Oral. Investig.* **2014**, *18*, 873–881. [CrossRef] [PubMed]

53. Aggeli, A.; Bell, M.; Boden, N.; Carrick, L.M.; Strong, A.E. Self-assembling peptide polyelectrolyte beta-sheet complexes form nematic hydrogels. *Angew. Chem.* **2003**, *42*, 5761–5764. [CrossRef]

54. Kirkham, J.; Firth, A.; Vernals, D.; Boden, N.; Robinson, C.; Shore, R.C.; Brookes, S.J.; Aggeli, A. Self-assembling peptide scaffolds promote enamel remineralization. *J. Dent. Res.* **2007**, *86*, 426–430. [CrossRef] [PubMed]

55. ten Cate, J.M. Remineralization of caries lesions extending into dentin. *J. Dent. Res.* **2001**, *80*, 1407–1411. [CrossRef] [PubMed]

56. Marshall, G.W., Jr.; Balooch, M.; Kinney, J.H.; Marshall, S.J. Atomic force microscopy of conditioning agents on dentin. *J. Biomed. Mater. Res.* **1995**, *29*, 1381–1387. [CrossRef] [PubMed]

57. Kato, M.T.; Bolanho, A.; Zarella, B.L.; Salo, T.; Tjäderhane, L.; Buzalaf, M.A. Sodium fluoride inhibits MMP-2 and MMP-9. *J. Dent. Res.* **2014**, *93*, 74–77. [CrossRef] [PubMed]

58. Altinci, P.; Mutluay, M.; Seseogullari-Dirihan, R.; Pashley, D.; Tjäderhane, L.; Tezvergil-Mutluay, A. NaF inhibits matrix-bound cathepsin-mediated dentin matrix degradation. *Caries Res.* **2016**, *50*, 124–132. [CrossRef] [PubMed]

59. Arends, J.; Christoffersen, J. Nature and role of loosely bound fluoride in dental caries. *J. Dent. Res.* **1990**, *69*, 601–605. [CrossRef] [PubMed]

60. Arends, J.; Christoffersen, J.; Ruben, J.; Jongebloed, W.L. Remineralization of bovine dentine in vitro. The influence of the F content in solution on mineral distribution. *Caries Res.* **1989**, *23*, 309–314. [CrossRef] [PubMed]

61. Prabhakar, A.R.; Manojkumar, A.J.; Basappa, N. In vitro remineralization of enamel subsurface lesions and assessment of dentine tubule occlusion from NaF dentifrices with and without calcium. *J. Indian Soc. Pedodontics Prev. Dent.* **2013**, *31*, 29–35. [CrossRef] [PubMed]

62. Comar, L.P.; Souza, B.M.; Gracindo, L.F.; Buzalaf, M.A.; Magalhães, A.C. Impact of experimental nano-HAP pastes on bovine enamel and dentin submitted to a pH cycling model. *Braz. Dent. J.* **2013**, *24*, 273–278. [CrossRef] [PubMed]

materials

MDPI

Article

Tuning Nano-Amorphous Calcium Phosphate Content in Novel Rechargeable Antibacterial Dental Sealant

Maria Salem Ibrahim [1,2], Faisal D. AlQarni [3], Yousif A. Al-Dulaijan [3], Michael D. Weir [4], Thomas W. Oates [4], Hockin H. K. Xu [4,5,6] and Mary Anne S. Melo [4,*]

[1]   Program in Dental Biomedical Sciences, University of Maryland School of Dentistry, Baltimore, MD 21201, USA; msyibrahim@gmail.com
[2]   Department of Preventive Dental Sciences, College of Dentistry, Imam Abdulrahman bin Faisal University, Dammam 34212, Saudi Arabia
[3]   Department of Substitutive Dental Sciences, College of Dentistry, Imam Abdulrahman bin Faisal University, Dammam 34212, Saudi Arabia; falqarni@icloud.com (F.D.A.); yaldulaijan@gmail.com (Y.A.A-D.)
[4]   Department of Advanced Oral Sciences and Therapeutics, University of Maryland School of Dentistry, Baltimore, MD 21201, USA; MWeir@umaryland.edu (M.D.W.); TOates@umaryland.edu (T.W.O.); HXu@umaryland.edu (H.H.K.X.)
[5]   Center for Stem Cell Biology & Regenerative Medicine, University of Maryland School of Medicine, Baltimore, MD 21201, USA
[6]   Marlene and Stewart Greenebaum Cancer Center, University of Maryland School of Medicine, Baltimore, MD 21201, USA
*   Correspondence: mmelo@umaryland.edu; Tel.: +1-410-706-8705

Received: 19 July 2018; Accepted: 20 August 2018; Published: 27 August 2018

Abstract: Dental sealants with antibacterial and remineralizing properties are promising for caries prevention among children and adolescents. The application of nanotechnology and polymer development have enabled nanoparticles of amorphous calcium phosphate (NACP) and dimethylaminohexadecyl methacrylate (DMAHDM) to emerge as anti-caries strategies via resin-based dental materials. Our objectives in this study were to (1) incorporate different mass fractions of NACP into a parental rechargeable and antibacterial sealant; (2) investigate the effects on mechanical performance, and (3) assess how the variations in NACP concentration would affect the calcium (Ca) and phosphate (PO_4) ion release and re-chargeability over time. NACP were synthesized using a spray-drying technique and incorporated at mass fractions of 0, 10, 20 and 30%. Flexural strength, flexural modulus, and flowability were assessed for mechanical and physical performance. Ca and PO_4 ion release were measured over 70 days, and three ion recharging cycles were performed for re-chargeability. The impact of the loading percentage of NACP upon the sealant's performance was evaluated, and the optimized formulation was eventually selected. The experimental sealant at 20% NACP had flexural strength and flexural modulus of 79.5 ± 8.4 MPa and 4.2 ± 0.4 GPa, respectively, while the flexural strength and flexural modulus of a commercial sealant control were 70.7 ± 5.5 MPa ($p > 0.05$) and 3.3 ± 0.5 GPa ($p < 0.05$), respectively. A significant reduction in flow was observed in the experimental sealant at 30% NACP ($p < 0.05$). Increasing the NACP mass fraction increased the ion release. The sealant formulation with NACP at 20% displayed desirable mechanical performance and ideal flow and handling properties, and also showed high levels of long-term Ca and PO_4 ion release and excellent recharge capabilities. The findings provide fundamental data for the development of a new generation of antibacterial and rechargeable Ca and PO_4 dental sealants to promote remineralization and inhibit caries.

Keywords: dental sealant; resin sealant; calcium phosphate nanoparticles; long-term ion release; remineralization; ion recharge

1. Introduction

Dental caries is still a highly prevalent oral disease worldwide despite the various approaches that have been used to prevent it [1,2]. These approaches include fluoride exposure, sugar intake control, brushing and dental sealants [3]. Dental sealants help prevent caries in pits and fissures of primary and permanent teeth, acting as a physical barrier for food accumulation and bacterial growth [4]. Accumulative evidence from epidemiologic findings has shown positive outcomes for caries prevention when the teeth are sealed, in comparison to non-sealed teeth in children and adolescents [4,5]. It was found that sealants on permanent molars may reduce dental caries for up to 24–48 months when compared to that of no sealant application [3,4]. In terms of sealants' retention, resin-based sealants are the materials with a higher success rate [6]. Even so, findings have shown an increase in dental resin-based sealants failures due to bacterial colonization under the restored fissures, thereby, initiating and progressing the carious lesion beneath the sealant [4].

Caries lesions at the sealed occlusal surfaces were initiated when the balance between the remineralization and demineralization of the tooth structure was adversely affected, and demineralization exceeded the remineralization abilities, resulting in the dissolution of hydroxyapatite crystals [7]. To allow the remineralization process to occur, adequate levels of calcium (Ca) and phosphate (PO_4) ions must be available [8]. Antibacterial and remineralizing resin-based sealants could be one of the most desirable approaches for management of dental caries in children and adolescents due to the potential of reduced bacteria and provide localized ion release near the tooth surface.

Recently, new fundamental research findings have highlighted the application of nanotechnology and polymer development in dental materials, enabling nanoparticles of amorphous calcium phosphate (NACP) and quaternary ammonium methacrylate such as dimethylaminohexadecyl methacrylate (DMAHDM) to emerge as anti-caries strategies via resin-based materials [9,10]. Amorphous calcium phosphate ($Ca_3[PO_4]_2$), as a precursor of the final crystalline hydroxyapatite, has been investigated as a suitable remineralizing agent. There is a growing body of evidence suggesting that NACP could enhance the remineralizing capacity due to a greater surface area-to-volume ratio [11,12]. NACP had a relatively high specific surface area of 17.76 m^2/g, compared to about 0.5 m^2/g of traditional micron-sized calcium phosphate particles used in dental resins [13,14]. Supported by the high performance of Ca and PO_4 ion release provided by a nanostructured compound, NACP has led to new possibilities for combating enamel demineralization [15].

Resin-based sealants with the ability to release Ca and PO_4 ions are expected to suppress the demineralization process and prevent dental caries. However, ion-depletion effect with loss of bioactivity over time (short-term ion release) has been a major drawback for calcium phosphate-containing resins [6,16]. Recently, significant levels of Ca and PO_4 ions were released from NACP-containing resins that were sustainable over long periods of time and with rechargeable capacity [17,18]. The rechargeable capability of NACP-containing dental materials have opened new horizons and are expected to lead to relevant changes in remineralizing approaches [19]. The repeatable recharge process to re-release Ca and PO_4 ions can lead to supersaturation into the surrounding microenvironments under acidic attacks, such as enamel areas located within deep occlusal pits and fissures with difficult access to brush. These ions can play a vital role in the precipitation of crystallites. This capability is highly desirable in a resin-based formulation for sealing the occlusal pits and fissures where high rates of demineralization happen and result in almost 50% of all caries in school children [2,4].

Many of the innovative bioactive strategies and technologies require new dental materials with new combinations of properties to meet the basic properties of the conventional polymeric materials [13]. This is true for resin-based formulations that are needed for dental sealant applications. For example, regarding the chemical and physical characteristics, the resin-based sealants must possess a high degree of wettability, and flow and viscosity that allow the penetration between the occlusal fissures and grooves of the teeth [5]. Another important characteristic is the resistance to abrasion and fracture, which would demonstrate the adequate mechanical performance of the material [6]. This is a challenging combination of characteristics when developing new resin-based formulations. Frequently, the incorporation of new agents in the resin-based system decreases the strength or bioactivity.

The present study reports the development of new antibacterial resin-based sealants that include NACP for Ca and PO_4 ion release and recharge properties. Our objectives were to (1) incorporate different mass fractions of NACP into parental rechargeable antibacterial sealant; (2) investigate the effects on mechanical performance; and (3) assess how the variations in NACP concentration would affect the Ca and PO_4 ions release and re-chargeability over time. It was hypothesized that adding an increased percentage of NACP would have acceptable mechanical and physical performances, while producing substantial initial ion release and a long-term repeated recharge capability.

2. Results

Illustration of the rechargeable NACP approach to dealing with the dissolution-diffusion process of enamel demineralization around dental sealants is shown schematically in Figure 1. The ion recharge cycle diagram displayed the ion re-release from the exhausted and recharged NACP-containing resin-based sealants in Figure 1A. Three recharge/re-release cycles were performed, and each re-release was measured for 14 days. The ion re-release increased with increasing the NACP filler level. Figure 1B shows the TEM image of NACP synthesized using the spray-drying technique having sizes of about 100–300 nm. The structural model of amorphous calcium phosphate is exemplified in this image, and the potential application for sealing the occlusal surface of the posterior teeth is illustrated.

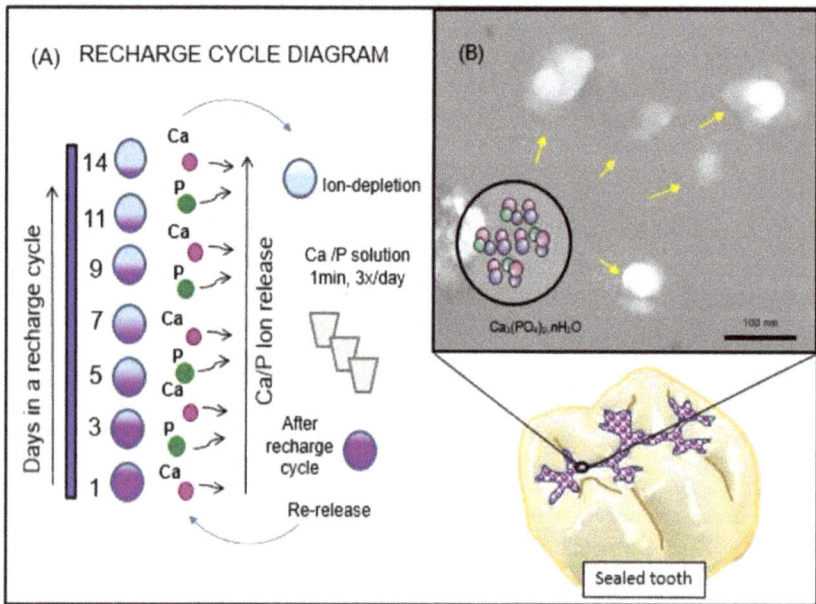

Figure 1. Schematic diagram of the rechargeable nanoparticles of amorphous calcium phosphate (NACP) sealant approach to deal with enamel demineralization around dental sealants: In (**A**), the recharge cycle diagram illustrates the re-release from the exhausted and recharged NACP sealants. Three recharge/re-release cycles were performed, and each re-release was measured for 14 days. The ion re-release increased with increasing the NACP filler level. In (**B**), the TEM image of NACP from the spray-drying technique having sizes of about 100–300 nm.

Figure 2 describes the flexural strength and modulus of the antibacterial and rechargeable resin-based sealants (Mean ± SD; $n = 8$). In Figure 2A, the flexural strength showed a decreasing trend with increasing NACP percentage, because the glass filler level was decreasing from 50% to 20%. The experimental sealant at 20% NACP had the flexural strength and flexural modulus of 79.5 ± 8.4 MPa and 4.2 ± 0.4 GPa, respectively. The commercial high viscosity sealant control showed flexural strength and flexural modulus of 70.7 ± 5.5 MPa ($p > 0.05$) and 3.3 ± 0.5 GPa ($p < 0.05$). These results demonstrated that the rechargeable antibacterial sealant at 20% mass fraction of NACP had mechanical properties similar to or higher than that of the commercial sealant/flowable composite control. Only the sealant with 30% NACP had a significantly lower strength that those other groups ($p < 0.05$).

(A)

(B)

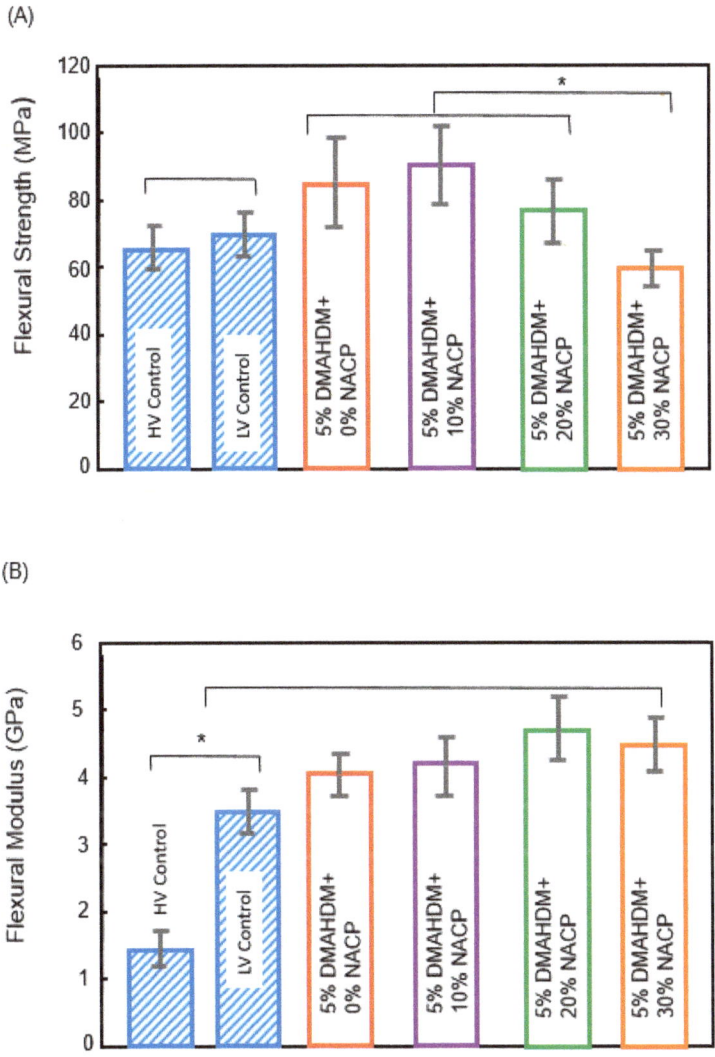

Figure 2. Bar graphs of (**A**) flexural strength and (**B**) flexural modulus (Mean ± SD; *n* = 8) of resin-based sealants. The asterisk means that there was a statistically significant difference between the groups.

The flow results for the experimental and control resin-based sealants (Mean $\pm$ SD; $n = 6$) are plotted in Figure 3. NACP at a mass fraction of 30% compromised the flow of the sealant ($p < 0.05$). Experimental sealant with a mass fraction of 20% NACP had a flow that was not significantly different from the control ($p > 0.05$).

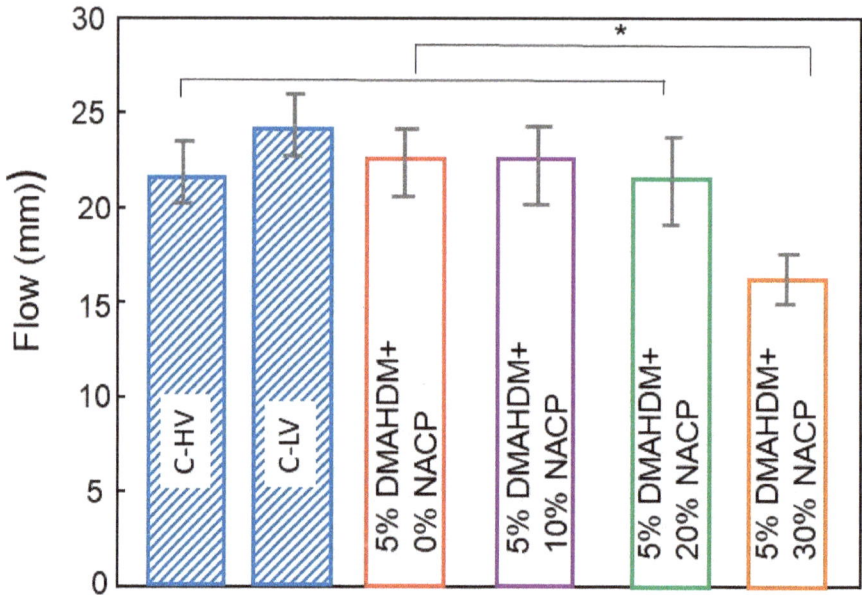

Figure 3. Means $\pm$ SD of flow analysis of the resin-based sealant formulations ($n = 6$). The asterisk means that there was a statistically significant difference between the groups.

Figure 4 plots the Ca and PO$_4$ initial ion release over time (Mean $\pm$ SD; $n = 6$). After 70 days of ion release, 30% NACP + 5% DMAHDM sealant had higher Ca ion release of 4.70 $\pm$ 0.95 mmol/L. This amount was significantly different from 20% NACP + 5% DMAHDM sealant that released 3.64 $\pm$ 0.11 mmol/L ($p < 0.05$). The PO$_4$ ion release had similar release behavior. The 30% NACP + 5% DMAHDM sealant showed 4.25 $\pm$ 0.12 mmol/L of PO$_4$ ion release, followed by the 20% NACP + 5% DMAHDM with 3.41 $\pm$ 0.10 mmol/L of PO$_4$ ion release ($p < 0.05$). During the 70 days, the pH increased from approximately 4.5 to 6.5 for the groups of the formulations containing 20% and 30% NACP. These changes represent the neutralizing capabilities of calcium and phosphate ion release.

(A)

(B)

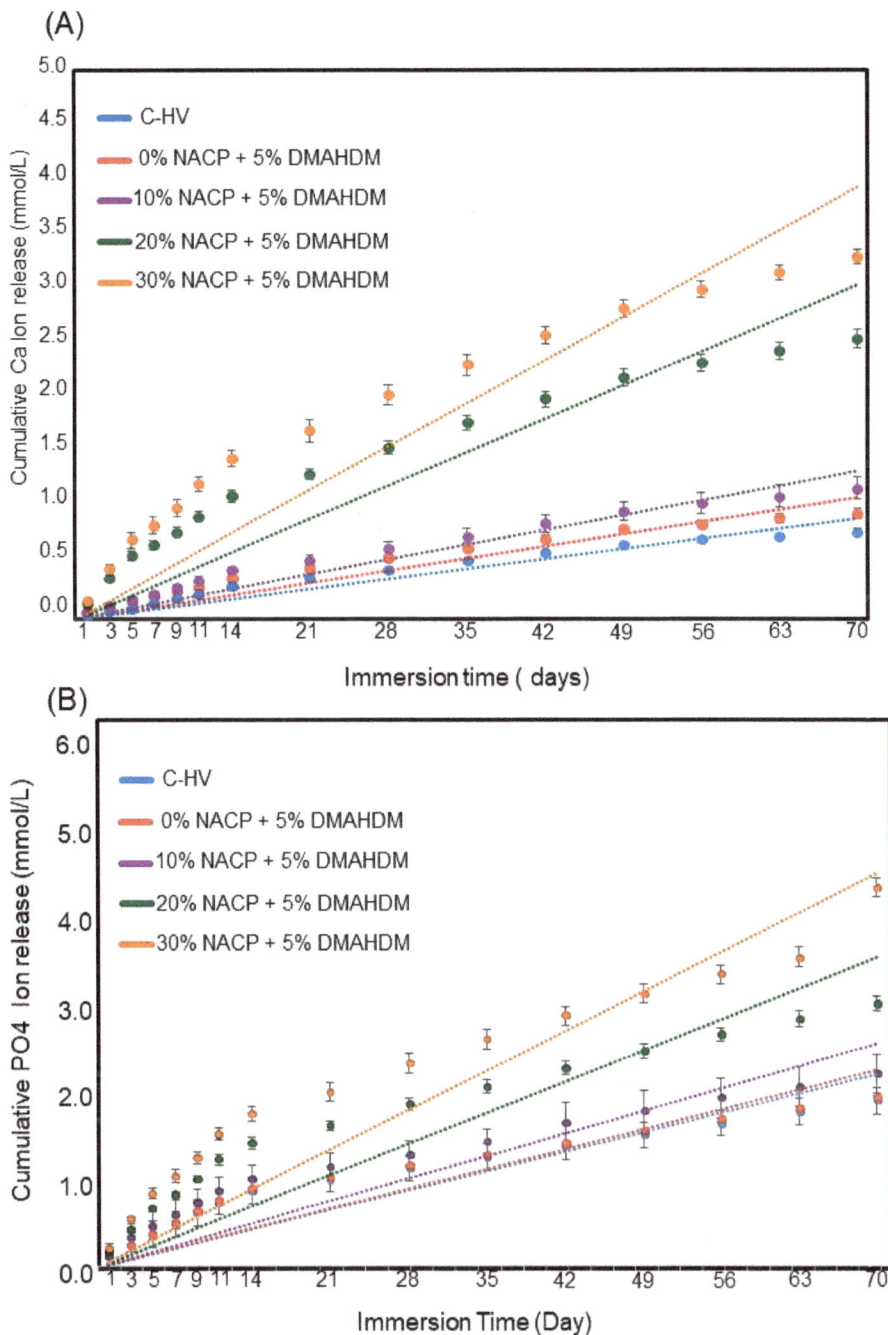

Figure 4. Cumulative initial ions release from sealants (Mean ± SD; *n* = 6). In (**A**), Calcium (Ca) ion and in (**B**) Phosphate ions. The dots for each group show the exact data and the dotted line is its approximate linear trend for each formulation.

The results of the three cycles of ion recharge and re-release are presented in Figure 5. The Ca ion re-releases after the first cycle were 0.46 ± 0.03, 0.73 ± 0.03, and 0.94 ± 0.04 for the experimental sealants with 10%, 20% and 30% NACP, respectively ($p < 0.05$ between the three groups). On the other hand, the PO_4 ion re-release after the first cycle were 1.10 ± 0.01, 1.22 ± 0.02, and 1.25 ± 0.03 for the experimental sealants with 10%, 20% and 30% NACP, respectively ($p > 0.05$ between the 20% and 30% NACP groups). All the experimental sealants showed good Ca and PO_4 ion recharging abilities in the three cycles tested.

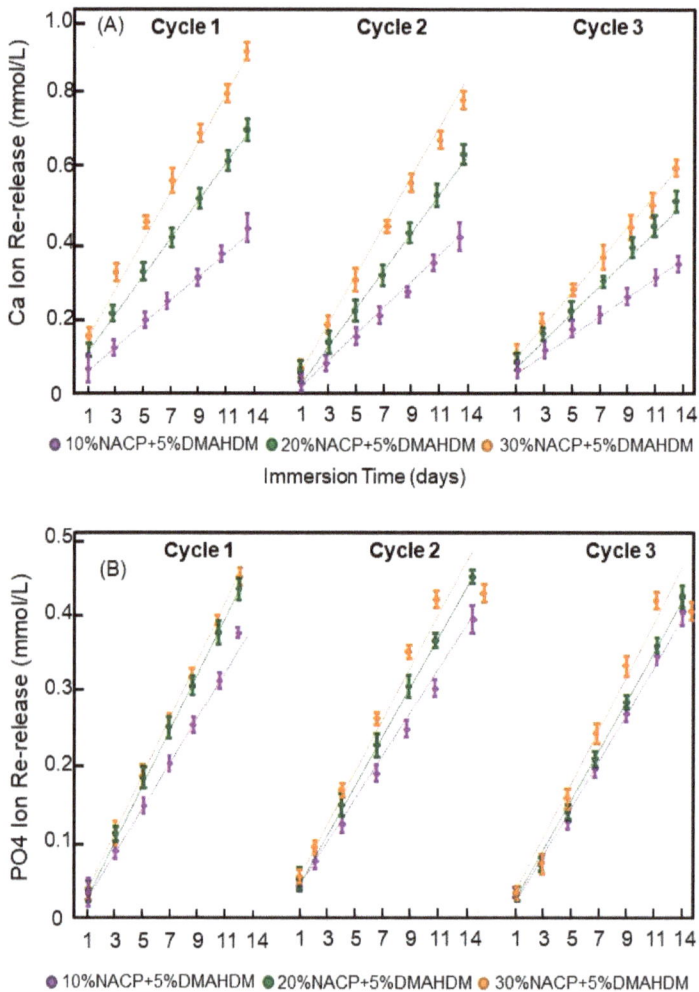

Figure 5. Cumulative ions re-release from the recharged resin dental sealants (Mean $\pm$ SD; $n = 3$) after three cycles of ions recharge and re-release. The dots for each group show the exact data and the dotted line is its approximate linear trend for each formulation. In (**A**), Calcium ion and in (**B**) Phosphate ions. There was no decrease in the ion re-release amounts with increasing the number of recharge and re-release cycles.

3. Discussion

Recent studies have developed rechargeable NACP resin-based materials for long-term Ca and PO_4 ion release to combat tooth decay [17,19], but the effects of NACP filler level on antibacterial sealant properties had not been reported. Developing a new dental material for sealant applications must meet the basic property requirements of clinicians. Thus, in the present study, an antibacterial and rechargeable Ca and PO_4 releasing sealant was developed for the first time. The effects of NACP mass fraction on mechanical and physical performance and Ca and PO_4 ions initial release and re-release were determined to allow the development of ideal formulation with bioavailable ions for potential enamel remineralization of occlusal pits and fissures.

In the case of the enlarged pit and fissure sealing, the mechanical properties of the material become more important since the material can be placed onto areas that encounter mechanical stresses during clenching [6]. In that situation, it appears clearly from the results that flowable resin composites have by far better flexural moduli than the pit and fissure sealants tested. Here, we assessed the flexural strength, flexural modulus and flow of different formulations and compared them to the commercial resin-based sealant and flowable composite that did not contain NACP. The percentage load of the glass filler ranged from 50% in 5% DMAHDM + 0% NACP sealant to 20% in 5% DMAHDM + 30% NACP formulation. Previous studies also revealed that the incorporation of a high content mass fraction of nanoparticles, such as 40% might have a negative impact on material properties [20,21]. Therefore, only three different mass fractions of NACP were used in this study: 5% DMAHDM + 10% NACP, 5% DMAHDM + 20% NACP, and 5% DMAHDM + 30% NACP. However, the results obtained in this study revealed a significant interference of the NACP at the 30% mass fraction with the flow of the experimental sealant. In deep fissures and grooves, the flow of the material is an important characteristic to help seal the total surface [4]. It is expected that the sealant's flow ability could be compromised when the experimental sealant incorporates the highest amount of NACP. Our findings suggest that the new formulation with 20% NACP did not negatively affect the mechanical and physical properties, while providing substantial Ca and PO_4 ions to inhibit caries.

Adding different caries-preventive measures and strategies to the sealant may increase its effect on caries prevention [5]. In previous studies, antibacterial resins containing quaternary ammonium methacrylates with an alkyl chain length of 16 (DMAHDM) were synthesized and assessed with a higher antibacterial potency against oral bacteria as an outcome. The mechanism of positively charged quaternary amine disrupting the negatively charged bacterial membranes supported the decrease in bacterial coverage nearly 90% for resin formulations at 5% DMAHDM [17,19]. The long-term antibacterial activity of DMAHDM was demonstrated by Zhang et al. [20] and attributed to the fact that the antibacterial monomer was copolymerized with the resin by forming a covalent bonding with the polymer network. Thus, the development of a resin-based dental sealant that also holds remineralizing properties could be essential to improve the function of the resin-based dental sealant on posterior teeth fissures and grooves for caries prevention.

Regarding recharge ability, this study builds upon the methodology provided by Zhang et al. [21]. Their report confirmed that the recharge of calcium and phosphate ions from resin-based formulations was achievable, while also establishing some of the variables important for ion release and recharge such as chemical composition of resin matrix and number of suitable cycles for recharge of the specimens. For the tested formulations, the 20% NACP sealant showed a release of 3.64 ± 0.11 mmol/L ($p < 0.05$) of Ca ions and 3.41 ± 0.10 mmol/L of PO_4 ions during the 70-day period. The effect of the NACP percentage loading within the resin-based antibacterial formulation containing 5% DMAHDM on the rate of Ca and PO_4 is reported in Figure 4A,B. When comparing the 20% NACP formulation and 10% NACP formulation, an approximate three times increase in Ca and PO_4 ions released was observed. However, as the NACP percentage load increased, the rate of diffusion of ions from the sealants increased approximately 5%. It appears that an increase in the NACP concentration by a factor of 1.5 does not provide the same corresponding increase in Ca and PO_4 ion concentration over 70 days, which makes the 20%NACP formulations suitable for the proposed application.

Referring to previous studies from our group, this release concentration of Ca and PO_4 ions was similar to release and recharge rates previously observed in a 30% NACP parental formulation with high inorganic filler content for reinforcement [17]. This variation could be attributed to the differences in the filler level in each formulation since in order for these ions to be available for remineralization, they must diffuse through the resin matrix, in this case, 50% PEHB.

The chemical composition of the formulation has a unique role for rechargeability. Since one of our objectives in this study was to develop ion-rechargeable resin-based sealants, PEHB resin matrix was chosen. The PEHB resin had shown high ion-recharging abilities in previous reports [19,22,23]. In the present study, the sealant showed promising ion recharging abilities because of the great amount of re-released Ca and PO_4 ions after each cycle of ion-recharge. This is likely due to the acidic adhesive monomer PMGDM [18], which consists of a large part of the parental resin matrix PEHB. The suggested recharge mechanism of NACP is based on the ability of PMGDM chelate with the recharging Ca ions during the recharging process, and release the ions when it is exposed to the acidic environment, such as pH 4. Further study is needed to investigate Ca and P ion recharge and re-release mechanisms in solutions that mimic the oral environment, such as artificial saliva.

4. Materials and Methods

4.1. Development of Dental Resin Sealants

The resin matrix consisted of 44.5% of pyromellitic glycerol dimethacrylate (PMGDM) (Hampford, Stratford, CT, USA), 39.5% of ethoxylated bisphenol a dimethacrylate (EBPADMA) (Sigma-Aldrich, St. Louis, MO, USA), 10% of 2-hydroxyethyl methacrylate (HEMA) (Esstech, Essington, PA, USA), and 5% of bisphenol a glycidyl dimethacrylate (Esstech) [17]. 1% of phenylbis (2,4,6-trimethylbenzoyl)-phosphine oxide (BAPO) (Sigma-Aldrich) was added as a photo-initiator. This resin matrix is referred to as PEHB. To formulate a sealant with antibacterial properties, dimethylaminohexadecyl methacrylate (DMAHDM) was synthesized via a modified Menschutkin reaction using a method as described previously [22].

NACP was synthesized by a spray-drying technique according to previous methodology [21]. Briefly, calcium carbonate and dicalcium phosphate were dissolved in acetic acid to produce Ca ions with a concentration of 8 mmol/L and PO_4 ions with a concentration of 5.3 mmol/L, then this solution was sprayed in a heated chamber using a spray-drying machine. An electrostatic precipitator was used to harvest the NACP with a particle size of average ± 116 nm. The silanized barium boroaluminosilicate glass particles were added for reinforcement purposes and had an average size of 1.4 μm (Caulk/Dentsply, Milford, DE, USA). The fillers were incorporated into the resin matrix at a filler level of 50%. Two main types of resin-based materials were available as pit and fissure sealants: filled flowable composite and unfilled resin-based sealants, both were used as commercial controls in this study. Hence, the following six sealants were tested.

1. Commercial Low-viscosity Sealant control termed "C-LV." (FluroShield, Dentsply Caulk, Milford, DE, USA)".
2. Commercial High-viscosity Sealant/Flowable Composite control termed "C-HV." (Virtuoso, Den-Mat Holdings, Lompoc, CA, USA).
3. Experimental Sealant termed "5% DMAHDM + 0% NACP" (45% PEHB + 5% DMAHDM + 50% Glass + 0% NACP).
4. Experimental Sealant termed "5% DMAHDM + 10% NACP" (45% PEHB + 5% DMAHDM + 40% Glass + 10% NACP).
5. Experimental Sealant termed "5% DMAHDM + 20% NACP" (45% PEHB + 5% DMAHDM + 30% Glass + 20% NACP).

6. Experimental Sealant termed "5% DMAHDM + 30% NACP" (45% PEHB + 5% DMAHDM + 20% Glass + 30% NACP).

4.2. Flexural Strength and Flexural Modulus

Samples for flexural strength and flexural modulus testing were prepared using $2 \times 2 \times 25$ mm stainless steel molds. Each paste was placed into the mold which was covered with Mylar strips and glass slides from both open sides of the mold then light-cured (500 mW/cm^2, 60 s, Triad 2000, Dentsply, York, PA, USA) for on each open side. Samples were stored at 37 °C for 24 h. Flexural strength and flexural modulus were measured using three-point flexure with a 10 mm span at a crosshead-speed of 1 mm/min on a computer-controlled universal testing machine (MTS 5500R, Cary, NC, USA) [24].

Flexural strength (F) was calculated by using the following formula:

$F = (3LS)/(2WH^2)$, where L is the maximum load; S is the span; W is the width of the specimen and H is the height.

Flexural Modulus (M) was determined as:

$M = (LS^33)/(4WH^33d)$, where L is the maximum load; S is the span; W is the width of the specimen, H is the height of the specimen, and d is the defluxion corresponding to the load L.

4.3. Flow Analysis

The recommendations of the ISO 6876/2012 and ANSI/ADA2000 standards were followed [25]. Briefly, two glass plates of 40 mm × 40 mm and approximately 5 mm thickness were used. The weight of one glass plate was approximately 20 g. The paste of each sealant was placed in the center of one of the glass plates using a graduated syringe. The amount of sealant was approximately 0.05 mL. The second glass plate was placed on top of the sealant; then a 100 g weight was used to make a total weight of approximately 120 g. After 10 minutes, the weight was removed, and the largest and smallest diameters of the discs formed by the compressed sealants were measured with the aid of a digital caliber (Mitutoyo MTI Corp., Huntersville, NC, USA) [26]. Six tests were done for each sealant.

4.4. Measurement of Initial Calcium and Phosphate Ions Release from NACP

Three specimens of approximately $2 \times 2 \times 12$ mm were immersed in 50 mL of sodium chloride (NaCl) solution (133 mmol/L). The NaCl solution was previously buffered to pH 4 with 50 mmol/L of lactic acid to simulate a cariogenic low pH condition [27]. The specimen volume per solution ratio was almost 2.9 mm^3/mL following previous study [22]. The tubes ($n = 6$) were kept in a 37 °C incubator during the experiment. Aliquots were collected at 1, 3, 5, 7, 9, 11, 14, 21, 28, 35, 42, 49, 56, 63 and 70 days. The Ca and PO$_4$ ions concentrations from collected aliquots were analyzed using SpectraMax$^{\circledR}$ M Series Multi-Mode Microplate Reader from Molecular Devices [28]. The absorbance was measured using known standards and calibration curves. After each collection, the NaCl solution was replaced by a fresh solution. All the groups were tested for the initial Ca and PO$_4$ ions release [29].

4.5. Calcium and Phosphate Ions Recharge and Re-Release

After the calcium and phosphate ions released for 70 days, the specimens were stored for almost 6 months before starting the ion recharging experiment to exhaust the ions in all the specimen and ensure there is no additional ion release. The pH of the immersion solutions was assessed by the same period of time. Then, these ion-exhausted specimens were used for the ion recharging experiment. The Ca ion recharging solution was made of 100 mmol/L CaCl$_2$ and 50 mmol/L HEPES buffer [18]. The PO$_4$ ion recharge solution was made of 60 mmol/L KHPO$_4$ and 50 mmol/L HEPES buffer. Both solutions were adjusted to pH 7 by the use of 1 mol/L KOH [19]. Three specimens of approx. 2 × 2 × 12 mm were immersed in 5 mL of the calcium ion or phosphate ion recharging solution and gently vortexed for 1 min using Analog Vortex Mixer, (Fisher Scientific, Waltham, MA, USA) to simulate the action of using mouthwash [30]. Specimens were immersed three times for 1 minute each time. This is for a total of 3 min of ion recharge. After that, the recharged specimens were immersed in a

50 mL NaCl solution, which was adjusted to pH 4 as described in Section 4.4. This immersion was to measure the Ca and PO$_4$ ion re-release on day 1, 2, 3, 5, 9, 11 and 14. After 14 days, the specimens were recharged again, and then the ion re-release was measured for 14 days at day 1, 2, 3, 5, 7, 9, 11 and 14 [31]. The same cycle or recharge and release were repeated for three times as described in Figure 1. The Ca and PO$_4$ ion measurements were assessed in the same way as mentioned in Section 4.4.

4.6. Statistical Analysis

Kolmogorov-Smirnov test and Levene test were performed to confirm the normality and equal variance of data. The results of flexural strength, flexural modulus, flow and Ca and PO$_4$ ion release and re-release were analyzed using one-way analysis of variance (ANOVA). Multiple comparisons between the different groups were conducted using Bonferroni's multiple comparison tests. All the statistical analyses were performed by SPSS 22.0 software (SPSS, Chicago, IL, USA) at an alpha of 0.05.

5. Conclusions

The effects of different percentage loading of the remineralizing agent, NACP in new dental material formulations have been studied in a thorough and systematic approach. Considering the clinically relevant properties for dental sealants, the formulation containing 20% of NACP was selected as the optimal composition. The material flow was highly related to the mass fraction of the filler in the resin. At 20% NACP, the PEHB-based resin sealant provides high levels of Ca and PO$_4$ ions release and durable repeated recharge capability, with no negative effect on the mechanical and physical properties of the sealant. Therefore, this is a promising approach to provide long-term ion release to promote remineralization and inhibit dental caries in occlusal surfaces of teeth in children and adolescents.

Author Contributions: Conceptualization, H.H.K.X., and M.A.S.M.; Methodology, H.H.K.X. and M.A.S.M.; Investigation, M.S.I., F.D.A., Y.A.A.-D.; Writing—M.S.I. and M.A.S.M.; Writing—Review & Editing, H.H.K.X.; Funding Acquisition, H.H.K.X., and M.A.S.M.; Resources, M.D.W., T.W.O.; Supervision, M.D.W. and T.W.O.; Project Administration, M.A.S.M. and M.D.W.

Funding: This study was supported by NIH R01DE17974 (HX) and a Seed Grant (HX, MM) from the University of Maryland School of Dentistry.

Acknowledgments: We thank Pei Feng, Associate Dean of Research, UMB School of Dentistry, and Nancy Lin, National Institute of Standards and Technology for comments that greatly improved the manuscript.

Conflicts of Interest: The authors declare no conflict of interest.

References

1. GBD 2015 Disease and Injury Incidence and Prevalence Collaborators. Global, regional, and national incidence, prevalence, and years lived with disability for 310 diseases and injuries, 1990–2015: A systematic analysis for the Global Burden of Disease Study 2015. *Lancet* **2017**, *388*, 1545–1602.
2. Benzian, H.; Hobdell, M.; Mackay, J. Putting teeth into chronic diseases. *Lancet* **2011**, *377*, 464. [CrossRef]
3. Skeie, M.S.; Klock, K.S. Dental caries prevention strategies among children and adolescents with immigrant-or low socioeconomic backgrounds- do they work? A systematic review. *BMC Oral Health* **2018**, *18*, 20. [CrossRef] [PubMed]
4. Ahovuo-Saloranta, A.; Forss, H.; Walsh, T.; Nordblad, A.; Mäkelä, M.; Worthington, H.V. Pit and fissure sealants for preventing dental decay in permanent teeth. *Cochrane Database Syst. Rev.* **2017**, *7*, 1–167. [CrossRef] [PubMed]
5. Beauchamp, J.; Caufield, P.W.; Crall, J.J.; Donly, K.; Feigal, R.; Gooch, B.; Ismail, A.; Kohn, W.; Siegal, M.; Simonsen, R. Evidence-based clinical recommendations for the use of pit-and-fissure sealants: A report of the American Dental Association Council on Scientific Affairs. *J. Am. Dent. Assoc.* **2008**, *139*, 257–268. [CrossRef] [PubMed]
6. Poulsen, S.; Beiruti, N.; Sadat, N. A comparison of retention and the effect on caries of fissure sealing with a glass-ionomer and a resin-based sealant. *Community Dent. Oral Epidemiol.* **2001**, *29*, 298–301. [CrossRef] [PubMed]

7. Cury, J.A.; Tenuta, L.M.A. Enamel remineralization: Controlling the caries disease or treating early caries lesions? *Braz. Oral Res.* **2009**, *23*, 23–30. [CrossRef] [PubMed]
8. Tenuta, L.M.; Del Bel Cury, A.A.; Bortolin, M.C.; Vogel, G.L.; Cury, J.A. Ca, Pi, and F in the fluid of biofilm formed under sucrose. *J. Dent. Res.* **2006**, *85*, 834–838. [CrossRef] [PubMed]
9. Melo, M.A.; Orrego, S.; Weir, M.D.; Xu, H.H.; Arola, D.D. Designing multi-agent dental materials for enhanced resistance to biofilm damage at the bonded interface. *ACS Appl. Mater. Interfaces* **2016**, *8*, 11779–11787. [CrossRef] [PubMed]
10. Feng, X.; Zhang, N.; Xu, H.H.; Weir, M.; Melo, M.A.; Bai, Y.; Zhang, K. Novel orthodontic cement containing dimethylaminohexadecyl methacrylate with strong antibacterial capability. *Dent. Mater. J.* **2017**, *36*, 669–676. [CrossRef] [PubMed]
11. Weir, M.D.; Chow, L.C.; Xu, H.H.K. Remineralization of demineralized enamel via calcium phosphate nanocomposite. *J. Dent. Res.* **2012**, *91*, 979–984. [CrossRef] [PubMed]
12. Melo, M.A.S.; Weir, M.D.; Rodrigues, L.K.A.; Xu, H.H.K. Novel calcium phosphate nanocomposite with caries-inhibition in a human in situ model. *Dent. Mater.* **2013**, *29*, 231–240. [CrossRef] [PubMed]
13. Melo, M.A.S.; Guedes, S.F.F.; Xu, H.H.K.; Rodrigues, L.K.A. Nanotechnology-based restorative materials for dental caries management. *Trends Biotechnol.* **2013**, *31*, 459–467. [CrossRef] [PubMed]
14. Cheng, L.; Zhang, K.; Weir, M.D.; Melo, M.A.; Zhou, X.; Xu, H.H. Nanotechnology strategies for antibacterial and remineralizing composites and adhesives to tackle dental caries. *Nanomedicine* **2015**, *10*, 627–641. [CrossRef] [PubMed]
15. Chole, D.; Lokhande, P.; Shashank, K.; Bakle, S.; Devagirkar, A.; Dhore, P. Comparative Evaluation of the Fluoride Release and Recharge through Four Different Types of Pit and Fissure Sealants: An In Vitro Study. *Int. J. Adv. Heal Sci.* **2015**, *2*, 1–6.
16. Melo, M.A.S.; Powers, M.; Passos, V.F.; Weir, M.D.; Xu, H.H. Ph-activated nano-amorphous calcium phosphate-based cement to reduce dental enamel demineralization. *Artif. Cells Nanomed. Biotechnol.* **2017**, *45*, 1778–1785. [CrossRef] [PubMed]
17. Al-Dulaijan, Y.A.; Cheng, L.; Weir, M.D.; Melo, M.A.S.; Liu, H.; Oates, T.W.; Wang, L.; Xu, H.H.K. Novel rechargeable calcium phosphate nanocomposite with antibacterial activity to suppress biofilm acids and dental caries. *J. Dent.* **2018**, *72*, 44–52. [CrossRef] [PubMed]
18. Melo, M.A.; Cheng, L.; Weir, M.D.; Hsia, R.C.; Rodrigues, L.K.; Xu, H.H. Novel dental adhesive containing antibacterial agents and calcium phosphate nanoparticles. *J. Biomed. Mater. Res. B Appl. Biomater.* **2013**, *101*, 620–629. [CrossRef] [PubMed]
19. Li, F.; Weir, M.D.; Chen, J.; Xu, H.H.K. Effect of charge density of bonding agent containing a new quaternary ammonium methacrylate on antibacterial and bonding properties. *Dent. Mater.* **2014**, *30*, 433–441. [CrossRef] [PubMed]
20. Zhang, K.; Cheng, L.; Wu, E.J.; Weir, M.D.; Bai, Y.; Xu, H.H. Effect of water-ageing on dentine bond strength and anti-biofilm activity of bonding agent containing new monomer dimethylaminododecyl methacrylate. *J. Dent.* **2013**, *41*, 504–513. [CrossRef] [PubMed]
21. Zhang, L.; Weir, M.D.; Chow, L.C.; Antonucci, J.M.; Chen, J.; Xu, H.H.K. Novel rechargeable calcium phosphate dental nanocomposite. *Dent. Mater.* **2016**, *32*, 285–293. [CrossRef] [PubMed]
22. Wu, J.; Zhou, H.; Weir, M.D.; Melo, M.A.; Levine, E.D.; Xu, H.H. Effect of the dimethylaminohexadecyl methacrylate mass fraction on fracture toughness and antibacterial properties of CaP nanocomposite. *J. Dent.* **2015**, *43*, 1539–1546. [CrossRef] [PubMed]
23. Zhang, N.; Melo, M.A.; Chen, C.; Liu, J.; Weir, M.D.; Bai, Y.; Xu, H.H. Development of a multifunctional adhesive system for prevention of root caries and secondary caries. *Dent. Mater.* **2015**, *31*, 1119–1131. [CrossRef] [PubMed]
24. Xie, X.J.; Xing, D.; Wang, L.; Zhou, H.; Weir, M.D.; Bai, Y.X.; Xu, H.H. Novel rechargeable calcium phosphate nanoparticle-containing orthodontic cement. *Int. J. Oral Sci.* **2017**, *9*, 24–32. [CrossRef] [PubMed]
25. Melo, M.A.; Wu, J.; Weir, M.D.; Xu, H.H. Novel antibacterial orthodontic cement containing quaternary ammonium monomer dimethylaminododecyl methacrylate. *J. Dent.* **2014**, *42*, 1193–1201. [CrossRef] [PubMed]
26. Zhang, N.; Melo, M.A.; Antonucci, J.; Lin, N.; Lin-Gibson, S.; Bai, Y.; Xu, H.H. Novel dental cement to combat biofilms and reduce acids for orthodontic applications to avoid enamel demineralization. *Materials* **2016**, *9*, 413. [CrossRef] [PubMed]

27. Zhang, N.; Weir, M.D.; Chen, C.; Melo, M.A.; Bai, Y.; Xu, H.H. Orthodontic cement with protein-repellent and antibacterial properties and the release of calcium and phosphate ions. *J. Dent.* **2016**, *50*, 51–59. [CrossRef] [PubMed]

28. *ISO 6876:2012. Dentistry–Root Canal Sealing Materials*; International Organization for Standardization: Geneva, Switzerland, 2012.

29. Xie, X.; Wang, L.; Xing, D.; Arola, D.D.; Weir, M.D.; Bai, Y.; Xu, H.H. Protein-repellent and antibacterial functions of a calcium phosphate rechargeable nanocomposite. *J. Dent.* **2016**, *52*, 15–22. [CrossRef] [PubMed]

30. Melo, M.A.; Cheng, L.; Zhang, K.; Weir, M.D.; Rodrigues, L.K.; Xu, H.H. Novel dental adhesives containing nanoparticles of silver and amorphous calcium phosphate. *Dent. Mater.* **2013**, *29*, 199–210. [CrossRef] [PubMed]

31. Zhang, L.; Weir, M.D.; Chow, L.C.; Reynolds, M.A.; Xu, H.H.K. Rechargeable calcium phosphate orthodontic cement with sustained ion release and re-release. *Sci. Rep.* **2016**, *6*, 36476. [CrossRef] [PubMed]

materials

MDPI

Article

Preclinical Studies of the Biosafety and Efficacy of Human Bone Marrow Mesenchymal Stem Cells Pre-Seeded into β-TCP Scaffolds after Transplantation

Mar Gonzálvez-García [1,2], Carlos M. Martinez [3], Victor Villanueva [2], Ana García-Hernández [1], Miguel Blanquer [1], Luis Meseguer-Olmo [1], Ricardo E. Oñate Sánchez [4], José M. Moraleda [1] and Francisco Javier Rodríguez-Lozano [1,4,5,*]

[1]   Cell Therapy Unit, IMIB-University Hospital Virgen de la Arrixaca, Faculty of Medicin, University of Murcia, 30120 Murcia, Spain; margonzalvez@yahoo.com (M.G.-G.); amgh8@hotmail.com (A.G.-H.); miguelblanquer@gmail.com (M.B.); l.meseguer.doc@gmail.com (L.M.-O.); jmoraled@um.es (J.M.M.)

[2]   Service of Oral and Maxillofacial Surgery, Clinical University Hospital Virgen de la Arrixaca, 30120 Murcia, Spain; villanuevasan@gmail.com

[3]   Inflammation and Experimental Surgery Unit, CIBERehd, Institute for Bio-Health Research of Murcia, Clinical University Hospital Virgen de la Arrixaca, 30120 Murcia, Spain; cmmarti@um.es

[4]   School of Dentistry, University of Murcia, 30003 Murcia, Spain; reosan@um.es

[5]   Unidad de Pacientes Especiales y Gerodontología, Universidad de Murcia, IMIB-Arrixaca, Hospital Morales Meseguer, 30008 Murcia, Spain

*   Correspondence: fcojavier@um.es; Tel.: +34-868-889518; Fax: +34-968-369-088

Received: 21 July 2018; Accepted: 31 July 2018; Published: 3 August 2018

Abstract: *Background*: Cell-Based Therapies (CBT) constitute a valid procedure for increasing the quantity and quality of bone in areas with an inadequate bone volume. However, safety and efficacy should be investigated prior to clinical application. The objective of this study was to evaluate the biodistribution, safety and osteogenic capacity of bone marrow-derived human mesenchymal stem cells (*h*BMMSCs) pre-seeded into β-tricalcium phosphate (TCP) and implanted into NOD/SCID mice at subcutaneous and intramuscular sites. *Methods*: *h*BMMSCs were isolated, characterized and then cultured in vitro on a porous β-TCP scaffold. Cell viability and attachment were analyzed and then *h*BMMSCs seeded constructs were surgically placed at subcutaneous and intramuscular dorsal sites into NOD/SCID mice. Acute and subchronic toxicity, cell biodistribution and efficacy were investigated. *Results*: There were no deaths or adverse events in treated mice during the 48-hour observation period, and no toxic response was observed in mice. In the 12-week subchronic toxicity study, no mortalities, abnormal behavioral symptoms or clinical signs were observed in the saline control mice or the *h*BMMSCs/β-TCP groups. Finally, our results showed the bone-forming capacity of *h*BMMSCs/β-TCP since immunohistochemical expression of human osteocalcin was detected from week 7. *Conclusions*: These results show that transplantation of *h*BMMSCs/β-TCP in NOD/SCID mice are safe and effective, and might be applied to human bone diseases in future clinical trials.

Keywords: preclinical biosafety; bone substitute; mesenchymal stem cells; β-tricalcium phosphate; tissue engineering

---

1. Introduction

Bone marrow is a source of mesenchymal stromal stem cells (MSCs) which have demonstrated in vivo and in vitro ability to differentiate into osteoblasts and chondrocytes, thus providing tissue repair capacities [1,2]. Their functional properties have been confirmed in several studies using autologous human bone marrow mesenchymal stem cells (*h*BMMSC) for bone repairing and tissue healing [3]. *h*BMMSC represent a cell type with a high potential for bone regeneration [4] as a result of their multipotential differentiation capacity, including differentiation into the osteogenic lineage, which constitutes a very valuable tool in medicine, specifically for tissue engineering in traumatology or maxillofacial applications [5,6]. When hBMMSCs are seeded into a scaffold, the final product brings together the osteoinductive and osteoconductive properties of the biomaterial and the regenerative and homeostatic properties of the cells. Therefore, this approach can provide an alternative to autogenous bone grafting that usually adds morbidity to the patients [7].

Cell-therapy approaches constitute one of the most promising instruments to enhance the reconstruction of both hard and soft tissues [8,9]. Nevertheless, cell dose and viability are always a problem when we move from the bench to preclinical, or even further, to the clinical setting. Therefore, this point remains to be optimized [10].

Cell-Based Therapies (CBT) are a promising approach to a wide variety of medical conditions that currently do not have satisfactory treatments. However, differentiation and proliferation potential of CBT involve new safety concerns that are not considered for conventional drug products [10]. Preclinical studies are needed to address the safety and efficacy of an investigational stem cell-based product before to move to the clinic. The development of new 3D scaffolds using advanced strategies [11,12], the mechanism of action of the mesenchymal stem cells and the most efficient route of administration have to be investigated in animal models that ideally should replicate human disease without compromising the ability of human cells to engraft and survive [13]. One step higher, MSCs from different sources are currently being tested as investigational medicinal products in several clinical trials (clinicaltrials.gov) [13]. However, many clinical trials have failed to demonstrate efficacy results because, as we have previously mentioned, critical aspects such as cell dose, homing, engraftment, and biodistribution in vivo of these "living drugs" are difficult to extrapolate from preclinical models [14]. In Europe, MSCs are somatic cell-therapy products, referred to as advanced therapy medicinal products (ATMPs) and are subject to European Regulation No. 1394/2007 [15].

The aim of this study was to test the biodistribution and security profile of *h*BMMSCs pre-seeded into β-tricalcium phosphate (TCP) after subcutaneous/intramuscular transplantation. In addition, the safety in terms of toxicity of the procedure and its capacity of osteocalcin production was evaluated.

2. Material and Methods

2.1. Isolation and Culture of Bone Marrow-Derived hBMMSCs

Multipotent *h*MSCs were isolated from bone marrow as described previously [16]. The study was approved by the Institutional Ethics Committee (Virgen de la Arrixaca University Hospital ID: 101212/1/AEMPS), while all patients signed an informed consent. For isolation, the aspirated material was transferred into transfer bags containing heparin. The mononuclear cell fraction was obtained using Ficoll density gradient media and a cell washing closed automated SEPAX™ System (Biosafe, Eysines, Switzerland). After estimating the viability with trypan blue staining, cells were plated out in 75 cm^2 culture flasks (Sarstedt, Nümbrecht, Germany) with 10 mL of basal culture growth medium (GM). The GM used was α-MEM (Minimum Essential Media) medium (Invitrogen, Carlsbad, CA, USA), supplemented with 15% fetal bovine serum (FBS, Invitrogen), 100 mM L-ascorbic acid phosphate (Sigma-Aldrich, Steinheim, Germany) and antibiotics/antimycotics before incubating at 37 °C in 5% CO_2. Cells in passage 3 were used for both in vitro and in vivo experiments.

2.2. Immunophenotypic Profiles of hBMMSC Cultures

hBMMSCs were analyzed by flow cytometry for mesenchymal (CD90, CD73), endothelial (CD105/endoglin,), hematopoietic (CD34, CD45) and HLA-DR stem cell (SC) markers, as previously described [17–19]. Single cell suspensions obtained by culture trypsinization were labelled or surface markers with fluorochrome-conjugated antibodies: CD73-PE, CD90-APC, CD105-FITC,

HLA-DR-FITC, CD34-APC and CD45-FITC (Human MSC Phenotyping Cocktail, Miltenyi Biotec, Bergisch Gladbach, Germany).

2.3. Human Bone Marrow-Derived Mesenchymal Stem Cells (hBMMSCs) Seeded into Scaffold (hBMMSCs/ β-TCP) Constructs Preparation

Synthetic β-Tricalcium phosphate (Cellplex™ TCP, Wright Medical Technology, Inc., Arlington, TN, USA) with size of 0.7–1.4 mm, a porosity of 60%, and a pore size of 100–400 pm was used as carrier. This dimension was appropriate for the specific application in the subcutaneous/intramuscular implantation. Prior to cell seeding, sterile β-TCP granules were pre-wetted for 1 h in complete medium. For cell seeding in the study group, *h*BMMSCs were trypsinized, centrifuged and resuspended in an appropriate volume; after cell counting, the density of cells in suspension was adjusted to about 1×10^6 cells. For the control group, β-TCP granules were pre-wetted with complete culture medium free of cells.

2.4. Cell Viability Assay

For this purpose, the MTT [3-(4,5-dimethylthiazol-2-yl)-2,5-diphenyltetrazolium bromide] assay was used, as previously described [20]. The cells/scaffold constructs were initially loaded with 1.0×10^4 cells/well in 96-well plates. After 1, 7 and 14 days, MTT (0.5 mg/mL in GM) was added to each cell/scaffold construct. Cells were seeded on β-TCP, as described above and 3 to 5 granules, depending on the granule size, and incubated for 4 h at 37 °C and 5% CO_2. The MTT insoluble formazan was then dissolved by means of DMSO (Dimethyl sulfoxide) that was applied for 2–4 h to the constructs at 37 °C. The optical density (OD) was measured against blank (DMSO) at a wavelength of 570 nm and a reference filter of 690 nm by an automatic microplate reader (ELx800; Bio-Tek Instruments, Winooski, VT, USA). Cell-free scaffolds incubated under the same conditions were used as reference controls and their OD values were subtracted from those obtained from the corresponding *h*BMMSCs /scaffold constructs. Population doubling number (PDN) was then calculated for the cells from days 1 to 14 using the cell number at day 1 as the seeding cell number (N0) and the day 14 as the harvested cell number (N1). The PDN was calculated with the following equation: Log10 (N1/N0) × 3.33 [21].

2.5. Scanning Electron Microscopy (SEM) Study of hBMMSCs Seeded on β-TCP

To evaluate the cell attachment of *h*BMMSCs adhered to β-TCP, study periods of 24 h, and 7 and 15 days were established. Then, *h*BMMSCs were directly seeded onto β-TCP granules at a density of 5×10^4 cells/mL. After 24 h, 7 and 15 days of culture, the samples seeded with *h*BMMSCs were primarily fixed in a solution of 3% glutaraldehyde, 0.1 M Sucrose, 0.1 M sodium cacodylate for 45 min at 4 °C. Then, they were rinsed again and dehydrated increasing concentrations (50–100% *v/v*) of ethanol and hexamethyldisilizane. The samples were dried in a critical point drier CPDO2 (Balzers Union, Liechtenstein, Germany) sputter-coated with a 20 nm thick layer of gold-palladium and observed under a SEM (JSM-6390 LV, JEOL, Tokyo, Japan).

2.6. In vivo hBMMSCs/β-TCP Constructs Transplantation

Thirty female NOD/SCID mice (Charles River Laboratories, Inc., Wilmington, MA, USA) with an average age of 6 weeks were used in this study. All animal experiments were conducted in accordance with the European Union guidelines for experimental animal use. The study protocol was approved by the Ethical Committee for Animal Care of the University of Murcia, Murcia, Spain (101212/1/AEMPS).

Mice were anesthetized intraperitoneally with a solution of ketamine (Renaudin, Aïnhoa, France, 100 mg/kg) and xylazine (Rompun, Bayer AG, Leverkusen, Germany, 10 mg/kg) and fixed on the board. After an aseptic preparation was applied to the skin. A subcutaneous incision was made at the middle of the dorsum.

The mice were randomly divided into two groups:

Group 1 formed by 25 NOD/SCID mice. A subcutaneous pocket was bluntly created in the left paravertebral area. 5 granules of hBMMSCs/β-TCP constructs were transplanted into the pockets, and the wound was suture closed. In addition, 5 granules of hBMMSCs/β-TCP construct was transplanted intramuscularly in the right paravertebral area.

Group 2 formed by 5 NOD/SCID mice. A subcutaneous pocket was bluntly created in the left paravertebral area. 5 constructs (1%PBS/β-TCP) were transplanted into the pockets, and the wound was suture closed. Also, 5 constructs (1%PBS/β-TCP) were transplanted intramuscularly in right paravertebral area.

Food and water was given *ad libitum* and the individuals' normal values for complications, abnormal locomotor activity, food and water consumption were recorded at different time points: 1 day, 2 days, 1 week, 2 weeks, 5 weeks, 7 weeks, 9 weeks and 12 weeks; 14 organs (lung, heart, kidney, spleen, tibialis anterior muscle, brain, inguinal fat pad, bone marrow, stomach, intestine, liver, ovary, blood, knee joint) were harvested and frozen at −80 °C.

2.7. Acute and Subchronic Toxicity Study

To assess the acute toxicity, the animals from both groups were observed continuously before surgery, at each hour for the first 4 hours and then at 6 hours interval for the next 48 hours after construct transplantation, to observe any deaths or abnormal locomotor activities. All mice were scored using a traditional welfare scoring system [22]. Values between 0–4 are considered a good welfare status, values of 5–9 indicate some kind of suffering, while 10–14 suggests that the mouse is in a state of considerable suffering. Finally, a score of between 15 and 19 (vocalization, self-mutilation, restlessness/stillness) is associated with intense pain and the animal should be sacrificed immediately. In addition, acute organ toxicity was evaluated by histological analysis 24 h and 48 h after surgery.

Subchronic toxicity was evaluated 1, 2, 5, 7, 9 and 12 weeks after surgery in all groups. The body weights and welfare status were recorded weekly. During the entire course of the study, animals were observed daily. In addition, subchronic organ toxicity was evaluated by histological analysis at the same time points.

2.8. Biodistribution

hBMMSCs were detected in mouse tissues using the quantitative polymerase chain reaction (qPCR) technique described by François et al. [23]. Genomic DNA from fresh tissues was prepared using the QIAamp DNA Mini Kit from Qiagen according to the manufacturer's instructions. The amount of human DNA in each sample was quantified by amplification of the human beta-globin gene, while endogenous mouse RAPSYN gene (Receptor-Associated Protein at the Synapse), served as internal control. Absolute standard curves were generated for the human beta-globin and mouse RAPSYN genes. One hundred nanograms of purified DNA from several tissues was amplified using Taqman Fast Advanced Master Mix and a Step-One Plus Real Time PCR (Polymerase Chain Reaction) system (Applied Biosytems, Foster City, CA, USA). The primers and probe for human beta-globin were: forward primer 5′GTGCACCTGACTCCTGAGGAGA3′ and reverse primer 5′CCTTGATACCAACCTGCCCAGG3′; the probe labelled with fluorescent reporter and quencher was 5′FAM-AAGGTGAACGTGGATGAAGTTGGTGG-TAMRA-3′. The primers and probe for mouse RAPSYN gene were forward primer 5′ACCCACCCATCCTGCAAAT3′ and reverse primer 5′ACCTGTCCGTGCTGCAGAA3′; the probe labelled with fluorescent reporter and quencher was 5′FAM-CGGTGCCAGTGATGAGGTTGGTC-TAMRA-3′. Likewise, human DNA was isolated from hMSC culture and used as a positive control [24].

2.9. Anatomic Pathology Examination

Representative samples from constructs and brain, lung, heart, liver, kidney, gut, spleen, lymph node, bone marrow were fixed in 4% buffered formalin (Panreac Quimica, Barcelona, Spain) for 48 h. Constructs and bone marrow were additionally decalcified in a formic-acid-based commercial solution (TBD-2, Thermo, Madrid, Spain) for 12–16 h. Samples were then washed, processed and paraffin-embedded. Sections were obtained and stained with hematoxylin and eosin (H&E) for standard histological analyses. To study the presence of human osteocalcin producer cells, a standard indirect ABC immunohistochemical staining was performed, using a specific polyclonal rabbit human anti-osteocalcin antibody (LsBio, Seattle, WA, USA) with a commercial kit EnVision FlexTM, (Dako, Carpinteria, CA, USA)). All samples were evaluated with a conventional light microscope (Axio Scope AX10, Zeiss, Oberkochen, Germany), with attached digital camera (Axio Cam Icc3, Carl Zeiss, Jenna, Germany).

2.10. Statistics

Data were analyzed using the SPSS software (version 19, SPSS, Inc., Chicago, IL, USA). Statistical analysis was conducted using the Mann-Whitney U-test or Student's t-test (others). $p < 0.05$ was interpreted as denoting statistical significance.

3. Results

3.1. Characterization of hBMMSCs In Vitro Experiments

The isolated *h*BMMSCs displayed a SC phenotype, and had a comparatively high purity; practically all cells showed a positive expression of the mesenchymal markers CD73, CD90 and CD105 (>95%) and lack expression of the hematopoietic markers, CD34, CD45 and HLA-DR (<5%) (Figure 1).

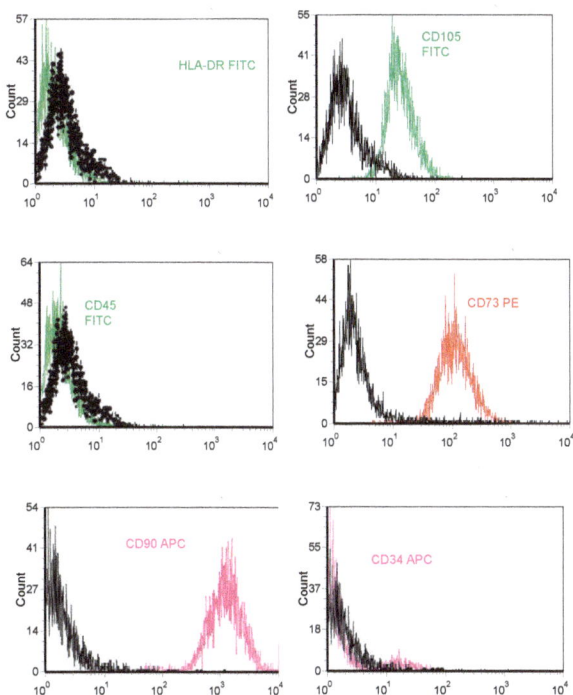

Figure 1. Immunophenotypic characterization of *h*BMMSCs by flow cytometry for the expression of mesenchymal (CD90, CD73, CD105/endoglin), hematopoietic (CD34, CD45) and HLA-DR markers (black line: unstained control; red, green and purple line: marker of interest). Results are means of triplicates (±SD) of three independent experiments.

3.2. Cell Proliferation and Attachment

Figure 2A shows the proliferation of *h*BMMSCs on β-TCP after 1, 7 and 14 days, as assessed by the MTT assay. *h*BMMSCs incubated in culture plates were monitored as positive control and cell-free scaffolds incubated under the same conditions were used as negative control. An MTT assay was performed at days 1, 7 and 14 after cell seeding into β-TCP to assess cell survival and proliferation. A significant increase in MTT reduction was seen at day 14 compared with days 1 and 7, indicating that *h*BMMSCs were able to survive and proliferate on β-TCP granules ($p < 0.01$). In addition, the PDN obtained with and without β-TCP was 2.22 ± 0.18 versus 2.09 ± 0.15, respectively.

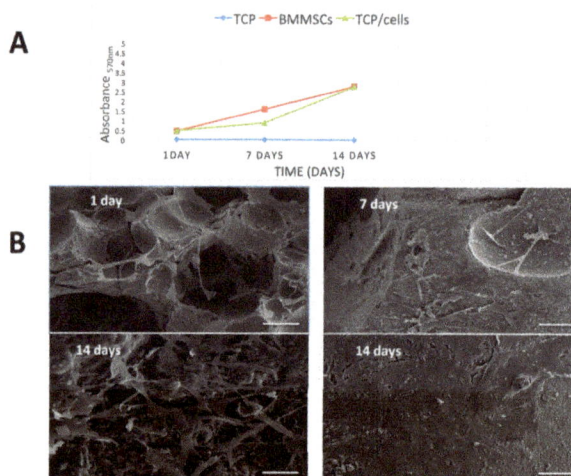

Figure 2. (**A**) MTT assay results of *h*BMMSCs and β-TCP/*h*BMMSCs construct. Results are expressed as relative MTT activity compared with the control. Data were shown as mean $\pm$ SD from three independent experiments; (**B**) Cellular shape and adherence of *h*BMMSCs onto β-TCP by scanning electron microscopy (SEM) 1, 7 and 14 days post-seeding on β-TCP. Scale bar: 100 μm.

SEM analyses revealed that small quantities of *h*BMMSCs were evenly attached to β-TCP granules after 24 h (Figure 2B). Importantly, at longer culture times (7 days) the hBMMSCs covered all the biomaterial, exhibiting a fibroblastoid morphology with several cytoplasmatic prolongations that allow the cells to anchor to β-TCP surface and establish intercellular connections. After 14 days of culture, large amounts of *h*BMMSCs adhered to the β-TCP granules, appearing as multilayered cultures. Moreover, calcified matrix deposition was detected on the surface of the cells.

3.3. Acute, Subchronic Toxicity Study

No death or clinical signs associated with toxicity occurred during the 48-hour observation period in animals. Mice exhibited normal behavior, without surgery complications or abnormal locomotor activities. No abnormal form or color was found in the animals' feces. Body weight changes were measured during this 2-day period. The welfare score of the 30 mice prior to and post-implantation was 0. According to Figure 3A, no statistically significant weight loss was observed between *h*BMMSCs/β-TCP group and the physiological saline control group ($p = 0.820$).

Local and subchronic toxicity of *h*BMMSCs/β-TCP constructs were assessed in a 12-week toxicity study. No mortalities or adverse clinical signs were found in both groups (Figure 3B). There was no significant difference in body weight between groups in each week. Dose-related change in mean daily food or water consumption was not observed between the negative control and the *h*BMMSCs/β-TCP groups throughout the experimental period. Additionally, no macroscopic findings were observed at necropsy, and microscopic analysis (Figure 3C) revealed no histopathological or tumor alterations in any of the paraffin-embedded tissues.

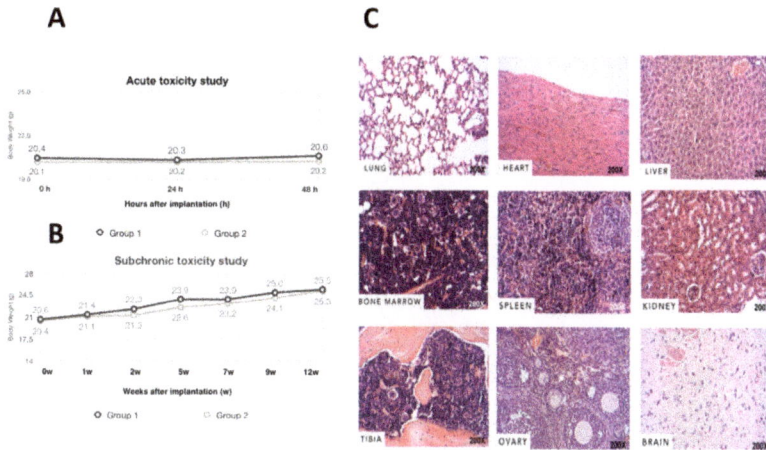

Figure 3. The body weight changes of the NOD/SCID mice after construct implantation for (**A**) 48 h (Acute Toxicity study) and (**B**) 12 weeks (Subchronic Toxicity study); (**C**) Histological analysis of various organs collected (lung, heart, liver, bone marrow, spleen, kidney, tibia, ovary and the brain). No structural changes or injuries were detected in theses organs.

3.4. Biodistribution

DNA extraction and PCR analysis were performed to detect the presence of human cells in mouse tissues. The results are expressed as the number of mice (or percentage) with PCR positive for human beta-globin gene. Our results showed the presence of human cells on scaffolds during all experiment (24 h to 12 weeks post-implantation) (Table 1). However, we did not detect human cells in lung, heart, kidney, spleen, tibialis anterior muscle, brain, inguinal fat pad, bone marrow, stomach, intestine, liver, ovary, blood, skin or the knee joint.

Table 1. Biodistribution of *h*BMMSCs using the quantitative polymerase chain reaction (qPCR) technique, for 12 weeks. Controls: cell-free scaffolds; (+), detection of human RNAse P gene; (−), no detection of human RNAse P gene.

Tissues/Organs	24 h	1 w	2 w	5 w	7 w	9 w	12 w	Controls
Scaffold	+	+	+	+	+	+	+	−
Lung	−	−	−	−	−	−	−	−
heart	−	−	−	−	−	−	−	−
kidney	−	−	−	−	−	−	−	−
spleen	−	−	−	−	−	−	−	−
tibialis anterior muscle	−	−	−	−	−	−	−	−
brain	−	−	−	−	−	−	−	−
inguinal fat pad	−	−	−	−	−	−	−	−
bone marrow	−	−	−	−	−	−	−	−
stomach	−	−	−	−	−	−	−	−
intestine	−	−	−	−	−	−	−	−
liver	−	−	−	−	−	−	−	−
ovary	−	−	−	−	−	−	−	−
blood	−	−	−	−	−	−	−	−
skin	−	−	−	−	−	−	−	−
knee joint	−	−	−	−	−	−	−	−

3.5. In Vivo Bone Formation

Next, the in vivo bone formation was analyzed. Histopathological analysis revealed signs of lamellar bone formation in both the subcutaneous (from week 7) and intramuscular (from week 9) constructs of Group 1 (*h*BMMSCs/β-TCP, Figure 4). No signs of lamellar bone neoformation were observed in the subcutaneous and intramuscular constructs of Group 2 at any time.

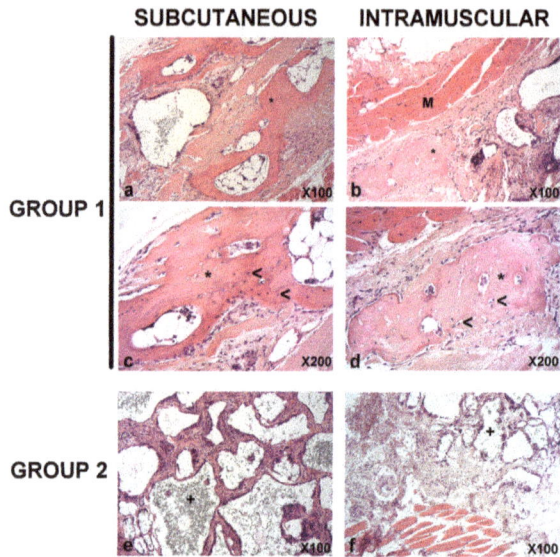

Figure 4. Representative images of subcutaneous and intramuscular constructs from Group 1 (**a–d**) and Group 2 (**e,f**) at week 9 after constructs implantation. While there was signs of formation of lamellar bone (asterisks) with signs of functional lacunae (presence of nucleus, head arrows) interspersed within the construct matrix in subcutaneous (**a,c**) and intramuscular (**b,d**) constructs from Group 1, in subcutaneous (**e**) and intramuscular (**f**) constructs from Group 2 there was a infiltration of connective tissue with trabecular disposition in which signs of a refringent material (+) could be identified within trabeculae. (M): Skeletal muscle. Hematoxylin and eosin (H&E) stain. Magnifications: $100 \times$ (**a,b,e,f**) and $200 \times$ (**c,d**).

Immunohistochemical expression of human osteocalcin was detected only in bone marrow from mice of Group 1 from week 7 onwards (Figure 5). On the other hand, no signs of positive immunoreaction were observed in subcutaneous and intramuscular constructs from Group 2.

Quantitative results (Table 2) exhibited a significant difference in the osteocalcin expression among the subcutaneous/intramuscular group (Group 1) and control group (Group 2) ($p < 0.05$). There was no significant difference between the subcutaneous and intramuscular localizations. Overall, the results indicated that the *h*BMMSCs/β-TCP group can promote the formation of calcified matrix and the osteocalcin expression compared with the control group. While there is a strong positive expression of osteocalcin in lacunae in human control bone, no positive reaction was observed either in bone or other tissues in the mouse (Figure 6).

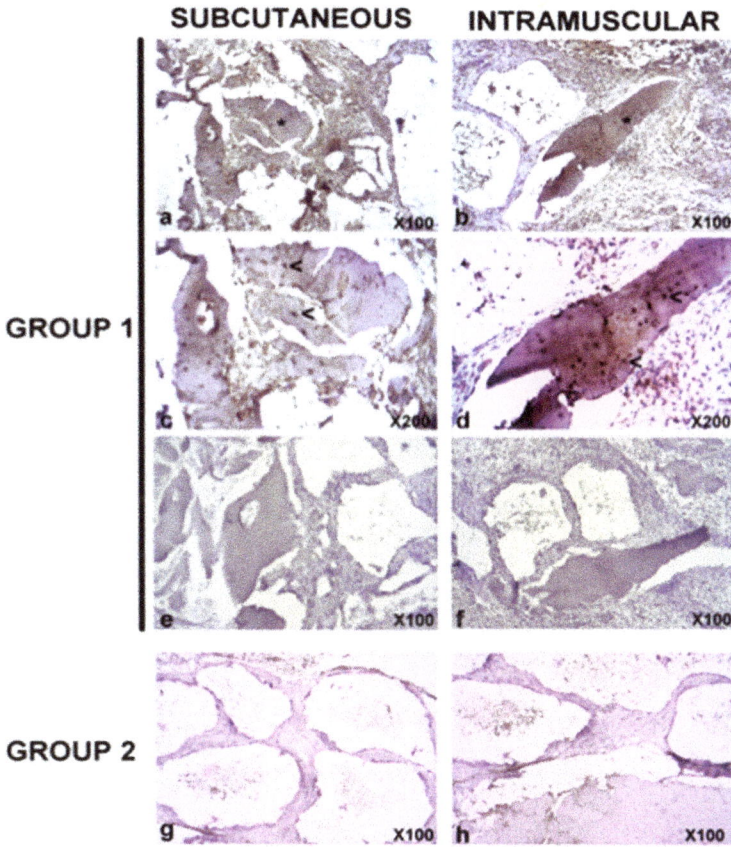

Figure 5. Representative images of human osteocalcin expression of subcutaneous and intramuscular constructs from Group 1 (**a–f**) and Group 2 (**g,h**) at week 9 after constructs implantation. There was positive expression of human osteocalcin in lamellar bone formations within subcutaneous or intramuscular constructs from Group 1 (**a,b** asterisks), particularly in functional lacunae (**c,d**, head arrows). Any sign of background was observed in negative controls of the same regions (**e,f**). On the other hand, no signs of positive immunoreaction were observed in subcutaneous and intramuscular constructs from Group 2 (**g,h**). ABC anti-human osteocalcin stain. Magnifications: $100 \times$ (**a,b,e–h**) and $200 \times$ (**c,d**).

Table 2. Frequency of human osteocalcin expression in mice after construct implantation compared with the control. ** $p < 0.01$.

	Intramuscular Implant			Subcutaneous Implant		
	Yes	No			Yes	No
Group I (hBMMSCs/TCP)	$n = 18$ **	$n = 7$		Group I (hBMMSCs/TCP)	$n = 17$ **	$n = 8$
Group II (Cell free TCP)	0	$n = 5$		Group II (Cell free TCP)	0	$n = 5$

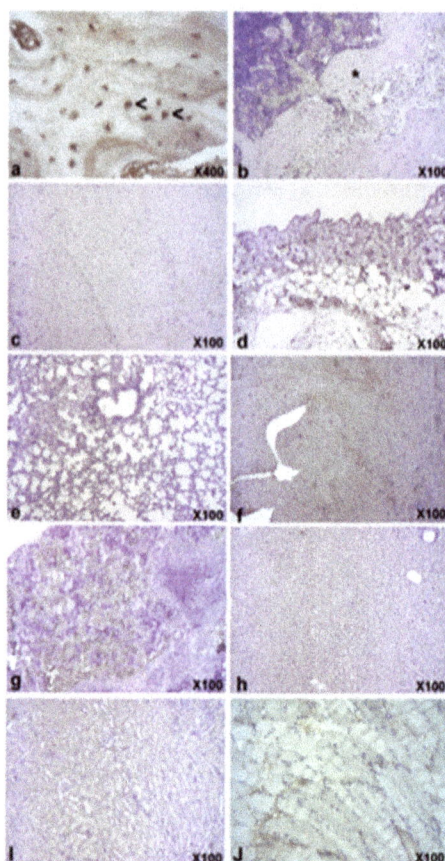

Figure 6. Representative images of expression of human osteocalcin in human bone (positive control, **a**); and in mouse bone (**b**); brain (**c**); skin (**d**); lung (**e**); heart (**f**); spleen (**g**); liver (**h**); kidney (**I**) and skeletal muscle (**j**). While there is a strong positive expression of osteocalcin in lacunae in human control bone (**a**, head arrows), no positive reaction was observed neither in bone (**b**), nor other tissues (**c**–**j**) in the mouse. ABC anti-human osteocalcin stain. Magnifications: 100 × (**a,b,e**–**j**).

4. Discussion

Preclinical studies of the products for use in new CBT need to be carried out in animal models in order to verify their biosecurity and efficacy [25]. In fact, determining the distributive fate and retention of CBT products after administration is key part of characterizing their mechanism of action and security profile [25,26]. The present study was prepared to analyze the biosafety of hBMMSCs pre-seeded into TCP scaffolds after subcutaneous/intramuscular transplantation.

We reported that (i) hBMMSCs/β-TCP constructs did not cause acute or subchronic toxicities to the mice (inspection of the health status of the operated mice and histologically analyses of several tissue samples); (ii) human cells do not migrated into tissues distant from the implantation sites (expression of human globin gene, by quantitative PCR, in several tissues); (iii) hBMMSCs/β-TCP constructs developed into bone tissue.

The limitation of this study was the animal model; immunocompetent animal model made the evaluation of the immune response of the implanted hBMMSCs under Good Laboratory Practice (GLP) conditions difficult and could be more significant by investigating the impact of SCs in larger animal models. In contrast, subcutaneous implantation is an easy and non-invasive technique, and allows performance of several test items in the same animal [27].

New materials must first manifest their biocompatibility before cells can proliferate and produce an extracellular mineralized matrix on a substrate [28]. For this purpose and to evaluate the possible cytotoxicity of the β-TCP, we investigated the viability and cell attachment of hBMMSCs cultured on β-TCP by MTT assay and SEM, respectively. A similar level of cell viability to the control was seen after 14 days of culture. Previous studies using colorimetric assays demonstrated good metabolic cell activity, cell adhesion and cell morphology promoted by β-TCP [29–31]. SEM is the most commonly used electron microscopy approach to analyze morphological appearance of cells seeded on certain biomaterials prior to implantation [32]. After 14 days of culture, we observed large amounts of hBMMSCs adhering to the β-TCP granules, giving the appearance of multilayered cultures. Arpornmaeklong et al. [33] showed that β-TCP stimulates the attachment and differentiation of human embryonic SCs (hESCs), especially the expression of genes related to neurogenesis (AP2a, FoxD3, HNK1, P75, Sox1, Sox10). Another recent study exhibited good morphology and cell attachment of dental pulp SCs into the β-TCP scaffolds [34].

Therapies based on SCs have shown great potential in many clinical studies. However, novel therapies using cell-based ATMPs require special safety testing strategies [27]. Thus, any additional information showing toxicity tests can help guide the design of clinical trials [35]. In our study, the local and systemic toxicity of hBMMSCs intramuscular and subcutaneous transplanted was monitored for 12 weeks. No mortality, morbidity or abnormal clinical symptoms were found. Moreover, no hBMMSC-related changes were observed in histopathological lesions. In a previous study involving mesenchymal progenitor cells derived from umbilical cord blood intravenously administered in mice, no toxicologically meaningful microscopic findings were observed in the animals [36]. Importantly we did not observe any tumors in the sacrificed animals.

Due to the cell migration after local administration, biodistribution studies are key elements for understanding the physiological or pathological behavior of the cells before clinical use [37]. Our biodistribution results did not show any hBMMSCs in the tested organs (lung, heart, kidney, spleen, tibialis anterior muscle, brain, inguinal pad, bone marrow, stomach, intestine, liver, ovary, blood, knee joint) 12 weeks after transplantation, suggesting that cells stay where they are placed and do not invade other tissues. These data were consistent with those of a previous study in which no hDNA was detected in such major organs as the brain, heart, lungs, kidneys, spleen or liver of animals after intramuscular administration of hMSCs [25]. In the same line, Choi et al. [38] reported that intracranially injected adipose mesenchymal SCs did not invade other tissues out of the brain in normal mice. However, after intra-articular injection, human Alu sequences were detected in several tissues and organs [39]. This suggests that the biodistribution potential of mesenchymal SCs could be influenced by the route of administration.

Regarding the efficacy of hBMMSCs, subcutaneous ectopic bone formation models are commonly used by CBT [40,41]. Previous reports have demonstrated that murine or human bone marrow stromal cells seeded on calcium phosphate (CaP) stimulate the bone formation implanted subcutaneously in immune-compromised mice [42]. Our results showed the presence of signs of lamellar bone formation in both subcutaneous and intramuscular constructs of group β-TCP + hBMMSCs from week 7 in those cases of subcutaneous implantation, and from week 9 in the intramuscular implants. While there was a strong positive expression of osteocalcin in lacunae in human control bone, a positive reaction was observed neither in bone, nor other tissues in the mouse. In this context, other authors have shown the therapeutic efficacy of BMMSCs/β-TCP in goat models of critical size bone defects [43].

5. Conclusions

Based on the data described in this work, it is concluded that transplantation of mesenchymal stem cell from bone marrow preseeded into β-TCP scaffolds in murine models is safe and effective. This results pave the way to perform "first in human" clinical trials to treat bone diseases in the future.

Author Contributions: M.G.-G., L.M.-O., A.G.H and F.J.R.L. designed the study. C.M.M, M.G.-G., and M.B. conducted the experiments; R.E.O.S. and F.J.R.-L. analysed the results. J.M.M. and F.J.R.L. contributed to preparation of the manuscript.

Funding: This work was supported by the Spanish Network of Cell Therapy (TerCel), RETICS subprogrammes of the I + D + I 2013–2016 Spanish National Plan. and projects 'RD12/0019/0001', 'RD16/0011/0001', 'PI13/02699' and 'EC11-009' funded by Instituto de Salud Carlos III and cofunded by European Regional Development Fund.

Conflicts of Interest: The authors declare no conflict of interest.

References

1. Delorme, B.; Chateauvieux, S.; Charbord, P. The concept of mesenchymal stem cells. *Regen. Med.* **2006**, *1*, 497–509. [CrossRef] [PubMed]
2. Moraleda, J.M.; Blanquer, M.; Bleda, P.; Iniesta, P.; Ruiz, F.; Bonilla, S.; Cabanes, C.; Tabares, L.; Martinez, S. Adult stem cell therapy: Dream or reality? *Transpl. Immunol.* **2006**, *17*, 74–77. [CrossRef] [PubMed]
3. Gonzálvez-García, M.; Rodríguez-Lozano, F.J.; Villanueva, V.; Segarra-Fenoll, D.; Rodríguez-González, M.A.; Oñate-Sánchez, R.; Blanquer, M.; Moraleda, J.M. Cell therapy in bisphosphonate-related osteonecrosis of the jaw. *J. Craniofac. Surg.* **2013**, *24*, 226–228. [CrossRef] [PubMed]
4. Qi, Y.; Niu, L.; Zhao, T.; Shi, Z.; Di, T.; Feng, G.; Li, J.; Huang, Z. Combining mesenchymal stem cell sheets with platelet-rich plasma gel/calcium phosphate particles: A novel strategy to promote bone regeneration. *Stem Cell Res. Ther.* **2015**, *6*, 256. [CrossRef] [PubMed]
5. Cella, L.; Oppici, A.; Arbasi, M.; Moretto, M.; Piepoli, M.; Vallisa, D.; Zangrandi, A.; Di Nunzio, C.; Cavanna, L. Autologous bone marrow stem cell intralesional transplantation repairing bisphosphonate related osteonecrosis of the jaw. *Head Face Med.* **2011**, *7*, 16. [CrossRef] [PubMed]
6. Suenaga, H.; Furukawa, K.S.; Suzuki, Y.; Takato, T.; Ushida, T. Bone regeneration in calvarial defects in a rat model by implantation of human bone marrow-derived mesenchymal stromal cell spheroids. *J. Mater. Sci. Mater. Med.* **2015**, *26*, 254. [CrossRef] [PubMed]
7. Shamsul, B.S.; Tan, K.K.; Chen, H.C.; Aminuddin, B.S.; Ruszymah, B.H. Posterolateral spinal fusion with ostegenesis induced BMSC seeded TCP/HA in a sheep model. *Tissue Cell* **2014**, *46*, 152–158. [CrossRef] [PubMed]
8. Jimi, E.; Hirata, S.; Osawa, K.; Terashita, M.; Kitamura, C.; Fukushima, H. The current and future therapies of bone regeneration to repair bone defects. *Int. J. Dent.* **2012**, *2012*, 148261. [CrossRef] [PubMed]
9. Sunil, P.; Manikandhan, R.; Muthu, M.; Abraham, S. Stem cell therapy in oral and maxillofacial region: An overview. *J. Oral Maxillofac. Pathol.* **2012**, *16*, 58–63. [CrossRef] [PubMed]
10. Basu, J.; Assaf, B.T.; Bertram, T.A.; Rao, M. Preclinical biosafety evaluation of cell-based therapies: Emerging global paradigms. *Toxicol. Pathol.* **2015**, *43*, 115–125. [CrossRef] [PubMed]
11. Patrício, T.; Domingos, M.; Gloria, A.; D'Amora, U.; Coelho, J.F.; Bártolo, P.J. Fabrication and characterisation of PCL and PCL/PLA scaffolds for tissue engineering. *Rapid Prototyping J.* **2014**, *20*, 145–156. [CrossRef]
12. Guarino, V.; Gloria, A.; Raucci, M.G.; De Santis, R.; Ambrosio, L. Bio-inspired composite and cell instructive platforms for bone regeneration. *Int. Mater. Rev.* **2013**, *57*, 256–275. [CrossRef]
13. Frey-Vasconcells, J.; Whittlesey, K.J.; Baum, E.; Feigal, E.G. Translation of stem cell research: Points to consider in designing preclinical animal studies. *Stem Cells Transl. Med.* **2012**, *1*, 353–358. [CrossRef] [PubMed]
14. Zhao, W.; Phinney, D.G.; Bonnet, D.; Dominici, M.; Krampera, M. Mesenchymal stem cell biodistribution, migration, and homing in vivo. *Stem Cells Int.* **2014**, *2014*, 292109. [CrossRef] [PubMed]
15. Rousseau, C.F.; Maciulaitis, R.; Sladowski, D.; Narayanan, G. Cell and Gene Therapies: European View on Challenges in Translation and How to Address Them. *Front. Med. (Lausanne)* **2018**, *5*, 158. [CrossRef] [PubMed]

16. De Aza, P.N.; Garcia-Bernal, D.; Cragnolini, F.; Velasquez, P.; Meseguer-Olmo, L. The effects of Ca_2SiO_4-$Ca_3(PO4)_2$ ceramics on adult human mesenchymal stem cell viability, adhesion, proliferation, differentiation and function. *Mater. Sci. Eng. C Mater. Biol. Appl.* **2013**, *33*, 4009–4020. [CrossRef] [PubMed]

17. Dominici, M.; Le Blanc, K.; Mueller, I.; Slaper-Cortenbach, I.; Marini, F.; Krause, D.; Deans, R.; Keating, A.; Prockop, D.; Horwitz, E. Minimal criteria for defining multipotent mesenchymal stromal cells. The International Society for Cellular Therapy position statement. *Cytotherapy* **2006**, *8*, 315–317. [CrossRef] [PubMed]

18. Horwitz, E.M.; Le Blanc, K.; Dominici, M.; Mueller, I.; Slaper-Cortenbach, I.; Marini, F.C.; Deans, R.J.; Krause, D.S.; Keating, A.; Therapy, I.S.f.C. Clarification of the nomenclature for MSC: The International Society for Cellular Therapy position statement. *Cytotherapy* **2005**, *7*, 393–395. [CrossRef] [PubMed]

19. Rodriguez-Lozano, F.J.; Garcia-Bernal, D.; Onate-Sanchez, R.E.; Ortolani-Seltenerich, P.S.; Forner, L.; Moraleda, J.M. Evaluation of cytocompatibility of calcium silicate-based endodontic sealers and their effects on the biological responses of mesenchymal dental stem cells. *Int. Endod. J.* **2017**, *50*, 67–76. [CrossRef] [PubMed]

20. Llena, C.; Collado-Gonzalez, M.; Tomas-Catala, C.J.; Garcia-Bernal, D.; Onate-Sanchez, R.E.; Rodriguez-Lozano, F.J.; Forner, L. Human Dental Pulp Stem Cells Exhibit Different Biological Behaviours in Response to Commercial Bleaching Products. *Materials (Basel)* **2018**, *11*, 1098. [CrossRef] [PubMed]

21. Eslaminejad, M.B.; Mirzadeh, H.; Nickmahzar, A.; Mohamadi, Y.; Mivehchi, H. Type I collagen gel in seeding medium improves murine mesencymal stem cell loading onto the scaffold, increases their subsequent proliferation, and enhances culture mineralization. *J. Biomed. Mater. Res. B Appl. Biomater.* **2009**, *90*, 659–667. [CrossRef] [PubMed]

22. Lloyd, M.H.; Foden, B.W.; Wolfensohn, S.E. Refinement: Promoting the three Rs in practice. *Lab. Anim.* **2008**, *42*, 284–293. [CrossRef] [PubMed]

23. Francois, S.; Bensidhoum, M.; Mouiseddine, M.; Mazurier, C.; Allenet, B.; Semont, A.; Frick, J.; Sache, A.; Bouchet, S.; Thierry, D.; et al. Local irradiation not only induces homing of human mesenchymal stem cells at exposed sites but promotes their widespread engraftment to multiple organs: A study of their quantitative distribution after irradiation damage. *Stem Cells* **2006**, *24*, 1020–1029. [CrossRef] [PubMed]

24. Francois, S.; Mouiseddine, M.; Allenet-Lepage, B.; Voswinkel, J.; Douay, L.; Benderitter, M.; Chapel, A. Human mesenchymal stem cells provide protection against radiation-induced liver injury by antioxidative process, vasculature protection, hepatocyte differentiation, and trophic effects. *Biomed. Res. Int.* **2013**, *2013*, 151679. [CrossRef] [PubMed]

25. Creane, M.; Howard, L.; O'Brien, T.; Coleman, C.M. Biodistribution and retention of locally administered human mesenchymal stromal cells: Quantitative polymerase chain reaction-based detection of human DNA in murine organs. *Cytotherapy* **2017**, *19*, 384–394. [CrossRef] [PubMed]

26. Bailey, A.M.; Mendicino, M.; Au, P. An FDA perspective on preclinical development of cell-based regenerative medicine products. *Nat. Biotechnol.* **2014**, *32*, 721–723. [CrossRef] [PubMed]

27. Zscharnack, M.; Krause, C.; Aust, G.; Thummler, C.; Peinemann, F.; Keller, T.; Smink, J.J.; Holland, H.; Somerson, J.S.; Knauer, J.; et al. Preclinical good laboratory practice-compliant safety study to evaluate biodistribution and tumorigenicity of a cartilage advanced therapy medicinal product (ATMP). *J. Transl. Med.* **2015**, *13*, 160. [CrossRef] [PubMed]

28. Stratton, S.; Shelke, N.B.; Hoshino, K.; Rudraiah, S.; Kumbar, S.G. Bioactive polymeric scaffolds for tissue engineering. *Bioact. Mater.* **2016**, *1*, 93–108. [CrossRef] [PubMed]

29. Seebach, C.; Schultheiss, J.; Wilhelm, K.; Frank, J.; Henrich, D. Comparison of six bone-graft substitutes regarding to cell seeding efficiency, metabolism and growth behaviour of human mesenchymal stem cells (MSC) in vitro. *Injury* **2010**, *41*, 731–738. [CrossRef] [PubMed]

30. Xu, L.; Lv, K.; Zhang, W.; Zhang, X.; Jiang, X.; Zhang, F. The healing of critical-size calvarial bone defects in rat with rhPDGF-BB, BMSCs, and beta-TCP scaffolds. *J. Mater. Sci. Mater. Med.* **2012**, *23*, 1073–1084. [CrossRef] [PubMed]

31. Xu, L.; Zhang, W.; Lv, K.; Yu, W.; Jiang, X.; Zhang, F. Peri-Implant Bone Regeneration Using rhPDGF-BB, BMSCs, and beta-TCP in a Canine Model. *Clin. Implant Dent. Relat. Res.* **2016**, *18*, 241–252. [CrossRef] [PubMed]

32. Wu, H.; Kang, N.; Wang, Q.; Dong, P.; Lv, X.; Cao, Y.; Xiao, R. The Dose-Effect Relationship Between the Seeding Quantity of Human Marrow Mesenchymal Stem Cells and In Vivo Tissue-Engineered Bone Yield. *Cell Transplant* **2015**, *24*, 1957–1968. [CrossRef] [PubMed]

33. Arpornmaeklong, P.; Pressler, M.J. Effects of ss-TCP Scaffolds on neurogenic and osteogenic differentiation of Human Embryonic Stem Cells. *Ann. Anat.* **2017**, *215*, 52–62. [CrossRef] [PubMed]

34. Vina-Almunia, J.; Mas-Bargues, C.; Borras, C.; Gambini, J.; El Alami, M.; Sanz-Ros, J.; Penarrocha, M.; Vina, J. Influence of Partial O($_2$) Pressure on the Adhesion, Proliferation, and Osteogenic Differentiation of Human Dental Pulp Stem Cells on beta-Tricalcium Phosphate Scaffold. *Int. J. Oral Maxillofac. Implants* **2017**, *32*, 1251–1256. [CrossRef] [PubMed]

35. He, J.; Ruan, G.P.; Yao, X.; Liu, J.F.; Zhu, X.Q.; Zhao, J.; Pang, R.Q.; Li, Z.A.; Pan, X.H. Chronic Toxicity Test in Cynomolgus Monkeys For 98 Days with Repeated Intravenous Infusion of Cynomolgus Umbilical Cord Mesenchymal Stem Cells. *Cell Physiol. Biochem.* **2017**, *43*, 891–904. [CrossRef] [PubMed]

36. Yun, J.W.; Ahn, J.H.; Kwon, E.; Kim, S.H.; Kim, H.; Jang, J.J.; Kim, W.H.; Kim, J.H.; Han, S.Y.; Kim, J.T.; et al. Human umbilical cord-derived mesenchymal stem cells in acute liver injury: Hepatoprotective efficacy, subchronic toxicity, tumorigenicity, and biodistribution. *Regul. Toxicol. Pharmacol.* **2016**, *81*, 437–447. [CrossRef] [PubMed]

37. Reyes, B.; Coca, M.I.; Codinach, M.; Lopez-Lucas, M.D.; Del Mazo-Barbara, A.; Caminal, M.; Oliver-Vila, I.; Cabanas, V.; Lope-Piedrafita, S.; Garcia-Lopez, J.; et al. Assessment of biodistribution using mesenchymal stromal cells: Algorithm for study design and challenges in detection methodologies. *Cytotherapy* **2017**, *19*, 1060–1069. [CrossRef] [PubMed]

38. Choi, S.A.; Yun, J.W.; Joo, K.M.; Lee, J.Y.; Kwak, P.A.; Lee, Y.E.; You, J.R.; Kwon, E.; Kim, W.H.; Wang, K.C.; et al. Preclinical Biosafety Evaluation of Genetically Modified Human Adipose Tissue-Derived Mesenchymal Stem Cells for Clinical Applications to Brainstem Glioma. *Stem Cells Dev.* **2016**, *25*, 897–908. [CrossRef] [PubMed]

39. Toupet, K.; Maumus, M.; Peyrafitte, J.A.; Bourin, P.; van Lent, P.L.; Ferreira, R.; Orsetti, B.; Pirot, N.; Casteilla, L.; Jorgensen, C.; et al. Long-term detection of human adipose-derived mesenchymal stem cells after intraarticular injection in SCID mice. *Arthritis Rheum.* **2013**, *65*, 1786–1794. [CrossRef] [PubMed]

40. Suzuki, K.; Nagata, K.; Yokota, T.; Honda, M.; Aizawa, M. Histological evaluations of apatite-fiber scaffold cultured with mesenchymal stem cells by implantation at rat subcutaneous tissue. *Biomed. Mater. Eng.* **2017**, *28*, 57–64. [CrossRef] [PubMed]

41. Ismail, T.; Osinga, R.; Todorov, A., Jr.; Haumer, A.; Tchang, L.A.; Epple, C.; Allafi, N.; Menzi, N.; Largo, R.D.; Kaempfen, A.; et al. Engineered, axially-vascularized osteogenic grafts from human adipose-derived cells to treat avascular necrosis of bone in a rat model. *Acta Biomater.* **2017**, *63*, 236–245. [CrossRef] [PubMed]

42. Bouvet-Gerbettaz, S.; Boukhechba, F.; Balaguer, T.; Schmid-Antomarchi, H.; Michiels, J.F.; Scimeca, J.C.; Rochet, N. Adaptive immune response inhibits ectopic mature bone formation induced by BMSCs/BCP/plasma composite in immune-competent mice. *Tissue Eng. Part A* **2014**, *20*, 2950–2962. [CrossRef] [PubMed]

43. Chu, W.; Gan, Y.; Zhuang, Y.; Wang, X.; Zhao, J.; Tang, T.; Dai, K. Mesenchymal stem cells and porous beta-tricalcium phosphate composites prepared through stem cell screen-enrich-combine(-biomaterials) circulating system for the repair of critical size bone defects in goat tibia. *Stem Cell Res. Ther.* **2018**, *9*, 157. [CrossRef] [PubMed]

materials

MDPI

Article

Calcium Charge and Release of Conventional Glass-Ionomer Cement Containing Nanoporous Silica

Koichi Nakamura [1,*], Shigeaki Abe [2], Hajime Minamikawa [3] and Yasutaka Yawaka [1]

[1] Department of Dentistry for Children and Disabled Person, Graduate School of Dental Medicine, Hokkaido University, Kita 13 Nishi 7, Kita-ku, Sapporo 060-8586, Hokkaido, Japan; yawaka@den.hokudai.ac.jp
[2] Department of Biomaterials and Bioengineering, Graduate School of Dental Medicine, Hokkaido University, Sapporo 060-8586, Hokkaido, Japan; sabe@den.hokudai.ac.jp
[3] Department of Dentistry for Molecular Cell Pharmacology, Graduate School of Dental Medicine, Hokkaido University, Sapporo 060-8586, Hokkaido, Japan; minami@den.hokudai.ac.jp
* Correspondence: pika@den.hokudai.ac.jp; Tel.: +81-11-706-4292

Received: 5 July 2018; Accepted: 23 July 2018; Published: 27 July 2018

Abstract: The aim of this study was to evaluate calcium charge and release of conventional glass-ionomer cement (GIC) containing nanoporous silica (NPS). Experimental specimens were divided into two groups: the control (GIC containing no NPS) and GIC-NPS (GIC containing 10 wt % NPS). The specimens were immersed in calcium chloride solutions of 5 wt % calcium concentration for 24 h at 37 °C, whereupon the calcium ion release of the specimens was measured. The calcium ion release behavior of GIC-NPS after immersion in the calcium solution was significantly greater than that of the control. Scanning electron microscopy and electron-dispersive X-ray spectroscopy results indicated that calcium penetrated inside the GIC-NPS specimen, while the calcium was primarily localized on the surface of the control specimen. It was demonstrated that NPS markedly improved the calcium charge and release property of GIC.

Keywords: nanoporous silica; glass-ionomer cement; calcium

1. Introduction

The overall incidence of dental caries has gradually decreased, but it still occurs quite frequently [1]. In the case of small dental carious lesions, the tooth is typically restored by a coronal restoration material, which often requires a retreatment owing to circumstances such as the progress of the dental caries around the restoration or the dental restoration falling out [2,3]. The main cause of coronal restoration treatment failure is secondary caries, which requires coronal restoration or pulp treatment [4,5]. To avoid tooth loss, therefore, it is necessary to prevent secondary caries, for which fluoride takes an important role [6,7]. Fluoride ions are included in dental items such as toothpaste and glass-ionomer cement (GIC) [8,9], and GIC is often used to prevent secondary caries in children [10,11]. GIC is widely used in dentistry such as the luting cement [12], temporary restoration [13], and adhering orthodontic band [14]. The GIC supplemented with TiO_2 nanoparticles also have the effect of antibacterial properties [15].

Calcium is an important component of the tooth. It is essential for the tooth mineralization that the calcium ion exists around the tooth. As one of the methods to supply calcium for teeth, some studies proposed the addition of casein phosphopeptide amorphous calcium phosphate (CPP-ACP) to GIC [16–18]. Addition of CPP-ACP to GIC increased the release of calcium ions. However, CPP-ACP is an ingredient derived from milk, and clinicians should consider potential side effects from ingestion of casein derivative protein in people with immunoglobulin E allergies to milk proteins [19].

Nanoporous silica (NPS) has a structure that possesses a uniform pore size of about 1 nm with a large specific surface area and is attractive in applications such as a sustained drug release carrier and catalyst support [20]. It is believed that NPS can adsorb formaldehyde gas and methylene blue pigments, and it has been confirmed that one of the properties of NPS is its ability to adsorb various ions and substances [21–23]. However, there are few studies on the application of NPS in dentistry.

The purpose of this study was to determine whether conventional GIC containing NPS was able to charge and release calcium ions. Furthermore, the cross-section of the GIC-NPS specimen was observed by scanning electron microscopy (SEM) and the compressive strength was measured. Our null hypotheses were that GIC containing the NPS did not charge and release calcium ions, and that there was no difference in the compressive strength of the GIC control and the GIC-NPS.

2. Materials and Methods

2.1. NPS Synthesis

The NPS was synthesized in accordance with the protocol described by Tagaya et al. [20]. In this process, 1.37 mmol of cetyltrimethylammonium bromide (CTAB: Wako, Osaka, Japan) was added to 120 mL of distilled water, to which 1.75 mL of 2.0 M sodium hydroxide (Wako) solution was added. The mixture was stirred for 30 min at 80 °C, whereupon 12.4 mmol of tetraethoxysilane (Wako) was added and the mixture stirred for 2 h at 80 °C. The resulting suspension was filtered and dried and, to remove the CTAB, the obtained particles were calcined at 550 °C for 4 h. The obtained white particles were subsequently observed using an SEM (S-4800, HITACHI, Tokyo, Japan) and a transmission electron microscope (TEM: JEM-2010, JEOL, Tokyo, Japan).

2.2. Preparation of Test Specimens

The obtained NPS was added to the powder component of the GIC (GC Fuji II, GC, Tokyo, Japan; lot 1606141) at 10 wt % concentration, whereupon the test specimens were prepared by mixing the powder component (with or without NPS) and the liquid component of the GIC (GC Fuji II, GC, Tokyo, Japan; lot 1607041) in accordance with the manufacturer specifications. The specimens were set in two patterns, where the first was 10 mm in diameter and 1 mm in height for the calcium ion release measurement and the SEM and electron-dispersive X-ray spectroscopy analysis, and the second was 4 mm in diameter and 6 mm in height for the compressive testing. The experimental specimens were divided into two groups comprising the control (no NPS) and the GIC-NPS (containing 10 wt % NPS). The specimens were wet ground via hand lapping using P400 grit silicon carbide abrasive papers (SANKYO, Saitama, Japan).

2.3. Calcium Charge and Release

Three 10 mm-diameter, 1 mm-height specimens were immersed in a calcium chloride solution with a 5 wt % calcium concentration (calcium 150 mg/3mL) for 24 h at 37 °C. After removal from the calcium solution, the specimens were rinsed with distilled water and subsequently immersed in 5 mL of distilled water at 37 °C for 7 d. The distilled water was changed every day and was further analyzed daily to determine the calcium ion concentration in the water. N = 6 samples (18 specimens) per group were used. The calcium ion concentration was measured using inductively coupled plasma atomic emission spectroscopy (ICPE-9000, SHIMADZU, Tokyo, Japan). The total weights of the calcium releasing from the specimens were calculated from the calcium concentration. The preparation of specimens was carried out according to Bando's conditions [24], but slightly modified.

2.4. Scanning Electron Microscopy (SEM) and Energy-Dispersive X-ray Spectroscopy (EDS)

Specimens containing calcium were analyzed using SEM and EDS (Genesis, EDAX Japan, Tokyo, Japan). The specimens were immersed in the 5 wt % calcium solution and dried. The specimens were cut perpendicularly, then the specimen surface and cross-section were observed after carbon coating.

2.5. Compressive Strength Test

The 4 mm-diameter, 6 mm-height test specimens were placed in a universal testing machine (Model 4204, Instron, Canton, OH, USA) with a cross-head speed of 1.0 mm/min. N = 12 specimens per group were used for this test. This procedure has been described in detail elsewhere [25].

2.6. Statistical Analysis

Statistical analysis was performed using IBM SPSS Statistics Version 21 (IBM Japan, Tokyo, Japan), and the results were analyzed statistically using the Mann-Whitney U test. The level of significance was set at $p < 0.05$.

3. Results

3.1. Morphological Characteristics of NPS

The SEM and TEM images of NPS particles are shown in Figure 1. The NPS particles were spherical and approximately 200–300 nm in diameter, exhibiting pores a few nanometers in diameter.

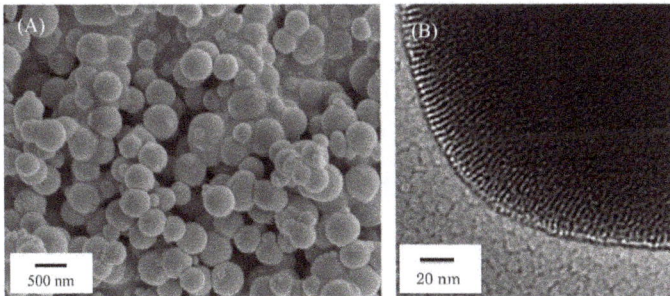

Figure 1. Typical (**A**) SEM and (**B**) TEM images of nanoporous silica particles. The nanoporous silica particles showed sizes approximately 200–300 nm in diameter (**A**), exhibiting pores a few nanometers in diameter (**B**).

3.2. Calcium Charge and Release

Figure 2 shows the time-profile of the release of calcium from the specimens after immersion in the 5 wt % calcium chloride solution. The calcium release behavior of GIC-NPS was significantly greater than that of the control ($p < 0.05$).

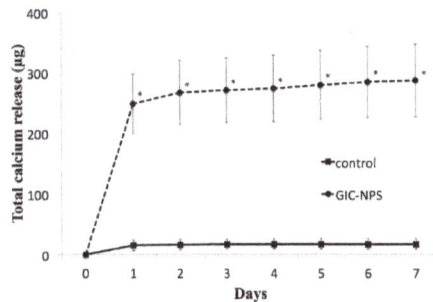

Figure 2. Time-profile of calcium release from the specimens after immersion in 5 wt % calcium chloride solution. Significance was determined using the Mann-Whitney U test ($p < 0.05$).

3.3. EDS Analysis

Figure 3A–C show the results of SEM and EDS for the surface and the cross-section of the specimens. As shown in Figure 3A, calcium was detected on the surfaces of both the control and GIC-NPS specimens; however, the specimen cross-sections exhibited significant amounts of calcium only for the GIC-NPS specimen. Although calcium was localized primarily on the surface of the control specimen, it was observed on the surface and throughout the inside of the GIC-NPS specimen (Figure 3B,C). In the surface of the control specimen (Figure 3B right and 3C right), calcium was distributed uniformly.

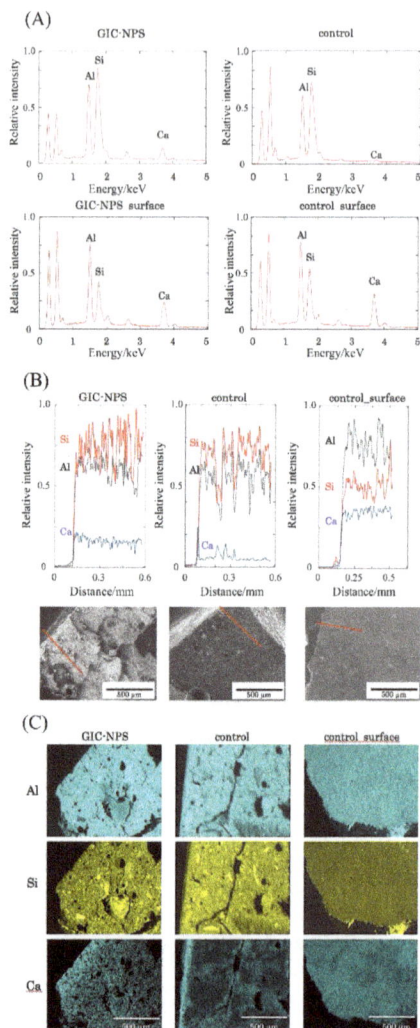

Figure 3. (**A**) Typical EDS spectra from the cross-section (upper) and surface (lower) of the GIC-NPS (left) and control (right) specimens, displaying the constitutive elements. (**B**) SEM images (lower) and EDS line analysis (upper) for the GIC-NPS cross-section (left), control cross-section (center), and control surface (right). (**C**) EDS element mapping of the Al (upper), Si (center), and Ca (lower) in the GIC-NPS cross-section (left), control cross-section (center), and control surface (right).

3.4. Compressive Strength

The compressive strength of the control and GIC-NPS were 111.37 ± 28.75 MPa and 100.32 ± 20.73 MPa (Mean ± SD, N = 12). There was no significant difference between the compressive strength of the control and GIC-NPS (p = 0.319).

4. Discussion

In the present study, the amount of the calcium release from the glass-ionomer cement containing nanoporous silica (GIC-NPS) specimen was determined. The results suggested the existence of an NPS-induced calcium ion charge and release property of the GIC-NPS. Even when a concentration of 10 wt % NPS was included in the GIC, the obtained specimens exhibited compressive strength, comparable to that of conventional GIC without NPS. It has been suggested that the possibility exists of NPS being used in dental materials as a calcium source, and it is therefore important that NPS did not reduce significantly the strength properties of the dental materials in the present study. However, other properties (solubility, adhesive strength, etc.) of GIC-NPS still have not been clarified, and so it is necessary for those concerned to hold further studies.

The nanopores in the NPS were approximately the size through which a calcium ion can pass [26]. The EDS images of the GIC-NPS specimens indicated that calcium ions penetrated into the GIC-NPS after immersion in the calcium solution. The NPS is well known as an adsorbent particle, and calcium might be adsorbed in the NPS pores in the same way other substances are adsorbed. It was not confirmed, however, which component of the NPS combined with the calcium, but this should be the subject of future research.

There were 150 mg of calcium in the immersed solution; nevertheless, the total amounts of the calcium released from specimen of control and GIC-NPS were 17.21 ± 8.66 μg and 287.71 ± 56.60 μg, respectively. Almost all of the calcium remained in the immersed solution or was washed away by rinsing with water. The percentage of total calcium released from the specimen in GIC-NPS on the first day, the second day, and the third day were 86.6%, 6.5%, and 1.5%, respectively. In contrast, those percentages for the control on the first day, the second day, and the third day were 87.7%, 6.6%, and 3.3%, respectively. Though the total amount of calcium was different for both, the percentage of calcium released had a similar tendency in both.

In this study, NPS was found to not only adsorb calcium ion but also to release adsorbed calcium ions when placed in an aqueous medium. The GIC-NPS could be expected to function as a source of calcium, and it was suggested that it could be useful to facilitate remineralization. Furthermore, the GIC used in this study was a fluoride-releasing material; thus, GIC-NPS may be useful for producing fluoroapatite by allowing the coexistence of calcium and fluoride ions [27]. This is significant because fluoroapatite possesses a higher acid resistance than hydroxyapatite, the main component of teeth. The amount of fluoride ions released from the GIC depends on the fluoroaluminosilicate glass composition included in the powder [28]. In this study, the inclusion of NPS decreases the relative amount of the GIC powder, so it is likely that the amount of fluoride ions released may be slightly decreased. This reduction may be negligible with a 10 wt % concentration of NPS, however, because of the amount of fluoride ions released from the GIC [29]. Furthermore, the specimens were immersed in calcium chloride solution in this study. Previous research has indicated that the surface hardness of GIC increases with immersion in the calcium chloride solution [30]. The surface property of GIC may have changed by immersion of the solution.

The SEM/EDS analysis indicated that calcium penetrated to the interior of the GIC-NPS specimen. After immersion in the calcium solution, calcium was distributed homogenously on the surfaces of the control specimen similar to the distribution of aluminum and silicon, which are the major constituents of GIC (Figure 3C, right column). However, the distribution behavior in the depth direction was clearly different in the two specimens. In the case of the control, calcium was localized on the surface of specimen (Figure 3C, middle column), while calcium was detected even inside of the GIC-NPS (Figure 3C, left column). The highest concentration of calcium was determined to be on the surface of the GIC-NPS specimen, similar to that found for the control specimen. The EDS mapping suggests that the calcium concentration slightly decreased as a function of depth but persisted throughout the interior of the GIC-NPS specimen.

The EDS spectral line analysis (Figure 3B) also revealed that the calcium concentration profile of the GIC-NPS specimen was clearly different than that of the control specimen. The latter exhibited a sharp peak only around the specimen surface, while the former exhibited a calcium concentration that slightly decreased in a path from the surface to a depth of 0.1 mm, then remained approximately constant up to 0.5 mm of depth.

Nanoporous silica is often applied as a film on the material surface and is rarely contained inside the material [31]. Nevertheless, it was found herein that NPS was capable of the charge and release of calcium ions through the matrix even when the NPS was incorporated throughout the material.

Previous studies of powdered inorganic additives have shown the compressive strength increase and decrease [25,32,33]. In this study, compressive strength has slightly decreased. It may be due to interference of the NPS with the normal GIC reaction. By increasing of NPS component, compressive strength might significantly decrease. However, under the conditions of the present study, there were no significant differences between the two groups.

The appearance of secondary caries requires a certain period time. In the present study, however, the calcium charge and release properties of GIC-NPS were observed over only a week. If it is a repeated charge and release of the calcium, secondary caries were prevented in the long term. To confirm this presumption, further study is needed. We also measured the amount of calcium charge and release; the study of the preventing effect for secondary caries may be needed in the model similar to the oral cavity.

In recent years, many studies of bioactive materials have been conducted [34–36]. The results of this study may be useful to the development of biomaterials. Firstly, the results herein may be extended to the development of remineralization-inducing materials by the uptake of phosphate ions, which are essential for remineralization, e.g., pit and fissure sealants and restoration materials. Also, this study may inform the drug delivery system by the uptake of antibacterial agents such as Cetylpyridinium Chloride (CPC). Furthermore, the NPS used in this study possesses a negative charge, and therefore cannot adsorb a negatively-charged fluoride ion. However, the NPS may be enabled to adsorb a fluoride ion by also causing it to retain a positive electric charge. This would give the ability of fluoride ion release to the material without also requiring a composite resin that exhibits fluoride ion release characteristics. We believe that it is necessary to study these possibilities in future work.

5. Conclusions

We demonstrated the capacity of GIC-NPS for calcium ion charge/release and contrasted it with that of conventional GIC. The presence of NPS was found to markedly improve the calcium ion charge/release property of GIC. Even for NPS concentrations up to 10 wt % in the GIC, the compressive strength of the GIC was not changed significantly.

Supplementary Materials: The following are available online at http://www.mdpi.com/1996-1944/11/8/1295/s1, Table S1: Total calcium release, Table S2: Compressive strength.

Author Contributions: K.N., S.A., and H.M. performed the experiments and analyzed the data; Y.Y. contributed as advisors; and K.N. wrote the paper. All authors read and approved the final manuscript.

Funding: This work was supported by JSPS KAKENHI Grant Number 15K20575.

Conflicts of Interest: The authors declare no conflict of interest.

References

1. Ramos-Jorge, J.; Alencar, B.M.; Pordeus, I.A.; Soares, M.E.; Marques, L.S.; Ramos-Jorge, M.L.; Paiva, S.M. Impact of dental caries on quality of life among preschool children: Emphasis on the type of tooth and stages of progression. *Eur. J. Oral Sci.* **2015**, *123*, 88–95. [CrossRef] [PubMed]

2. Metz, I.; Rothmaier, K.; Pitchika, V.; Crispin, A.; Hickel, R.; Garcia-Godoy, F.; Bücher, K.; Kühnisch, J. Risk factors for secondary caries in direct composite restorations in primary teeth. *Int. J. Paediatr. Dent.* **2015**, *25*, 451–461. [CrossRef] [PubMed]

3. Pallesen, U.; Van Dijken, J.W. A randomized controlled 27 years follow up of three resin composites in Class II restorations. *J. Dent.* **2015**, *43*, 1547–1558. [CrossRef] [PubMed]

4. Brantley, C.F.; Bader, J.D.; Shugars, D.A.; Nesbit, S.P. Does the cycle of rerestoration lead to larger restorations? *J. Am. Dent. Assoc.* **1995**, *126*, 1407–1413. [CrossRef] [PubMed]

5. Hsu, C.Y.; Donly, K.; Drake, D.; Wefel, J. Effects of aged fluoride-containing restorative materials on recurrent root caries. *J. Dent. Res.* **1998**, *77*, 418–425. [CrossRef] [PubMed]

6. Comar, L.P.; Wiegand, A.; Moron, B.M.; Rios, D.; Buzalaf, M.A.; Buchalla, W.; Magalhães, A.C. In situ effect of sodium fluoride or titanium tetrafluoride varnish and solution on carious demineralization of enamel. *Eur. J. Oral Sci.* **2012**, *120*, 342–348. [CrossRef] [PubMed]

7. Bonetti, D.; Clarkson, J.E. Fluoride Varnish for Caries Prevention: Efficacy and Implementation. *Caries Res.* **2016**, *50*, 45–49. [CrossRef] [PubMed]

8. Kucukyilmaz, E.; Savas, S. Evaluation of shear bond strength, penetration ability, microleakage and remineralisation capacity of glass-ionomer-based fissure sealants. *Eur. J. Paediatr. Dent.* **2016**, *17*, 17–23. [PubMed]

9. Yönel, N.; Bikker, F.J.; Lagerweij, M.D.; Kleverlaan, C.J.; Van Loveren, C.; Özen, B.; Çetiner, S.; Van Strijp, A.J. Anti-erosive effects of fluoride and phytosphingosine: An in vitro study. *Eur. J. Oral Sci.* **2016**, *124*, 396–402. [CrossRef] [PubMed]

10. Goldman, A.S.; Chen, X.; Fan, M.; Frencken, J.E. Cost-effectiveness, in a randomized trial, of glass-ionomer-based and resin sealant materials after 4 yr. *Eur. J. Oral Sci.* **2016**, *124*, 472–479. [CrossRef] [PubMed]

11. Zhao, I.S.; Mei, M.L.; Burrow, M.F.; Lo, E.C.; Chu, C.H. Prevention of secondary caries using silver diamine fluoride treatment and casein phosphopeptide-amorphous calcium phosphate modified glass-ionomer cement. *J. Dent.* **2017**, *57*, 38–44. [CrossRef] [PubMed]

12. Heintze, S.D. Crown pull-off test (crown retention test) to evaluate the bonding effectiveness of luting agents. *Dent. Mater.* **2010**, *26*, 193–206. [CrossRef] [PubMed]

13. Coll, J.A.; Campbell, A.; Chalmers, N.I. Effects of glass ionomer temporary restorations on pulpal diagnosis and treatment outcomes in primary molars. *Pediatr. Dent.* **2013**, *35*, 416–421. [PubMed]

14. Millett, D.T.; Glenny, A.M.; Mattick, R.C.; Hickman, J.; Mandall, N.A. Adhesives for fixed orthodontic bands. *Cochrane Database Syst. Rev.* **2016**, *10*, CD004485. [CrossRef] [PubMed]

15. Garcia-Conreras, R.; Scougall-Vilchis, R.J.; Contreras-Bulner, R.; Sakagami, H.; Morales-Luckie, R.A.; Nakajima, H. Mechanical, antibacterial and bond strength properties of nano-titanium-enriched glass ionomer cement. *J. Appl. Oral Sci.* **2015**, *23*, 321–328. [CrossRef] [PubMed]

16. Mazzaoui, S.A.; Burrow, M.F.; Tyas, M.J.; Dashper, S.G.; Eakins, D.; Reynolds, E.C. Incorporation of casein phosphopeptide–amorphous calcium phosphate into a glass-ionomer cement. *J. Dent. Res.* **2003**, *82*, 914–918. [CrossRef] [PubMed]

17. Al Zraikat, H.; Palamara, J.E.; Messer, H.H.; Burrow, M.F.; Reynolds, E.C. The incorporation of casein phosphopeptide–amorphous calcium phosphate into a glass ionomer cement. *Dent. Mater.* **2011**, *27*, 235–243. [CrossRef] [PubMed]

18. Zalizniak, I.; Palamara, J.E.; Wong, R.H.; Cochrane, N.J.; Burrow, M.F.; Reynolds, E.C. Ion release and physical properties of CPP-ACP modified GIC in acid solutions. *J. Dent.* **2013**, *41*, 449–454. [CrossRef] [PubMed]

19. Azarpazhooh, A.; Limeback, H. Clinical efficacy of casein derivatives: A systematic review of the literature. *J. Am. Dent. Assoc.* **2008**, *139*, 915–924. [CrossRef] [PubMed]

20. Tagaya, M.; Ikoma, T.; Yoshioka, T.; Motozuka, S.; Xu, Z.; Minami, F.; Tanaka, J. Synthesis and luminescence properties of Eu(III)-doped nanoporous silica spheres. *J. Colloid Interface Sci.* **1999**, *218*, 462–467. [CrossRef] [PubMed]

21. Barbé, C.; Bartlett, J.; Kong, L.; Finnie, K.; Lin, H.Q.; Larkin, M.; Calleja, S.; Bush, A.; Calleja, G. Silica Particles: A Novel Drug-Delivery System. *Adv. Mater.* **2004**, *16*, 1959–1966. [CrossRef]

22. Fazaeli, Y.; Feizi, S.; Jalilian, A.R.; Hejrani, A. Grafting of [(64)Cu]-TPPF20 porphyrin complex on Functionalized nano-porous MCM-41 silica as a potential cancer imaging agent. *Appl. Radiat. Isotopes* **2016**, *112*, 13–19. [CrossRef] [PubMed]

23. Wang, L.; Chen, Q.; Li, C.; Fang, F. Nano-Web Cobalt Modified Silica Nanoparticles Catalysts for Water Oxidation and MB Oxidative Degradation. *J. Nanosci. Nanotechnol.* **2016**, *16*, 5364–5368. [CrossRef] [PubMed]

24. Bando, Y.; Nakanishi, K.; Abe, S.; Yamagata, S.; Yoshida, Y.; Iida, J. Electric charge dependence of controlled dye-release behavior in glass ionomer cement containing nano-poruos silica particles. *J. Nanosci. Nanotechnol.* **2018**, *18*, 75–79. [CrossRef] [PubMed]

25. Elsaka, S.E.; Hamouda, I.M.; Swain, M.V. Titanium dioxide nanoparticles addition to a conventional glass-ionomer restorative: Influence on physical and antibacterial properties. *J. Dent.* **2011**, *39*, 589–598. [CrossRef] [PubMed]

26. Hille, B. *Ion Channels of Excitable Membranes*, 3rd ed.; Sinauer Associates: Sunderland, MA, USA, 2001.

27. Wegehaupt, F.J.; Tauböck, T.T.; Sener, B.; Attin, T. Long-term protective effect of surface sealants against erosive wear by intrinsic and extrinsic acids. *J. Dent.* **2012**, *40*, 416–422. [CrossRef] [PubMed]

28. Dennis, C.S. Development of glass-ionomer cement systems. *Biomaterials* **1998**, *19*, 467–478.

29. Okte, Z.; Bayrak, S.; Fidanci, U.R.; Sel, T. Fluoride and aluminum release from restorative materials using ion chromatography. *J. Appl. Oral Sci.* **2012**, *20*, 27–31. [CrossRef] [PubMed]

30. Shiozawa, M.; Takahashi, H.; Iwasaki, N.; Wada, T.; Uo, M. Effect of immersion time of restorative glass ionomer cements and immersion duration in calcium chloride solution on surface hardness. *Dent. Mater.* **2014**, *30*, 377–383. [CrossRef] [PubMed]

31. Song, D.P.; Naik, A.; Li, S.; Ribbe, A.; Watkins, J.J. Rapid, Large-Area Synthesis of Hierarchical Nanoporous Silica Hybrid Films on Flexible Substrates. *J. Am. Chem. Soc.* **2016**, *138*, 13473–13476. [CrossRef] [PubMed]

32. Xie, D.; Weng, Y.; Guo, X.; Zhao, J.; Gregory, R.L.; Zheng, C. Preparation and evaluation of a novel glass-ionomer cement with antibacterial functions. *Dent. Mater.* **2011**, *27*, 487–496. [CrossRef] [PubMed]

33. Bellis, C.A.; Nobbs, A.H.; O'sullivan, D.J.; Holder, J.A.; Barbour, M.E. Glass ionomer cements functionalised with a concentrated paste of chlorhexidine hexametaphosphate provides dose-dependent chlorhexidine release over at least 14 months. *J. Dent.* **2016**, *45*, 53–58. [CrossRef] [PubMed]

34. Lee, J.H.; Seo, S.J.; Kim, H.W. Bioactive glass-based nanocomposites for personalized dental tissue regeneration. *Dent. Mater. J.* **2016**, *35*, 710–720. [CrossRef] [PubMed]

35. Uo, M.; Wada, T.; Asakura, K. Structural analysis of strontium in human teeth treated with surface pre-reacted glass-ionomer filler eluate by using extended X-ray absorption fine structure analysis. *Dent. Mater. J.* **2017**, *36*, 214–221. [CrossRef] [PubMed]

36. Gandolfi, M.G.; Iezzi, G.; Piattelli, A.; Prati, C.; Scarano, A. Osteoinductive potential and bone-bonding ability of ProRoot MTA, MTA Plus and Biodentine in rabbit intramedullary model: Microchemical characterization and histological analysis. *Dent. Mater.* **2017**, *33*, 221–238. [CrossRef] [PubMed]

materials

MDPI

Article

Human Dental Pulp Stem Cells Exhibit Different Biological Behaviours in Response to Commercial Bleaching Products

Carmen Llena [1], Mar Collado-González [2], Christopher Joseph Tomás-Catalá [2], David García-Bernal [2], Ricardo Elías Oñate-Sánchez [2], Francisco Javier Rodríguez-Lozano [2,3,]* and Leopoldo Forner [1]

[1]   Department of Stomatology, University de Valencia, 46010 Valencia, Spain; llena@uv.es (C.L.); forner@uv.es (L.F.)
[2]   School of Dentistry/Cellular Therapy and Hematopoietic Transplant Unit, Hematology Department, Virgen de la Arrixaca Clinical University Hospital, IMIB-Arrixaca, University of Murcia, 30120 Murcia, Spain; mdmcg1@um.es (M.C.-G.); ctc20203@um.es (C.J.T.-C.); redond@gmail.com (D.G.-B.); reosan@um.es (R.E.O.-S.)
[3]   School of Dentistry, Hospital Morales Meseguer 2pl. Av. Marqués de los Vélez s/n, University of Murcia, 30008 Murcia, Spain
*   Correspondence: fcojavier@um.es; Tel.: +34-86-888-9518

Received: 29 May 2018; Accepted: 22 June 2018; Published: 27 June 2018

Abstract: The purpose of this study was to evaluate the diffusion capacity and the biological effects of different bleaching products on human dental pulp stem cells (hDPSCs). The bleaching gel was applied for 90, 30 or 15 min to enamel/dentine discs that adapted in an artificial chamber. The diffusion of hydrogen peroxide (HP) was analysed by fluorometry and the diffusion products were applied to hDPSCs. Cell viability, cell migration and cell morphology assays were performed using the eluates of diffusion products. Finally, cell apoptosis and the expression of mesenchymal stem cell markers were analysed by flow cytometry. Statistical analysis was performed using analysis of variance and Kruskal–Wallis or Mann–Whitney tests ($\alpha < 0.05$). Significant reductions of approximately 95% in cell viability were observed for the 3×15 min groups ($p < 0.001$), while 1×30 min of PerfectBleach and 1×90 min of PolaNight resulted in reductions of 50% and 60% in cell viability, respectively ($p < 0.001$). Similar results were obtained in the migration assay. Moreover, the 3×15 min group was associated with cell morphology alterations and reductions of >70% in cell live. Finally, hDPSCs maintained their mesenchymal phenotype in all conditions. Similar concentrations of carbamide peroxide (CP) and HP in different commercial products exhibited different biological effects on hDPSCs.

Keywords: bleaching products; diffusion; cytotoxicity; dental pulp; stem cells

1. Introduction

The first study examining the chemical mechanism of hydrogen peroxide (HP) when coming into contact with enamel was published in 1970, which demonstrated that HP reached the dentin structure to produce a whitening effect [1]. The permeability of hard dental tissues and the low molecular weight of HP explain HP diffusion [2,3]. The bleaching process occurs when the low molecular weight of HP diffuses through enamel and dentin as it releases reactive oxygen species (ROS) that react with other free or weakly bound substances before regaining molecular stability. This oxidant phenomenon may explain the complex mechanism of dental bleaching [4]. The presence of oxidising agents and the penetration capacity of HP are closely correlated with each other [5].

HP reaches the pulp chamber via the dentinal tubules, decreases cell metabolism and viability [6,7] and induces vascular permeability changes [8], DNA modifications and pulpal necrosis [9–12].

Dental pulp is a loose connective tissue that occupies the pulp chamber of the tooth, which originates from the embryonic dental papilla (ectomesenchymal tissue) [13]. It has a high capacity for repair since it exhibits a significant regenerative response in reaction to injury or trauma, which causes the secretion of tertiary dentine and promotion of the differentiation of dental pulp stem cells (*h*DPSCs) into odontoblast-like cells. In addition, *h*DPSCs are able to differentiate toward osteogenic, myogenic and adipogenic lineages, melanocytes, Schwann cells and neurons [14].

Previous studies have analysed the biocompatibility of several bleaching agents, although there are fewer investigations studying specific commercial products and the effect of different times of applications [15–19]. Furthermore, the different compositions of commercial bleaching products are often unknown and commercial bleaching products have provided differing results for their biocompatibility in vitro and in vivo.

Cell cultures are suitable for evaluating the effects of different dental products and materials on pulp tissues [20] and the biological response of human pulp cells to these products and materials. The results of these in vitro studies are not directly comparable to those in humans, but these studies provide a good model with which to analyse different products and techniques and their potential risks [21].

The literature has questioned whether the use of bleaching products with a high concentration of HP is necessary or even safe. However, these products have been applied and reapplied multiple times in the same clinical session in order to increase the speed of changing the colour of the teeth. Although the whitening effect is known to be related to the diffusion of peroxide through the dental tissues, studies suggest that the bleaching effect is not related to the constant reapplication of the gel because good results have been obtained with the technique of a single clinical application [22]. Thus, this finding might support the adoption of a new dosage that is based on a single application of the bleaching product. Thus, given that high concentrations of peroxide are potentially harmful to the pulp cells [7,23], a focus on the posology that is guided by the adoption of milder protocols is both appealing and justifiable in an effort to find safe alternatives to bleaching.

In addition to the discrepancy in the clinical application protocols, the diffusion of and pulp damage caused by different commercial products with similar HP concentrations is still not clear.

The present study evaluated the diffusion capacity and cytotoxicity of different high-concentration commercial bleaching products and different application protocols in vitro. The null hypothesis was that different commercial bleaching products with equal or similar concentrations of HP would not promote different responses in dental pulp stem cells. This study was conducted in order to determine if other substances contained in the product are responsible for the cytotoxicity of the commercial bleaching product.

There are no studies that have determined the toxicity of the excipients contained in different commercial bleaching products with equal or similar HP concentrations.

2. Materials and Methods

2.1. Specimen Selection and Preparation

We selected freshly impacted mandibular third molars ($n = 25$) from patients aged 18 to 35 years to obtain a homogeneous sample. Selected teeth were not subjected to the oral environment, occlusal loads or other functional aggressions. The study was approved by the Ethics Committee of the University of València (registration number H1443515306255).

Organic remains were removed from the teeth and the integrity of the dental structure was confirmed. Selected teeth were immersed in a 0.1% thymol solution and preserved at 4 °C for 48 h. Teeth were removed from the thymol solution and immersed in deionized water at 4 °C until use.

Twenty-five samples that contained 2 mm of enamel and 2 mm of dentin were prepared as follows. Each tooth was cut mesio-distally with a diamond disk mounted on a hand piece with water-cooling. Roots were discarded. The dentin surface was reduced with 400- and 600-grit silicon carbide paper Sof-lex™ discs (3M Dental Products, St. Paul, MN, USA) until the dentin thickness was 2 mm. The sample dimensions were in the range of 0.5–0.7 cm × 0.4–0.6 cm. The dentin surface was treated with 0.5 M ethylene diamine tetra-acetic acid (EDTA) at a pH of 7.4 for 30 s, which was subsequently rinsed with deionized water to remove the smear layer and stored in deionized water at 4 °C until use.

An artificial pulp chamber was prepared with heavy silicone (Panasil R Putty, Kettenbach, Huntington Beach, CA, USA) with a capacity of 100 μL. To stabilize the sample, a heavy silicone ring was fabricated to anchor the sample, which allowed the dentin to be in contact with the buffer and had an upper window for placing the study gel (Scheme 1). This ring fit perfectly in the artificial pulp chamber, which maintained the fixation of the sample, while additional sealing was performed between the sample and the ring by using wax.

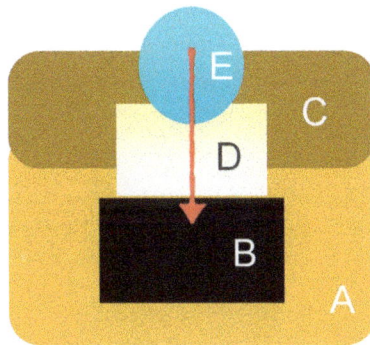

Scheme 1. Schematic representation of the artificial pulp chamber. Heavy silicone (A,C) was utilized to stabilize the sample (D), maintain the buffer (B) and prepare the artificial pulp chamber.

2.2. Diffusion Evaluation

The experimental specimens were randomized into five different groups (n = 5 each): group 1 was exposed to a neutral pH gel of 37.5% HP -Pola Office + (PO)- (SDI, Bayswater, VIC, Australia) for 30 min (PO30); group 2 was exposed to the same product as group 1 but with 3 applications of 15 min (PO3x15); group 3 was exposed to neutral pH gel of 35% HP -PerfectBleach (PB)- (Voco, Cuxhaven, Germany) for 30 min (PB30); group 4 was exposed to the same product as group 3 but with 3 applications of 15 min (PB3x15); and the group 5 was exposed to 16% carbamide peroxide (CP) -PolaNight (PN)- (SDI) for 90 min (PN90).

The diffusion of HP from different bleaching products was analysed using fluorimetry. The production of a homovanillic acid dimer from a reaction catalysed by peroxidase using HP as a substrate was measured as described by Barja [24]. Samples that had reacted with the peroxidase were measured using a fluorimeter (model F-4500 fluorescence spectrophotometer, Hitachi, Japan). A standard fluorimetry signal curve of HP (H1009-100ML, Sigma-Aldrich, St. Louis, MO, USA) was generated. The same dilutions were measured in a Helios Alpha UV–vis spectrophotometer (Thermo Fisher Scientific Inc., Waltman, MA, USA). Therefore, the fluorimetry signal was related to the concentration of the fluorescent dimer.

Phosphate-buffered saline (PBS; 400 μL) was placed in the buffer in the reservoir. Gel (0.5 μL) was applied to the external enamel surface and the film was coated (Parafilm M, Sigma-Aldrich, St. Louis, MO, USA). The HP in the PO15 and PB15 groups was replaced every 15 min without removal of the PBS from the reservoir. The HP remained for 30 min in the PO30 and PB30 groups and 90 min in the

PN90 group. The PBS containing the diffused HP from the reservoir was removed. The bleaching agent application process was performed at 37 °C.

A volume of 100 µL of the sample containing the diffused HP was removed and added to 1400 µL of the reaction buffer. Glycine−EDTA buffer (500 µL) was added after 15 min. Fluorescence intensity was measured, before the concentration value was extrapolated from the standard curve from the controls.

2.3. Biological Assays

2.3.1. Isolation and Culture of hDPSCs

hDPSCs were isolated and characterised as described previously [25]. Human DPSCs (hDPSCs) were obtained from impacted third molars collected from healthy subjects (*n* = 15). Donors provided written informed consent according to the guidelines of the Ethics Committee of our Institution. Briefly, the pulp chamber was exposed after removal of the tooth and the pulp tissue was retrieved, thoroughly minced and digested by means of a collagenase I (3 mg/mL) and dispase II (4 mg/mL) buffer (Invitrogen, Karlsruhe, Germany) for 45 min at 37 °C. The cells were cultured with a-MEM (Minimum Essential Media) medium (Invitrogen), supplemented with 15% fetal bovine serum (FBS, Invitrogen), 100 mM L-ascorbic acid phosphate (Sigma-Aldrich, Steinheim, Germany) and antibiotics/antimycotics (Complete Culture Medium, CCM) before incubating at 37 °C in 5% CO_2. This study was performed using hDPSCs from 4th passage of the culture.

2.3.2. Characterisation Assay

The mesenchymal immunophenotype of cultured human DPSCs was analyzed by flow cytometry. Single cell suspensions obtained by culture trypsinization were labelled or surface markers with fluorochrome-conjugated antibodies: CD73 (clone AD2), CD90 (clone DG3), CD105 (clone 43A4E1), CD14 (clone TÜK4), CD20 (clone LT20.B4), CD34 (clone AC136) and CD45 (clone 5B1) (Human Mesenchymal (MSC) Stem Cell Phenotyping Cocktail, Miltenyi Biotec, Bergisch-Gladbach, Germany), following the recommendations of the International Society of Cellular Therapy (ISCT) [26]. Flow cytometry analyses were performed using a FACSCanto II flow cytometer (BD Biosciences, San José, CA, USA).

2.3.3. Conditioned Medium

Culture medium conditioned with the bleaching products (BPs) was obtained as described by Cavalcanti et al. [27] in compliance with the ISO 10993-12 [28] proceedings. Briefly, 100 µL of the sample containing the diffused HP was removed and was collected into aliquots, before dilutions of 1%, 0.5% and 0.25% were created using cell culture medium. The conditioned medium was filtered in a sterile environment immediately and was used for cell culture exposure.

2.3.4. Cell Viability Assay

The metabolic activity of hDPSCs growing in the presence of different bleaching extracts was evaluated using the MTT assay (MTT Cell Growth Kit, Chemicon, Rosemont, IL, USA). hDPSCs were initially loaded with 1×10^3 cells per well and a volume of 180 µL in 96-well plates. Cells were starved for 24 h in serum-free medium at 37 °C in a humidified 5% CO_2 atmosphere prior to testing. The serum-free medium was replaced with different material elutes. Cells cultured in α-MEM medium containing 10% FBS were the negative control. After 24, 48 and 72 h, MTT (1 mg/mL in CCM) was added after cell seeding and incubated for 4 h at 37 °C and 5% CO_2. The MTT insoluble formazan was then dissolved by means of dimethyl sulfoxide DMSO that was applied for 2–4 h at 37 °C. The optical density (OD) was measured against blank (DMSO) at a wavelength of 570 nm (Abs570) and a reference filter of 690 nm by an automatic microplate reader (ELx800; Bio-Tek Instruments, Winooski, VT, USA)

2.3.5. Cell Migration

To study the chemotactic effect of bleaching extracts, hDPSCs were seeded on 12-well plastic dishes and incubated in culture medium for 24 h until until a monolayer formed. A pipette tip was used to generate a cross-shaped scratch or "wound" and the cell debris was removed with PBS. Plated cells were incubated with various dilutions of the eluates up to 48 h to allow cell migration back into the wounded area. Images of the scratched areas were captured at 0, 24 and 48 h using a phase-contrast microscope (Nikon, Tokyo, Japan). Two images photomicrographs per well were takenat each indicated time. ImageJ software (NIH, Bethesda, MD, USA) was used to evaluate the wound closure area.

2.3.6. Cell Morphology in Presence of Extracts

hDPSCs were plated on the glass coverslips at a low density and cultured in culture medium containing different material extracts to investigate morphological changes. The cells were fixed with 4% paraformaldehyde in phosphate-buffered saline (PBS) for 20 min at 2537 °C, permeabilized with 0.25% Triton X-100 (Sigma-Aldrich, St. Louis, MO, USA) for 3 min, and washed 3 times with PBS. For staining filamentous actin (F-actin), cells were incubated with CruzFluor594-conjugated phalloidin (Santa Cruz Biotechnology, Dallas, TX, USA). Nuclei were counterstained with 4,6-diamidino-2-phenylindole dihydrochloride (DAPI; Sigma-Aldrich). Fluorescent images were obtained using a confocal microscopy (Zeiss, Oberkochen, Germany).

2.3.7. Analysis of the Expression of Mesenchymal Stem Cell Surface Markers on hDPSCs Exposed to Bleaching Products Using Flow Cytometry

The expression of mesenchymal stem cell surface molecules was analysed in cultures of hDPSCs in 2nd–4th passages using flow cytometry. Briefly, cells were seeded at a density of 3×10^4 cells/cm^2 in 48-well plates and treated with the different eluates for 72 h at 37 °C. Cells were detached using a 0.25% w/v trypsin-EDTA solution. This was then rinsed twice with PBS and incubated in the dark at 4 °C for 30 min with fluorescence-conjugated specific monoclonal antibodies for human CD73, CD90 and CD105 (MiltenyiBiotec, BergischGladbach, Germany), as recommended by the International Society of Cellular Therapy (ISCT), to confirm the mesenchymal phenotype of the cells (Dominici et al., 2006). The lack of expression of the hematopoietic markers CD14, CD20, CD34 and CD45 was also analysed. Non-specific fluorescence was measured using specific-isotype monoclonal antibodies. Cells were acquired using a BD FACS Canto flow cytometer (BD Biosciences) and analysed using Kaluza analysis software (Beckman Coulter, Inc., Brea, CA, USA).

2.3.8. Detection of Apoptosis and Necrosis Using Flow Cytometry (Annexin-V/7-AAD Staining)

hDPSCs were cultured in different bleaching extracts for 72 h, which was followed by double staining with PE-conjugated Annexin-V and 7-AAD (BD Biosciences, Pharmingen). Percentages of live (Annexin-V$^-$/7-AAD$^-$), early apoptotic (Annexin-V$^+$/7-AAD$^-$) or late apoptotic and necrotic cells (Annexin-V$^+$/7-AAD$^+$ and Annexin-V$^-$/7-AAD$^+$) were determined using flow cytometry, before the percentages of each population were calculated.

2.4. Statistical Analysis

Two independent experiments were performed for all protocols in this study. Data were compiled and subjected to Levene's test to verify the homogeneity of variance. Data data were compiled and subjected to one-way ANOVA and Tukey's test (* $p < 0.05$, ** $p < 0.01$ and *** $p < 0.001$.)

3. Results

3.1. HP/CP Diffusion

All bleached groups differed in HP diffusion. We created a 1:2 dilution of the productions of one application, while we created 1:4 dilutions of the products of 3 applications to obtain unsaturated measures of fluorometry. Our results demonstrated variation in the quantification of homovanillic dimer (HD) between bleaching products (Figure 1). Table 1 shows that HD concentrations with PN were lower than those with the other products despite the longer exposure time (90 min). Pola Office exhibited no differences with different numbers of applications. Perfect Bleach exhibited great diffusion with three applications.

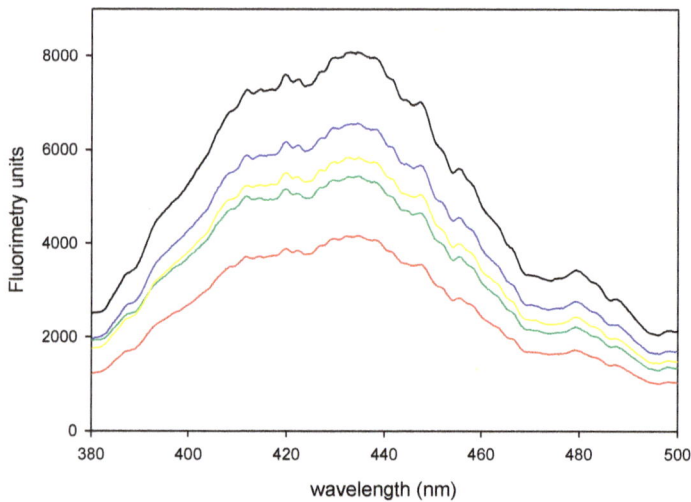

Figure 1. Fluorometric spectrum of samples of PO30 diluted 1:2 (black), PO3x15 diluted 1:4 (red), PB30 diluted 1:2 (green), PB3x15 diluted 1:4 (yellow) and PN90 diluted 1:2 (blue).

Table 1. Bleaching products diffusion.

Bleaching Products	Applications	Diffusion (mM)
Pola Office	1×30	38.4 ± 8.2
Pola Office	3×15	39.3 ± 1.2
Perfect Bleach	1×30	28.2 ± 6.5
Perfect Bleach	3×15	48.6 ± 1.7
Polanight	1×90	27.7 ± 0.4

3.2. Isolation and Characterisation of hDPSCs

Positive cell surface marker expression of the mesenchymal markers CD73, CD90 and CD105 and negative expression of the hematopoietic markers CD14, CD20, CD34 and CD45 were evaluated on *h*DPSCs using flow cytometry. More than 95% of viable *h*DPSCs were positive for the mesenchymal markers CD73, CD90 and CD105 and negative for the hematopoietic markers CD14, CD20, CD34 and CD45 (Figure 2).

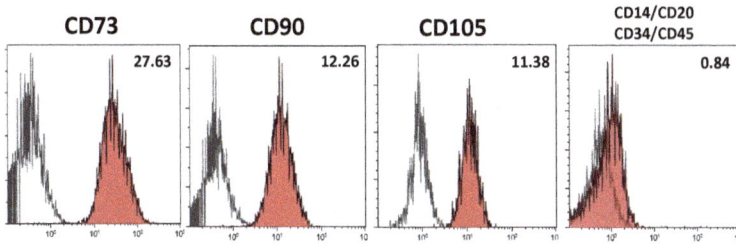

Figure 2. Immunophenotypic characterization of *h*DPSCs by flow cytometry for the expression of mesenchymal (CD73, CD90 and CD105) and hematopoietic (CD14, CD20, CD34 and CD45) stem cell (SC) markers (grey histogram: unstained control, red histogram: marker of interest). Results are means of triplicates of three independent experiments.

3.3. MTT Assays

MTT assays were performed to evaluate the effects of the eluates of different bleaching agents on *h*DPSC viability (Figure 3). The incubation of *h*DPSCs with 1% of each bleaching product yielded a significant reduction in cell viability rates compared with the control group at the indicated times (*** $p < 0.001$). In contrast, the 1%, 0.5% and 0.25% PB30 or PN allowed a slight but significant recovery of cell viability compared to PO30, PO3x15 or PB3x15 ($^{\Delta\Delta\Delta}$ $p < 0.001$). There was not difference in cell viability between 1 and 3 applications of the bleaching agents, except PO as there were significant differences in cell viability between 0.25% PO30 and PO3x15 (** $p < 0.01$). Our results revealed differences between bleaching agents with one application and PB30 produced better viability rates than PO30 and PN (*** $p < 0.001$).

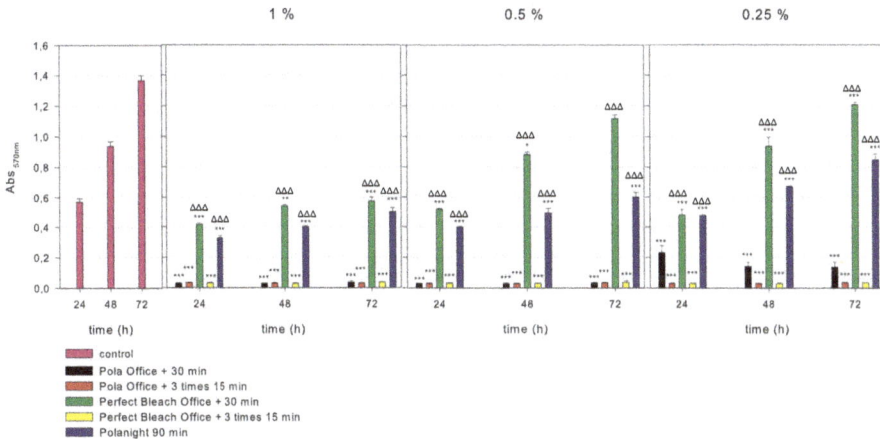

Figure 3. Cell viability was determined using the 3-(4,5-dimethyl-thiazol)-2,5-diphenyl-tetrazolium bromide (MTT) assay. The incubation of hDPSCs with 1% of each bleaching product revealed a significant reduction in cell viability rates compared with those in the control group at the indicated times of culture (*** $p < 0.001$). In contrast, a slight but significant recovery of cell viability was observed with 1%, 0.5% and 0.25% PB30 or PN compared to the others ($^{\Delta\Delta\Delta}$ $p < 0.001$).

3.4. Cell Migration

Compared to treatment with the complete medium (control), treatment with the different dilutions of PB3x15, PB30 and PO3x15 (except 0.25%) for 24 h induced a significantly lower cell migration rate

(** $p < 0.01$ and *** $p < 0.001$; Figure 4). Cell migration with PN and PO30 was still impaired but slightly higher at 24 h. Extracts of PO30 significantly promoted wound closure after 48 h of treatment and achieved comparable levels to control extracts. The PN group exhibited a slightly faster cell migration than the PB30 group, but both groups exhibited larger wound openings than the control group. Taken together, our results demonstrated that PO30 exerted a stronger effect on the migration ability of *h*DPSCs, while PN and PB30 produced values that were similar to those of the controls.

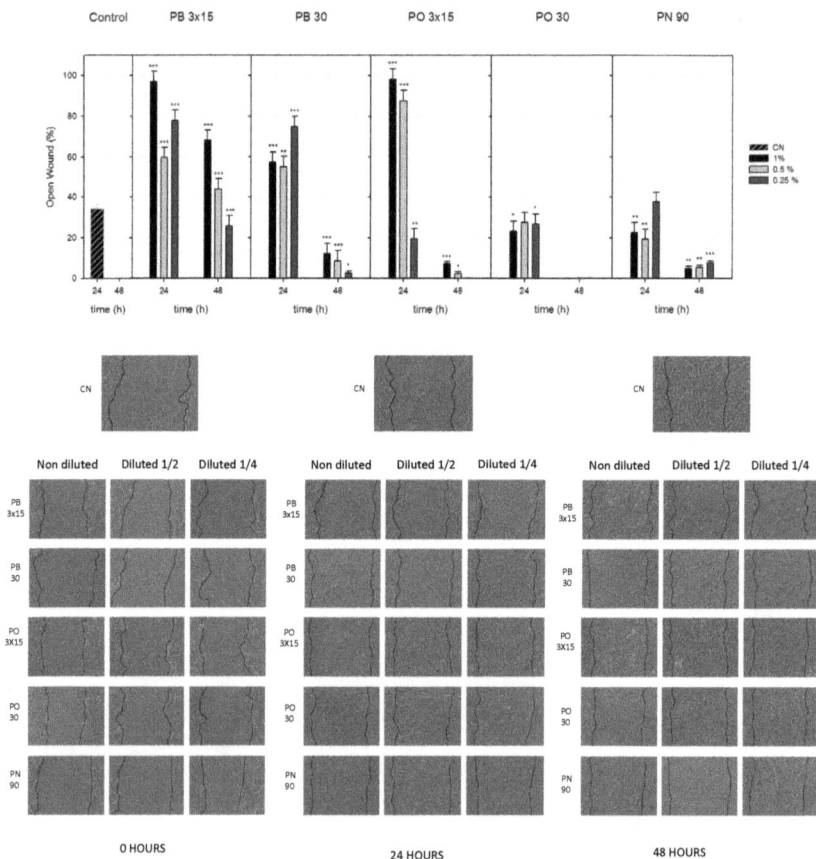

Figure 4. Wound-healing assay. Observation of wound closure change. Confluent hDPSCs were wounded and stimulated with bleaching extracts for up to 48 h. Cell migration is represented as the percentage of the open wound area under each condition compared with the control (* $p < 0.05$; ** $p < 0.01$; *** $p < 0.001$), which was analysed using one-way ANOVA.

3.5. Cell Morphology

The immunofluorescence assay was performed to detect changes in cellular morphology and cytoskeletal organisation of hDPSCs by phalloidin (red fluorescence) and DAPI (blue fluorescence). hDPSCs without bleaching extracts (control) showed a gradual increase in cell proliferation with fibroblastic morphology, a high expression of F-actin and confluency after 72 h of culture (Figure 5). Notably, treatment with extracts of PB3x15, PB30, PO3x15 and PO30 resulted in a small proportion of cells in both groups that lacked F-actin cytoskeleton staining and exhibited condensed or fragmented nuclei, which is typical of apoptotic cells. PN, especially at 0.25% and 0.5% concentrations, exhibited a similarly organised and stretched stress assembly of fibers compared with controls, which provides evidence for the optimal status of *h*DPSCs.

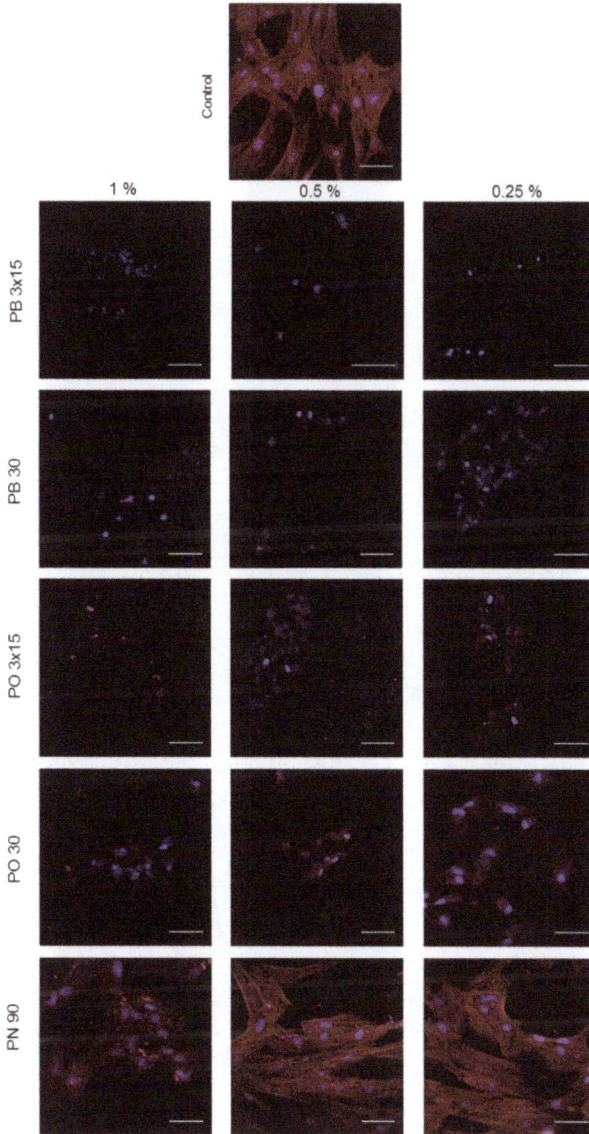

Figure 5. Representative immunofluorescence micrographs revealing the cytoskeletal organisation of *h*DPSCs exposed to control and bleaching products. Cultured *h*DPSCs were treated with undiluted extracts of bleaching products for 24 h. For staining filamentous actin (F-actin), cells were incubated with CruzFluor594-conjugated phalloidin (Santa Cruz Biotechnology, Dallas, TX, USA). Nuclei were counterstained with 4,6-diamidino-2-phenylindole dihydrochloride (DAPI) (blue). Scale bar = 150 μm.

3.6. Analysis of the Expression of Mesenchymal Stem Cell Surface Markers on hDPSCs Exposed to Bleaching Extracts Using Flow Cytometry

Flow cytometry studies were performed to determine possible phenotypic changes in viable *h*DPSCs after culture with the eluates of the different bleaching agents. MSC surface molecules CD73, CD90 and CD105 were expressed at levels greater than 95%, while the expression level of the hematopoietic markers CD14, CD20, CD34 and CD45 was lower than 5% (Figure 6). Notably, the incubation of *h*DPSCs with different dilutions of the extracts (1%, 0.5%, or 0.25%) did not significantly alter the percentage of positive expression of these mesenchymal markers compared to that in untreated *h*DPSCs (controls).

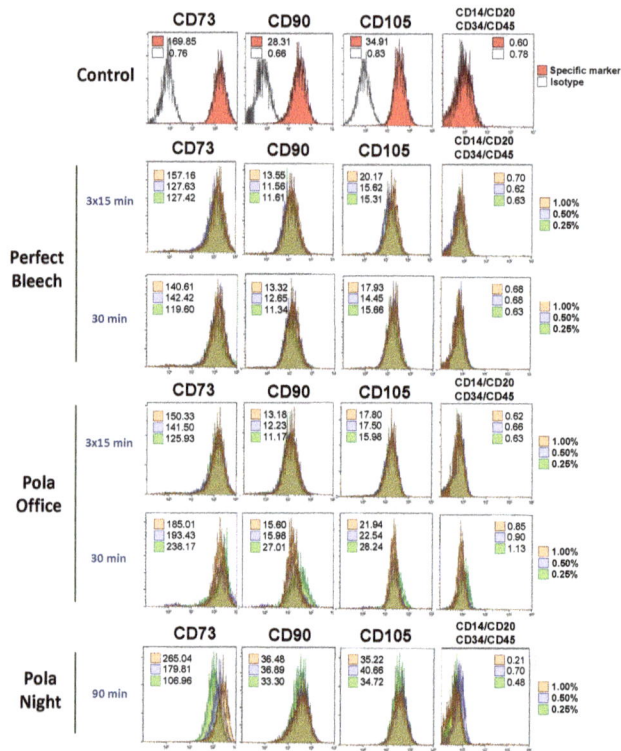

Figure 6. Mesenchymal phenotype analysis of *h*DPSCs after culture with bleaching products using flow cytometry. Cells were cultured for 72 h, detached and labelled with fluorescence-conjugated specific antibodies for the mesenchymal surface markers CD73, CD90 and CD105 and the hematopoietic markers CD14, CD20, CD34 and CD45. Inset numbers represent the mean fluorescence intensity values of viable cells. Histograms show representative flow cytometry results from three independent experiments.

3.7. Apoptosis/Necrosis of hDPSCs in the Presence of Bleaching Extracts

Figure 7 shows the representative 2D dot plots of the distribution of viable (Annexin-V$^-$/7-AAD$^-$), early apoptotic (Annexin-V$^+$/7-AAD$^-$) or late apoptotic/necrotic (Annexin-V$^+$/7-AAD$^+$ and Annexin-V$^-$/7-AAD$^+$) cells among the untreated *h*DPSCs (controls) or cells exposed to different dilutions of the bleaching extracts. The PN extract was associated with more than 90% of the cells remaining viable after 72 h. In contrast, the more concentrated eluates (1% and

0.5%) of the other bleaching agents decreased cell viability. This cytotoxic effect was minor in the most diluted extracts (0.25%).

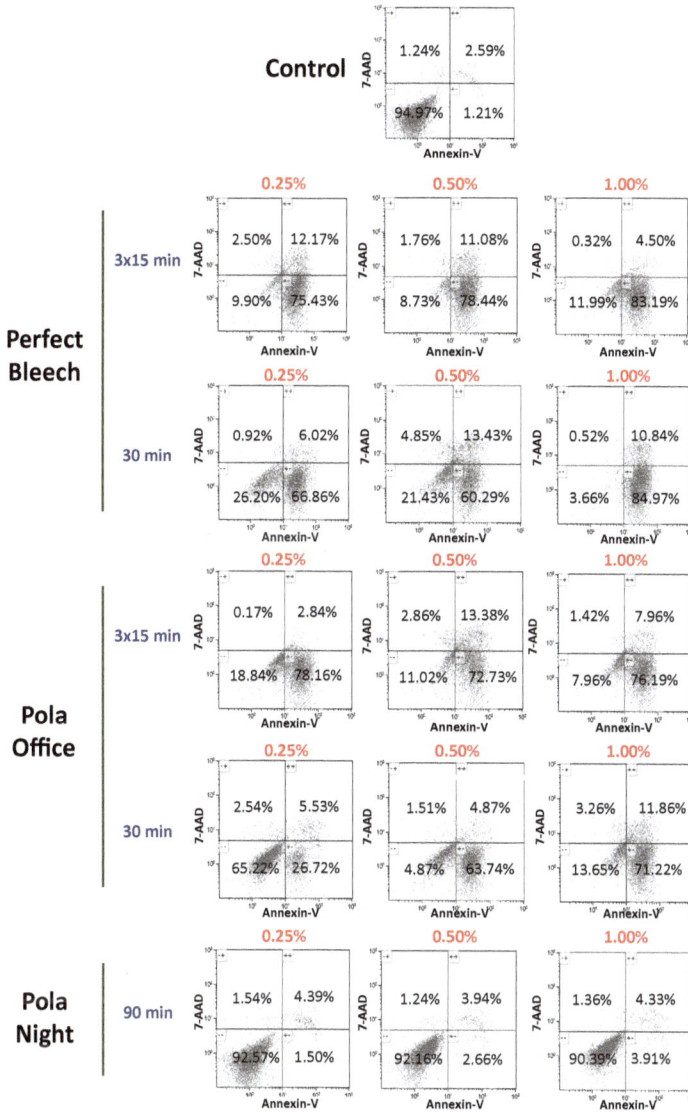

Figure 7. *h*DPSCs were cultured with bleaching extracts and plastic (control) for 72 h, before being labelled with Annexin-V and 7-AAD and analysed using flow cytometry. Numbers within the different quadrants represent the percentages of live (Annexin-V⁻/7-AAD⁻), early apoptotic (Annexin-V⁺/7-AAD⁻) or late apoptotic and necrotic (Annexin-V⁺/7-AAD⁺ and Annexin-V⁻/7-AAD⁺) cells. Dot-plots display representative flow cytometry results from three independent experiments. comma in the figure should be decimal point.

4. Discussion

To our knowledge, this is the first study to compare commercial bleaching products and their biological effects on stem cells from dental pulp. Little information on the diffusion capacity of commercial bleaching products and their effects on dental pulp tissue is available. Therefore, the present study quantified the diffusion of commercial bleaching products and their influence on the biological responses of hDPSCs.

The trans-enamel and trans-dentinal diffusion of bleaching products was based on a previously reported method [24], which is more sensitive than spectrophotometry. Our results demonstrated a higher diffusion of bleaching products after 3 applications. These results corroborate previous studies, which revealed that three applications of gels with greater HP concentrations produced greater diffusion of HP to the pulp chamber and increased damage to the dental pulp [11,12]. Therefore, the exposure time was less important than the number of applications. Our results demonstrated that PN (90 min) exhibited lower rates of diffusion than the other bleaching products. This is the less concentrated product as 16% CP is equivalent to 5–6% HP, but its diffusion was similar to that obtained with PB30.

Pulpal inflammation associated with local tissue necrosis was demonstrated in vivo by the application of high-concentration HP bleaching gels (35–38%) to human teeth [7,29]. hDPSCs are used as the cell line model to simulate clinical conditions because these cells play a major role in dental pulp inflammation and tooth hypersensitivity [25,30].

We analysed the biological cell responses of hDPSCs with respect to cell viability, cell migration, phenotype, apoptosis and morphology in the presence of three commercial bleaching products. The present investigation demonstrated that the PB3x15 and PO3x15 bleaching gel applications were cytotoxic to hDPSCs because cell viability was reduced. These results are clinically interesting because the number of applications may be involved with the occurrence of tooth sensitivity. In fact, Benetti et al. [31] demonstrated that major damage to rat dental pulp occurred after three applications of an HP gel, while de Oliveira et al. [6] also observed a significant reduction in cell viability following three applications of 10% HP gel for 15 min.

Previous studies demonstrated the healing of damaged pulp via the formation of a hard-tissue barrier or reparative dentine. The migration of pulp stem cells to substitute for irreversibly damaged odontoblasts after differentiation is an important step in this process [25]. In general, three applications of the bleaching gel produced lower cell migration than PN or PO30. However, we observed better cell migration in Pola Office groups than in Perfect bleach groups. Notably, Vaz et al. [32] reported that pulp inflammation was related to increased macrophage migration. These authors observed higher macrophage migration with in-office bleaching compared to at-home bleaching.

Optimising the state of dental pulp stem cells contributes to cell migration and dental pulp repair [21]. Immunofluorescence staining showed that undiluted extracts of cells treated with PN produced no morphological alterations and no changes in cytoskeletal organisation patterns compared to cells treated with the other bleaching products. Again, we observed better cell morphology in PO30 than PB30.

Analysis of the expression of mesenchymal stem cell markers is important in regenerative dentistry [33], which was performed in this study. The ISCT states that multipotent mesenchymal stromal cells must express CD105, CD73 and CD90 and should be devoid of the expression of hematopoietic markers, such as CD45, CD34, CD14 and CD11b [26]. We characterised the surface expression pattern of these markers using flow cytometry to evaluate possible changes in the expression. With commercial bleaching products, the MSC surface molecules CD73, CD90 and CD105 were expressed at levels greater than 95% following commercial bleaching, while the expression level of the hematopoietic markers CD34 and CD45 was less than 5%. Therefore, the bleaching products used in this study maintained the mesenchymal phenotype of hDPSCs.

Apoptosis plays an important role in the formation of reparative dentin by providing room for new dentin and preventing inflammation. Apoptosis also leads to the formation of mature dental pulp when exposed to extraneous stimuli [34]. Our results revealed that PN did not induce apoptosis, which is consistent with the results from Benneti et al. [31] as they demonstrated that the effects on pulp tissue varied with the HP concentration. These authors observed that higher HP concentrations produced pulp necrosis and had a prolonged effect on the apoptotic process, while lower HP concentrations induced moderate inflammation, cell proliferation and apoptosis. Nevertheless, in the present study, also in the apoptosis experiment, bleaching products with 35% HP concentration were more cytotoxic than 37% HP and caused membrane permeability-related apoptosis and necrosis. Thus, this effect is more dependent on the commercial product than on the composition.

5. Conclusions

In general, a low concentration of bleaching products, such as PN applied for 90 min, was less cytotoxic than other commercial bleaching products with 35–37.5% HP concentration. This effect occurred independently of the application protocol. For high concentration products, bleaching products with 35% HP concentration were more cytotoxic than 37% HP, which suggests that unknown agents in bleaching products could play an important role in determining their level of toxicity.

Author Contributions: C.L., M.C.-G., C.J.T.-C., D.G.-B., R.O.-S., F.J.R.-L., and L.F. conceived of and designed the experiments; M.C.-G., C.J.T.-C., and D.G.-B. performed the experiments; C.L. and F.J.R.-L. analysed the data; R.O.-S., F.J.R.-L., D.G.-B., and L.F. contributed materials/analysis tools; C.L., F.J.R.-L., and L.F. wrote the manuscript.

Funding: Research was funded by by the Spanish Net of Cell Therapy (TerCel), RETICS subprograms of the I+D+I 2013–2016 Spanish National Plan, and projects "RD12/0019/0001" and "RD16/0011/0001" funded by the Instituto de Salud Carlos III to JMM and co-funded by the European Regional Development Fund.

Conflicts of Interest: The authors explicitly state that there are no conflicts of interest related to this article.

References

1. Cohen, S.C. Human pulpal response to bleaching procedures on vital teeth. *J. Endod.* **1979**, *5*, 134–138. [CrossRef]
2. Pugh, G., Jr.; Zaidel, L.; Lin, N.; Stranick, M.; Bagley, D. High levels of hydrogen peroxide in overnight tooth-whitening formulas: Effects on enamel and pulp. *J. Esthet. Restor. Dent.* **2005**, *17*, 40–45, discussion 46–47. [PubMed]
3. Kwon, S.R.; Li, Y.; Oyoyo, U.; Aprecio, R.M. Dynamic model of hydrogen peroxide diffusion kinetics into the pulp cavity. *J. Contemp. Dent. Pract.* **2012**, *13*, 440–445. [PubMed]
4. Azer, S.S.; Machado, C.; Sanchez, E.; Rashid, R. Effect of home bleaching systems on enamel nanohardness and elastic modulus. *J. Dent.* **2009**, *37*, 185–190. [CrossRef] [PubMed]
5. Sulieman, M. An overview of bleaching techniques: I. History, chemistry, safety and legal aspects. *Dent. Update* **2004**, *31*, 608–610, 612–604, 616. [CrossRef] [PubMed]
6. De Oliveira Duque, C.C.; Soares, D.G.; Basso, F.G.; Hebling, J.; de Souza Costa, C.A. Influence of enamel/dentin thickness on the toxic and esthetic effects of experimental in-office bleaching protocols. *Clin. Oral Investig.* **2017**, *21*, 2509–2520. [CrossRef] [PubMed]
7. Soares, D.G.; Basso, F.G.; Hebling, J.; de Souza Costa, C.A. Immediate and late analysis of dental pulp stem cells viability after indirect exposition to alternative in-office bleaching strategies. *Clin. Oral Investig.* **2015**, *19*, 1013–1020. [CrossRef] [PubMed]
8. Ferreira, V.G.; Nabeshima, C.K.; Marques, M.M.; Paris, A.F.; Gioso, M.A.; dos Reis, R.S.; Machado, M.E. Tooth bleaching induces changes in the vascular permeability of rat incisor pulps. *Am. J. Dent.* **2013**, *26*, 298–300. [PubMed]
9. Lima, A.F.; Marques, M.R.; Soares, D.G.; Hebling, J.; Marchi, G.M.; de Souza Costa, C.A. Antioxidant therapy enhances pulpal healing in bleached teeth. *Restor. Dent. Endod.* **2016**, *41*, 44–54. [CrossRef] [PubMed]
10. Costa, C.A.; Riehl, H.; Kina, J.F.; Sacono, N.T.; Hebling, J. Human pulp responses to in-office tooth bleaching. *Oral Surg. Oral Med. Oral Pathol. Oral Radiol. Endod.* **2010**, *109*, e59–e64. [CrossRef] [PubMed]

11. Cintra, L.T.A.; Ferreira, L.L.; Benetti, F.; Gastelum, A.A.; Gomes-Filho, J.E.; Ervolino, E.; Briso, A.L.F. The effect of dental bleaching on pulpal tissue response in a diabetic animal model. *Int. Endod. J.* **2017**, *50*, 790–798. [CrossRef] [PubMed]

12. Ferreira, L.L.; Gomes-Filho, J.E.; Benetti, F.; Carminatti, M.; Ervolino, E.; Briso, A.L.F.; Cintra, L.T.A. The effect of dental bleaching on pulpal tissue response in a diabetic animal model: A study of immunoregulatory cytokines. *Int. Endod. J.* **2018**, *51*, 347–356. [CrossRef] [PubMed]

13. Rodríguez-Lozano, F.J.; Insausti, C.L.; Iniesta, F.; Blanquer, M.; Ramírez, M.D.; Meseguer, L.; Meseguer-Henarejos, A.B.; Marín, N.; Martínez, S.; Moraleda, J.M. Mesenchymal dental stem cells in regenerative dentistry. *Med. Oral Patol. Oral Cir. Bucal* **2012**, *17*, e1062–e1067. [CrossRef] [PubMed]

14. Pisciotta, A.; Bertoni, L.; Riccio, M.; Mapelli, J.; Bigiani, A.; La Noce, M.; Orciani, M.; de Pol, A.; Carnevale, G. Use of a 3D Floating Sphere Culture System to Maintain the Neural Crest-Related Properties of Human Dental Pulp Stem Cells. *Front. Physiol.* **2018**, *9*, 547. [CrossRef] [PubMed]

15. Cintra, L.T.; Benetti, F.; Ferreira, L.L.; Rahal, V.; Ervolino, E.; Jacinto Rde, C.; Gomes Filho, J.E.; Briso, A.L. Evaluation of an experimental rat model for comparative studies of bleaching agents. *J. Appl. Oral Sci.* **2016**, *24*, 171–180. [CrossRef] [PubMed]

16. De Almeida, L.C.; Soares, D.G.; Gallinari, M.O.; de Souza Costa, C.A.; Dos Santos, P.H.; Briso, A.L. Color alteration, hydrogen peroxide diffusion, and cytotoxicity caused by in-office bleaching protocols. *Clin. Oral Investig.* **2015**, *19*, 673–680. [CrossRef] [PubMed]

17. Dias Ribeiro, A.P.; Sacono, N.T.; Lessa, F.C.; Nogueira, I.; Coldebella, C.R.; Hebling, J.; de Souza Costa, C.A. Cytotoxic effect of a 35% hydrogen peroxide bleaching gel on odontoblast-like mdpc-23 cells. *Oral Surg. Oral Med. Oral Pathol. Oral Radiol. Endod.* **2009**, *108*, 458–464. [CrossRef] [PubMed]

18. Hanks, C.T.; Fat, J.C.; Wataha, J.C.; Corcoran, J.F. Cytotoxicity and dentin permeability of carbamide peroxide and hydrogen peroxide vital bleaching materials, in vitro. *J. Dent. Res.* **1993**, *72*, 931–938. [CrossRef] [PubMed]

19. Kinomoto, Y.; Carnes, D.L., Jr.; Ebisu, S. Cytotoxicity of intracanal bleaching agents on periodontal ligament cells in vitro. *J. Endod.* **2001**, *27*, 574–577. [CrossRef] [PubMed]

20. Peters, O.A. Research that matters—Biocompatibility and cytotoxicity screening. *Int. Endod. J.* **2013**, *46*, 195–197. [CrossRef] [PubMed]

21. Tomas-Catala, C.J.; Collado-Gonzalez, M.; Garcia-Bernal, D.; Onate-Sanchez, R.E.; Forner, L.; Llena, C.; Lozano, A.; Moraleda, J.M.; Rodriguez-Lozano, F.J. Biocompatibility of new pulp-capping materials neomta plus, mta repair hp, and biodentine on human dental pulp stem cells. *J. Endod.* **2018**, *44*, 126–132. [CrossRef] [PubMed]

22. Marson, F.C.; Sensi, L.G.; Vieira, L.C.; Araujo, E. Clinical evaluation of in-office dental bleaching treatments with and without the use of light-activation sources. *Oper. Dent.* **2008**, *33*, 15–22. [CrossRef] [PubMed]

23. Soares, D.G.; Ribeiro, A.P.; Lima, A.F.; Sacono, N.T.; Hebling, J.; de Souza Costa, C.A. Effect of fluoride-treated enamel on indirect cytotoxicity of a 16% carbamide peroxide bleaching gel to pulp cells. *Braz. Dent. J.* **2013**, *24*, 121–127. [CrossRef] [PubMed]

24. Barja, G. The quantitative measurement of H_2O_2 generation in isolated mitochondria. *J. Bioenerg. Biomembr.* **2002**, *34*, 227–233. [CrossRef] [PubMed]

25. Collado-Gonzalez, M.; Pecci-Lloret, M.R.; Tomas-Catala, C.J.; Garcia-Bernal, D.; Onate-Sanchez, R.E.; Llena, C.; Forner, L.; Rosa, V.; Rodriguez-Lozano, F.J. Thermo-setting glass ionomer cements promote variable biological responses of human dental pulp stem cells. *Dent. Mater.* **2018**, *34*, 932–943. [CrossRef] [PubMed]

26. Dominici, M.; Le Blanc, K.; Mueller, I.; Slaper-Cortenbach, I.; Marini, F.; Krause, D.; Deans, R.; Keating, A.; Prockop, D.; Horwitz, E. Minimal criteria for defining multipotent mesenchymal stromal cells. The international society for cellular therapy position statement. *Cytotherapy* **2006**, *8*, 315–317. [CrossRef] [PubMed]

27. Cavalcanti, B.N.; Rode, S.M.; Marques, M.M. Cytotoxicity of substances leached or dissolved from pulp capping materials. *Int. Endod. J.* **2005**, *38*, 505–509. [CrossRef] [PubMed]

28. International Organization for Standardization. *10993-12. Biological evaluation of Medical Devices—Part 12: Sample Preparation and Reference Materials*; International Organization for Standardization: Geneva, Switzerland, 2007.

29. De Almeida, L.C.; Soares, D.G.; Azevedo, F.A.; Gallinari Mde, O.; Costa, C.A.; dos Santos, P.H.; Briso, A.L. At-home bleaching: Color alteration, hydrogen peroxide diffusion and cytotoxicity. *Braz. Dent. J.* **2015**, *26*, 378–383. [CrossRef] [PubMed]

30. Jha, N.; Ryu, J.J.; Choi, E.H.; Kaushik, N.K. Generation and role of reactive oxygen and nitrogen species induced by plasma, lasers, chemical agents, and other systems in dentistry. *Oxid. Med. Cell. Longev.* **2017**, *2017*, 7542540. [CrossRef] [PubMed]

31. Benetti, F.; Gomes-Filho, J.E.; Ferreira, L.L.; Ervolino, E.; Briso, A.L.F.; Sivieri-Araujo, G.; Dezan-Junior, E.; Cintra, L.T.A. Hydrogen peroxide induces cell proliferation and apoptosis in pulp of rats after dental bleaching in vivo: Effects of the dental bleaching in pulp. *Arch. Oral Biol.* **2017**, *81*, 103–109. [CrossRef] [PubMed]

32. Vaz, M.M.; Lopes, L.G.; Cardoso, P.C.; Souza, J.B.; Batista, A.C.; Costa, N.L.; Torres, E.M.; Estrela, C. Inflammatory response of human dental pulp to at-home and in-office tooth bleaching. *J. Appl. Oral Sci.* **2016**, *24*, 509–517. [CrossRef] [PubMed]

33. Rodríguez-Lozano, F.J.; Bueno, C.; Insausti, C.L.; Meseguer, L.; Ramírez, M.C.; Blanquer, M.; Marín, N.; Martínez, S.; Moraleda, J.M. Mesenchymal stem cells derived from dental tissues. *Int. Endod. J.* **2011**, *44*, 800–806. [CrossRef] [PubMed]

34. Wu, T.T.; Li, L.F.; Du, R.; Jiang, L.; Zhu, Y.Q. Hydrogen peroxide induces apoptosis in human dental pulp cells via caspase-9 dependent pathway. *J. Endod.* **2013**, *39*, 1151–1155. [CrossRef] [PubMed]

materials

MDPI

Article

Physicochemical and Microbiological Assessment of an Experimental Composite Doped with Triclosan-Loaded Halloysite Nanotubes

Diana A. Cunha [1], Nara S. Rodrigues [1], Lidiane C. Souza [1], Diego Lomonaco [1,2], Flávia P. Rodrigues [1,3], Felipe W. Degrazia [4], Fabrício M. Collares [4], Salvatore Sauro [5,6,*] and Vicente P. A. Saboia [1,7]

[1] Post-Graduate Programme in Dentistry, Federal University of Ceará, Rua Monsenhor Furtado S/N, Rodolfo Teófilo, Fortaleza 60430-355, Ceará, Brazil; araujo.diana@gmail.com (D.A.C.); nara.sousa.rodrigues@gmail.com (N.S.R.); lidiane_costa26@hotmail.com (L.C.S.); lomonaco@ufc.br (D.L.); flapiro@gmail.com (F.P.R); vpsaboia@yahoo.com (V.P.A.S.)
[2] Department of Organic and Inorganic Chemistry, Federal University of Ceará, Fortaleza 60440-900, Ceará, Brazil
[3] School of Dentistry, Paulista University—UNIP, R. Dr. Bacelar 1212, Vila Clementino, São Paulo 04026-002, SP, Brazil
[4] Laboratório de Materiais Dentários, Faculdade de Odontologia, Universidade Federal do Rio Grande do Sul, Rua Ramiro Barcelos, 2492, Rio Branco, Porto Alegre 90035-003, Rio Grande do Sul, Brazil; fdegrazia@hotmail.com (F.W.D.); fabricio.collares@ufrgs.br (F.M.C.)
[5] Departamento de Odontología, Facultad de Ciencias de la Salud, Universidad CEU-Cardenal Herrera, C/Del Pozos/n, Alfara del Patriarca, 46115 Valencia, Spain
[6] Tissue Engineering and Biophotonics Research Division King's College London Dental Institute (KCLDI), London SE1 9RT, UK
[7] Department of Restorative Dentistry, School of Dentistry, of Ceará, Fortaleza 60430-355, Ceará, Brazil
*   Correspondence: salvatore.sauro@uchceu.es; Tel.: +34-6348-68517

Received: 6 May 2018; Accepted: 22 June 2018; Published: 25 June 2018

Abstract: This study is aimed at evaluating the effects of triclosan-encapsulated halloysite nanotubes (HNT/TCN) on the physicochemical and microbiological properties of an experimental dental composite. A resin composite doped with HNT/TCN (8% w/w), a control resin composite without nanotubes (HNT/TCN-0%) and a commercial nanofilled resin (CN) were assessed for degree of conversion (DC), flexural strength (FS), flexural modulus (FM), polymerization stress (PS), dynamic thermomechanical (DMA) and thermogravimetric analysis (TGA). The antibacterial properties (M) were also evaluated using a 5-day biofilm assay (CFU/mL). Data was submitted to one-way ANOVA and Tukey tests. There was no significant statistical difference in DC, FM and RU between the tested composites ($p > 0.05$). The FS and CN values attained with the HNT/TCN composite were higher ($p < 0.05$) than those obtained with the HNT/TCN-0%. The DMA analysis showed significant differences in the TAN δ ($p = 0.006$) and Tg ($p = 0$) between the groups. TGA curves showed significant differences between the groups in terms of degradation ($p = 0.046$) and weight loss ($p = 0.317$). The addition of HNT/TCN induced higher PS, although no significant antimicrobial effect was observed ($p = 0.977$) between the groups for CFUs and ($p = 0.557$) dry weight. The incorporation of HNT/TCN showed improvements in physicochemical and mechanical properties of resin composites. Such material may represent an alternative choice for therapeutic restorative treatments, although no significance was found in terms of antibacterial properties. However, it is possible that current antibacterial tests, as the one used in this laboratory study, may not be totally appropriate for the evaluation of resin composites, unless accompanied with aging protocols (e.g., thermocycling and load cycling) that allow the release of therapeutic agents incorporated in such materials.

Keywords: mechanical properties; nanotubes; resin composite; *Streptococcus mutans*; triclosan

1. Introduction

Resin dental composites (RDCs) have been widely modified in the last few decades. Indeed, in order to increase their clinical performance, especially in terms of wear resistance and lower polymerization shrinkage, modern RDCs are formulated with a high amount of glass/ceramic fillers (60–80 wt.%) [1]. Moreover, inorganic nanoparticles and nanofibers have also been incorporated within the composition of RDCs to advance their mechanical and esthetic properties [2], as well as their biological and bioactive properties [3].

Nowadays, the formation of secondary caries remains one of the main reasons for the replacement of resin composite restorations [4–6]. Carious lesions along the margins of our restorations present an important causal relationship with the accumulation of a cariogenic biofilm; this is probably facilitated by gaps formed at the tooth-restoration interface, as well as by excessive roughness of resin composite [7]. *Streptococcus mutans* is one of the main species of a cariogenic biofilm responsible for secondary caries [8], hence, restorative materials with bioactive and antimicrobial properties are needed in order to improve the clinical outcome in daily practice [9].

Triclosan (TCN) is a well-known antibacterial agent used in a wide range of products such as toothpastes and mouthwashes. Several studies demonstrated the efficiency of such therapeutic substances against gram-positive microorganisms, (e.g., *Streptococcus mutans* [10,11], *Staphylococcus aureus* [12], *Lactobacillus* spp. [13], and *Actinomyces* spp. [14]). Indeed, due to such antimicrobial potential, along with its low molecular weight, uses and applications of TCN have radically increased in the last 30 years [12]. However, neat TCN incorporation in resin-based materials could lead to a high leachability that can cause a rapid decrease of the antimicrobial properties of such materials. To overcome this issue, TCN was previously incorporated in specific "vehicles" known as nanotubes, in order to achieve a slower and more controlled release of such an antibacterial agent [3,11].

Halloysite nanotubes (HNT) are natural aluminosilicates with a hollow tubular structure [3], which are typically used as a reinforcing nano-filler to improve some mechanical properties of resin-composites [15] such as tensile strength, flexural strength, storage modulus [16], as well as microhardness and bond strength [17]. HNT is a "green" biocompatible nanomaterial characterized by very low cytotoxicity [18]. Furthermore, it acts as "biologically-safe" reservoirs for the encapsulation and controlled release of a variety of therapeutic drugs [15,19], bioactive molecules [20] and matrix metalloproteinase inhibitors [21,22]. Active principles released by halloysite may last 30 to 100 times more than when these were incorporated alone or using some different nanocarriers in polymer nanocomposite [23,24].

Thus, the aim of the present study was to evaluate in vitro the effects of triclosan-encapsulated halloysite nanotubes (HNT/TCN) on the physical-chemical and microbiological properties of an experimental micro-hybrid resin dental composite.

The null hypothesis tested in this study was that the inclusion of HNT-TCN would not influence the physicochemical and microbiological properties of such an experimental resin composite.

2. Material and Methods

2.1. Material

Halloysite nanotubes ($Al_2Si_2O_5(OH)_4 \cdot 2H_2O$) with a diameter of 30–70 nm and length of 1–3 μm (Sigma-Aldrich, St. Louis, MO, USA) were treated with a silane solution (5 wt.% of 3-metacryloxypropyltrimetoxysilane and 95 wt.% acetone) at 110 °C for 24 h. Subsequently, these were mixed [1:1 ratio] with 2,4,4-Trichloro-2-hydroxydiphenyl ether (TCN: Triclosan, Fagron, Rotterdam, SH, The Netherlands) under constant shaking for 1 h [3]. The mixture was then dispersed in 95 wt.% pure ethanol (0.03 mg/mL^{-1}) and ultasonicated for 1 h. The nanoparticles were finally desiccated for 10 days at 30 °C to ensure complete evaporation of the residual solvents. The HNT/TCN nanoparticles obtained after such processing method were finally prepared and characterized using a Transmission Electron Microscope (TEM) JEM 120 Exll (JEOL, Tokyo, Japan) at 80 kV at a magnification X 300,000.

An experimental resin composite was created by mixing 75 wt.% Bis-GMA (2,2-bis-[4-(hydroxyl-3-methacryloxy-propyloxy)phenyl]propane) and 25 wt.% triethylene glycol dimethacrylate (TEGDMA) (Sigma-Aldrich, St. Louis, MO, USA) under continuous agitation and sonication for 30 min. Camphorquinone (CQ), ethyl4-dimethylaminobenzoate (EDAB), and diphenyliodoniumhexafluorophosphate (DPIHFP; Milwaukee, MI, USA) were also added at 1 mol % to obtain a light-curable resin-based material. Incorporation of 8 wt.% of HNT/TCN and 72 wt.% silica micro-hybrid filler was performed by stirring for 12 h under continuous sonication of the experimental resin composite.

The control experimental resin composite was formulated with the same organic matrix and silica micro-hybrid filler (80 wt.%), but without the use of the HNT/TCN nanotubes (HNT/TCN-0%) (Table 1).

Table 1. Composition of the resin composite used in this study.

Name of the Composite	Type of Composite	Manufacturer/Lot No.	Composition
HNT/TCN-0%	Experimental resin-composite	—	**Organic matrix:** Bis-GMA, TEGDMA.
			Filler type: Silica micro-hybrid filler 80 wt.%
			Filler content: 80 wt.%
HNT/TCN	Experimental resin-composite	—	**Organic matrix:** Bis-GMA, TEGDMA.
			Filler type: Silica micro-hydrid filler 72 wt.% and halloysite nanotubes 8 wt.%.
			Filler content: 80 wt.%
Filtek Z-350XT (Shade A1D)	Commercial nano-filled composite	3M ESPE (St Paul, MN, USA)/N702257	**Organic matrix:** Bis-GMA, Bis-EMA, UDMA, TEGDMA
			Filler type: Silica and zirconia nanofillers, agglomerated zirconia-silica nanoclusters
			Filler content: 82 wt.%

A commercial nano-filled resin composite was used in this study as a control group. According to the manufacturer (3M ESPE), this material has nanoclusters of zirconia (4–11 nm) and silica (20 nm) nanoparticles, along with micro-filled silica/zirconia particles (0.6 mm) (Table 1).

2.2. Degree of Conversion (DC)

Three discs were created for each composite used in this study using teflon molds (6 mm in diameter × 2 mm thick) and light-cured in the absence of oxygen for 40 s under an acetate transparent strip using a light-curing system (1200 mW/cm^2, Bluephase, Ivoclar Vivadent, Schaan, Liechtenstein), at a standardized 2-mm distance. The degree of conversion (DC) of each material tested in this study was evaluated through micro-Raman spectroscopy (Xplora Horiba, Paris, France) in the range between 1590 and 1670 cm^{-1} using the 638 nm laser emission wavelength, 5 s acquisition time and 10 accumulations [11,25]. Three specimens for each group were analyzed at a standardized room temperature of 23 ± 1 °C. DC was calculated as described in a previous study by Rodrigues et al. [25] on the intensity of the C=C stretching vibrations (peakheight) at 1635 cm^{-1} and using the symmetric ring stretching at 1608 cm^{-1} from the polymerized and non-polymerized specimens.

$$DC\% = \left(1 - \left(\frac{R\ cured}{R\ uncured}\right)\right) \times 100 \qquad (1)$$

2.3. Flexural Strength and Flexural Modulus

The flexural strength (FS) and flexural modulus (FM) ($n = 5$) evaluation was performed according to ISO 4049/2000 [26] using a universal mechanical testing machine (Instron 3345, Canton, MA, EUA). Specimens with standard dimensions of $25 \times 2 \times 2$ mm were prepared using a Teflon split mold. A polyester strip and a glass slide covered the resin-composite, and the light tip guide was placed over the center of the mold, to light-cure the specimens for 40 s as described in Section 2.2. After irradiation, the specimens were removed from the molds and carefully polished using a 320 grit SiC abrasive paper. All of the specimens were stored in water at 37 °C for 24 h. The specimens were positioned in a 3-point bending apparatus with 2 parallel supports with a distance of 20 mm. The specimens were loaded until fracture with a 500 Kgf load cell at a cross-head speed of 0.05 mm/min. The flexural strength (MPa) was calculated using the following formula:

$$\sigma = \frac{3L \times F_{max}}{2w \times h^2} \tag{2}$$

L is the distance between the parallel supports (mm); F_{max} is the load at fracture (N); w is the width (mm), and h is the height (mm).

The flexural modulus (GPa) was calculated using the following formula:

$$E_f = \frac{F \times L^3}{4wh^3 d} \tag{3}$$

F and d stands for the load and deflection increment, respectively, between 2 specific points in the elastic portion of the curves (N and mm); L is the distance between the parallel supports (mm); w is the width (mm), h is the height (mm).

2.4. Dynamic Thermomechanical Analysis (DMA)

Three specimens (8 mm $\times$ 2 mm $\times$ 2 mm) were prepared for each group (Table 1) and light-cured as previously described. A DMA system (Mettler Toledo, Columbus, OH, USA), equipped with a single bending cantilever was used to determine the mechanical properties in clamped mode. The viscoelastic properties were characterized by applying a sinusoidal deformation force to the material under dynamic conditions: Temperature, time, frequency, stress, or a combination of these parameters. The storage modulus (E'), glass transition temperature (Tg) and tangent delta (TAN-δ) of the tested materials were evaluated at different temperatures under cyclic stress (frequency of 2.0 Hz and amplitude of 10 μm) and from 50 to 800 °C at the heating rate of 2 °C min^{-1}. The TAN-δ value represents the damping properties of the material, serving as an indicator of all types of molecular motions and phase transitions.

2.5. Thermogravimetric Analysis (TGA)

A further three specimens were prepared for each group (Table 1) (mass of 10 mg). A thermogravimetric analysis was performed to determine the thermal degradation and the weight percentage of fillers resin-composites. A thermal program from 30 to 193 °C at the heating rate of 2 °C min^{-1} in nitrogen atmosphere determined the weight changes as a function of time and temperature. Thermogravimetric analysis was performed using the Pyris 1 TGA (SDTA851—Mettler Toledo) thermal analyzer.

2.6. Polymerization Stress Measurements (PS)

Poly(methyl methacrylate) rods, 5 mm in diameter and 13 or 28 mm in length, had one of their flat surfaces sandblasted with 250 μm alumina. On the shorter rod, to allow for the highest possible light transmission during photoactivation, the opposite surface was polished with silicone SiC papers (600, 1200, and 2000 grit) followed by felt disks with 1μm alumina paste (Alumina 3, ATM, Altenkirchen, Germany). The sandblasted surfaces received a layer of methylmethacrylate

(JET Acrilico Auto Polimerizante, Sao Paulo, Brazil), followed by two thin layers of unfilled resin (Scotchbond Multi-Purpose Plus, bottle 3, 3M ESPE).

The resin composite was light-cured as previously described for 40 s. The rods were attached to the opposing clamps of a universal testing machine (Instron 5565, Canton, MA, USA) with the treated surfaces, facing each other with a 1-mm gap. The resin composite was inserted into the gap and shaped into a cylinder following the perimeter of the rods. An extensometer (0.1 μm resolution), attached to the rods (Instron 2630-101, Bucks, UK) in order to assess the height of the specimen, provided the feedback to the testing machine to keep the height constant. Therefore, the force registered by the load cell was necessary to counteract the polymerization shrinkage to maintain the specimen's initial height. A hollow stainless steel fixture with a lateral slot attached the short rod to the testing machine, allowing the tip of the light guide to be positioned in contact with the polished surface of the rod. Force development was monitored for 10 min from the beginning of the photoactivation; the nominal stress was calculated by dividing the maximum force value by the cross-section area of the rod. Five specimens were tested for each tested material [27].

2.7. Microbiology Assay

Streptococcus mutans (*S. mutans*) UA159 (ATTCC) was obtained from single colonies isolated on blood agar plates, inoculated in Tryptone yeast-extract broth containing 1% glucose (*w/v*) and incubated for 18 h at 37 °C under micro-aerophilic conditions in partial atmosphere of 5% CO_2.

To analyze antimicrobial effects, blocks (4 × 4 × 2 mm) of each group were produced (Table 1). Materials were dispensed in a silicone mold, covered with a polyester tape and then submitted to digital pressure for 2 s to better accommodate the material, with curing light being activated for 40 s. Specimens were sterilized by exposure to Plasma Hydrogen Peroxide before starting biofilm formation. Mono-species *S. mutans* biofilms were formed on blocks placed in bath cultures at 37 °C in 5 % CO_2 up to 5 days in 24-well polystyrene plates. The biofilms grew in tryptone yeast-extract broth containing 1% sucrose (*w/w*) and were kept undisturbed for 24 h to allow an initial biofilm formation. During the biofilm formation period, once daily the discs were dip-washed three times in a plate containing NaCl 0.89% solution to remove the loosely bound biofilm and they were transferred to new 24-well plates with sterile medium. The blocks of each experimental group were removed after 5 days of initial biofilm formation and transferred to pre-weighed microtubes containing 1 mL of NaCl 0.89 % solution. Biofilms were then dispersed with 3 pulses of 15 s with 15 s of interval at a 7-W output (Branson Sonifier 150; Branson Ultrasonics, Danbury, CT, USA). An aliquot (0.05 mL) of the homogenized biofilm was serially diluted (10^{-1}–10^{-7}) and plated onto blood agar plates. Plates were then incubated at 37 °C, 5% CO_2 for 48 h, before enumerating viable microorganisms. Results were expressed as colony forming units (CFU)/mL and transformed in $\log_{10}$ CFU to reduce variance heterogeneity [28].

To determine the biofilm dry weight, 200 μL aliquots of the initial biofilm suspension were transferred to pre-weighed tubes and dehydrated with ethanol solutions (99 %). The tubes were centrifuged, and the supernatants were discarded before the pellet was dried into a desiccator (P_2O_5) for 24 h and weighted (±0.00001 mg). The dry weight of the biofilm was determined by calculating the weight in the tube (initial weight − final weight) and in the original suspension (dry weight in 1 mL = dry weight in 200 μL × 5) [29].

2.8. Statistical Analysis

Physicochemical properties data was submitted to analysis of variance with one factor (One way-ANOVA), followed by Tukey test. For analyzing antimicrobial effects was performed analysis of variance with one factor (One way-ANOVA). Significance level was set at 5%. The program used to perform the analyses was IBM SPSS Statistics Version 20.0 (Armonk, NY, USA).

3. Results

TEM analysis showed that TCN was successfully deposited inside the lumen of the HNTs (Figure 1A,B). Means and standard deviations of physicochemical properties of the tested materials are presented in Tables 2 and 3. To summarize, the degree of conversion (DC) test showed that there was no significant difference between the tested materials ($p = 0.879$). The flexural strength of HNT/TCN and that of Z350XT was greater than attained with the control HNT/TCN-0% ($p = 0.005$) composite. However, no significant difference was encountered between the storage modulus (E_f) of the three tested groups (Figure 2). The maximum polymerization stress obtained in the specimens created with HNT/TCN composite was significantly ($p < 0.05$) greater than that observed in the control HNT/TCN-free resin composite. The DMA assessment showed no significant differences between the three tested composites for the TAN δ at Tg ($p = 0.006$) and Tg ($p = 0$) (Table 3). The TGA curves (Figure 3) obtained in nitrogen atmosphere showed that there was no significant difference between the three tested composites on the first degradation step ($p > 0.05$): HNT/TCN (296 °C), HNT/TCN-0% (301 °C) and the commercial resin composite (286 °C). Moreover, on the second degradation step, no significant difference was observed: HNT/TCN (419 °C), HNT/TCN-0% (426 °C) and the commercial resin composite (415 °C). There was no difference ($p = 0.317$) between the weight loss of the composites HNT/TCN (25.7%), HNT/TCN (24.5%) and the commercial resin composite (30%) (Table 3).

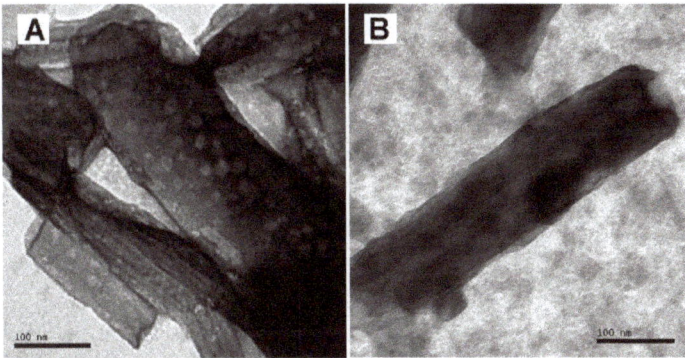

Figure 1. (**A**) TEM image that shows the presence of TCN nanoparticles with diameter of 5–10 nm. (**B**) TEM image of a nanotube with its inner-surface of 40–50 nm diameter range and outer-surface of 90 nm diameter.

Table 2. Results Degree of Conversion (DC), Flexural Modulus (E) and Flexural Strength (FS), Maximum polymerization stress (PS).

Composite	DC (%)	E (GPa)	FS (MPa)	PS (MPa)
HNT/TCN-0%	75.9 (5.4) [A]	6.8 (0.9) [A]	75.9 (10.1) [B]	3.6 (0.3) [B]
HNT/TCN	78.5 (2.2) [A]	7.5 (0.2) [A]	107.2 (6.6) [A]	5.4 (0.9) [A]
Z350XT	72.5 (10.6) [A]	6.8 (0.4) [A]	101.4 (18.4) [A]	3.6 (0.3) [B]

* Different capital letters in column indicate statistical difference ($p < 0.05$).

Table 3. Results Glass Transition Temperature (Tg), TAN-δ and Thermogravimetric analysis (TGA).

Composite	Tg (°C)	Tanδ ($\times 10^3$) at Tg	TGA Weight Loss (%)	TGA Temperature of the First Degradation Step (°C)	TGA Temperature of the Second Degradation Step (°C)
HNT/TCN-0%	102 (6.56) [B]	156.7 (0.15) [A]	27.3 (0.58) [A]	301 (1.73) [A]	426 (3.6) [A]
HNT/TCN	154 (4.36) [A]	106.7 (0.11) [B]	26.3 (0.58) [A]	296 (1.0) [B]	419 (1.15) [A,B]
Z350XT	105.3 (3.51) [B]	103.3 (0.15) [B]	27 (1.0) [A]	286 (2.64) [C]	415 (5.72) [B]

* Different capital letters in column indicate statistical difference ($p < 0.05$).

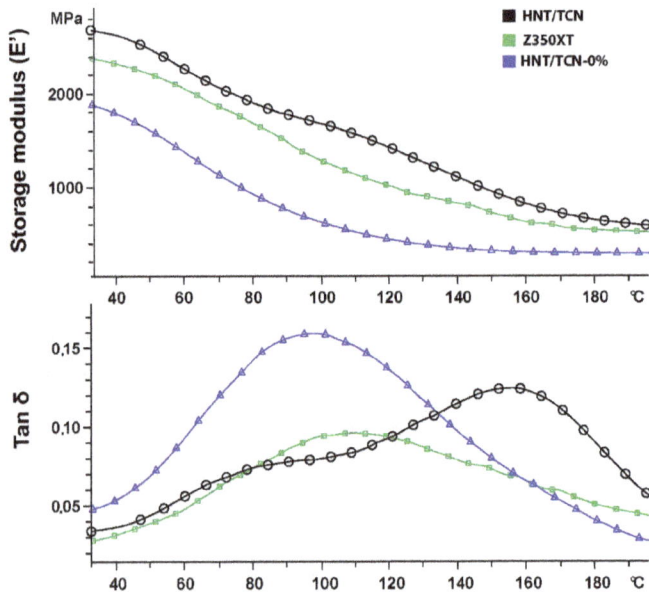

Figure 2. DMA curves of the specimens showed that the MPa, TAN δ at Tg for HNT/TCN resin composite presented a lower TAN δ at Tg when compared to the control composite containing no HNT/TCN-0%. The glass transition temperature (Tg) of HNT/TCN is higher than the other resin composite.

The results of the microbiological test are presented in Table 4. There were no statistically significant differences ($p = 0.977$) between the experimental groups for CFUs and ($p = 0.557$) for dry weight.

Table 4. Results microbiological tests, as colony forming units after 5 days (CFU)/mL/mm² and biofilm dry weight. This test showed no statistical difference between experimental groups and commercial resin composite ($p = 0.977$).

Composite	(CFU)/mL/mm²	Dry Weight (g)
HNT/TCN-0%	6.9267 (0.35) [A]	0.0004 (0.00025) [A]
HNT/TCN	6.8733 (0.28) [A]	0.0008 (0.00082) [A]
Z350XT	6.8867 (0.28) [A]	0.0003 (0.00006) [A]

* The same capital letters indicate absence of statistical difference ($p < 0.05$).

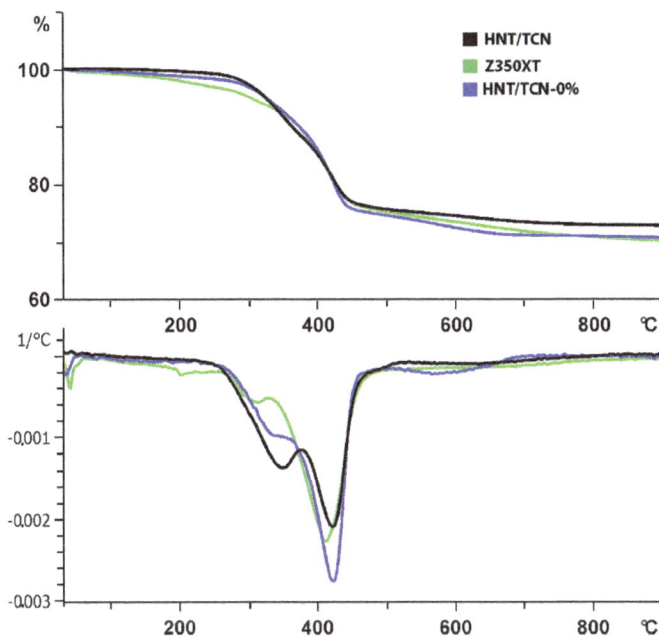

Figure 3. TGA curves of the specimens. Curve of the weight loss (%) showing that HNT/TCN resin composite presented a lower weight loss than other groups. The second curve below shows that there were two steps of degradation for each group.

4. Discussion

Recent studies have demonstrated that the incorporation of HNTs (5 wt.%) in resin-based materials (i.e., dental adhesive systems and enamel infiltrants) could improve their micro-hardness, flexural strength [15] and maximum polymerization rate [11]. However, if the concentration of HNTs is higher than 10 wt.% it is likely to attain a decrease of both flexural strength [15] and maximum polymerization rate [11]. The reason of such outcomes has been attributed to the behavior of such nanotubes to agglomerate in micro-cluster; this causes interference in the mechanism of interaction between nanotubes and the polymer matrix [30]. The null hypothesis that the inclusion of HNT-TCN at concentration of 8% into an experimental resin composite would have increased the physicochemical and microbiological properties must be partially accepted since some physicochemical properties were improved, although no significant differences were attained in terms of antibacterial activity (CFUs) and dry weight after 5 days of initial biofilm formation.

The physicochemical results obtained in this study are in agreement with those recently presented by Degrazia et al. [11]. Indeed, the organic matrix structure and the characteristic of fillers employed in the formulation of composite materials exert a direct influence on the surface roughness, degree of conversion, finishing, and polishing procedures; this may influence the surface quality of resin composite when applied in clinical scenarios [31]. Moreover, it is believed that if a higher degree of conversion is achieved, it is possible to extend the long-term stability and longevity of resin composites, especially in those cases when relatively "short" light-curing periods are performed [32]. However, despite the absence of statistical significance, the HNT/TCN composite presented greater numerical degree of conversion compared to the two control composites tested in this study. It is important to consider that for monomers with the same functionality, the higher the conversion of double bond monomers, the greater the mechanical strength of the cured resin [33]. Indeed, it has been already demonstrated that during the light-curing process HNT/TCN nanotubes may increase the intermolecular interactions between monomers during polymerization [30], due to the C=O and Al-O-H groups present on the inner and outer surface of the HNTs [34]. In general, in case of greater cross-link conversion degree, the conversion of monomers may advance and increase the density of the polymer network [35].

The dynamic mechanical analysis is usually employed for the evaluation of the glass transition temperature (Tg) of resin-based materials, since this allows a more "in-depth" knowledge of the network homogeneity and cross-linking density [36]. As previously stated, the addition of nanotubes in resin-based materials may cause the formation of intermolecular interactions (hydrogen bond formation between hydroxyl groups) between the outer surface of HNTs and Bis-GMA molecule by hydroxyl groups [30].

It has been reported that, low values of TAN δ at specific Tg values indicate a better interfacial adhesion between the organic matrix and the filler [35]. In the present study, the glass transition temperature of the tested materials showed that the HNT/TCN composite presented a lower TAN δ at Tg when compared to the control composite containing no HNT. This outcome indicates that the nanotubes may enhance the chemical interaction during polymerization reaction between the organic matrix and the inorganic fillers (Figures 2 and 3).

Furthermore, the strength, modulus and impact resistance of resin-based materials can be increased when these are doped with HNTs, even at relative low concentration (5 wt.%) [23]. According to the ISO 4049/2000 standard [26], the flexural strengths of universal resin-based restorative materials should be higher than 80 MPa. The Flexural strength test (Table 2) showed that FS of the commercial nanofilled composite and that of the experimental composite HNT/TCN were statistically higher than FS obtained with the control composite HNT/TCN-0%; the latter did not meet the minimal required values of ISO 4049/2000. High flexural strength is necessary to prevent cohesive fractures within the bulk material, especially when considering posterior restorations [37,38]. However, HNT/TCN presented FS values similar to that of the modern commercial composite used in this study. However, although the similarity in flexural strength, the elastic modulus of the Filtek Z350 XT attained in our study was lower than that reported by Rosa et al., 2012 [39]. The main reason for such a difference can be attributed to the different light-curing protocol used in our study.

A further important characteristic that may influence the stress development is the flexural modulus (FM); this is often associated with the composition of the tested material. In this study, the FM tests showed no significant difference between the tested groups (Table 2). Correlation between the FM and the polymerization stress values is a valid simplified approach [38]. However, the limitations of this in vitro test are related to the fact that the light needed to pass through the acrylic rod, as well as the 1 mm thick material before reaching the upper rod in order to create the bonding between the materials and the rods [40]. Indeed, in this study the high standard deviation of polymerization stress test obtained in the HNT/TCN resin composite may be related to the compliance of the test. Tensilometer test configurations can also present some limitations when considering a clinical scenario and the configuration of the cavities [41]. Nonetheless, it was previously reported that it is a valid method to simulate the conditions present in small cavities, where the volume of the resin composite is restricted; this may represent the clinical situation in minimally invasive restorative procedures [42].

Resin composites with lower values of the polymerization stress are those having the lowest elastic modulus; this is probably due to the greater deformation capacity of these latter materials [43]. The addition of HNT/TCN in the experimental resin composite resulted in a significant increase in PS (Table 2). It is known that resin-composites with high modulus of elasticity may generate stiffer restorations; such conditions may increase the effect of the polymerization stress on the tooth-composite interfaces [44]. Conversely, resin composites with a lower elastic modulus create less stress on the interface, although these may lack full dimensional recovery to withstand the masticatory load [44]. Indeed, it can be observed that despite the higher polymerization stress value of HNT/TCN, FM values had the tendency to be higher than the other two materials tested in this study. Another reason HNT/TCN composite showed higher polymerization stress may be associated with the lower micro-hybrid silica filler content (72%) compared to the HNT-free control composite, which had 80 wt.% of micro-hybrid filler.

To test the antibacterial effect, a test was carried out with 5 days of biofilm growth. Degrazia and collaborators [11], showed that their resin-based material doped with HNT/TCN had antibacterial effect up to 72h; antibacterial properties can be also evaluated after 24 h of biofilm growth. It is suggested that the antibacterial effect can be obtained by direct contact of *S. mutans* with the inhibitory agent present in the resin composite (TCN in case of the present work). In agreement with this idea, Feitosa and collaborators [22] demonstrated strong antibacterial activity when *S. mutans* was in direct contact with doxycycline-encapsulated nanotube-modified dentin adhesive after 24 h. In the present study, no statistically significant difference was found between the CFUs values of the 3 tested composites biofilm formed over a 5-day period ($p = 0.977$) (Table 4). It seems that the decrease in the inhibitory antibacterial effect overtime might be related to the inability of such agent from the resin composite to reach the whole thickness of the plaque, in particular the outer layers of bacteria. Probably, the absence in degradation and wearing of the material seems to be considered as an advantage. Due to the interestingly low CFU of the HNT/TCN, in the present work, we believe that the daily removal of the biofilm can reduce its thickness and allow the direct contact of the agent with the bacteria and promoting a more effective antibacterial action. Furthermore, it is possible that current antibacterial tests performed in a laboratory may not be totally appropriate for the evaluation of resin composites, unless accompanied with aging protocols that allow the release of therapeutic agents such as the TCN.

Future studies are suggested to improve the development of antimicrobial materials and the understanding of the relationship between their formulations, morphology and properties, to promote the longevity (shelf-life and ageing) of resin-based materials restorations. Moreover, studies including its rheological properties with alternative methods for synthesis and nanotubes can extend the range of application of such materials as well the satisfaction of as patients and clinicians.

5. Conclusions

It can be concluded that:

- Incorporation of 8 wt.% seems to be a satisfactory formulation of halloysite nanotube for achieving appropriate mechanical properties for an experimental resin micro-hybrid composite without affecting the viscosity and the material and increase the risk for phase separation;
- The experimental resin composite containing 8 wt.% halloysite nanotube doped with triclosan, from a physicochemical point of view, seems to be a suitable restorative material such as the current commercial nano-filled resin-composite;
- Incorporation of triclosan seems to be test-dependent, since it showed no response in mature biofilms as used in this study.

Author Contributions: D.A.C. performed the experiments and wrote the manuscript; N.S.R. and L.C.S. performed the experiments; D.L. analyzed the data and contributed materials/analysis tools; F.P.R. wrote part of the manuscript and revised the manuscript; F.W.D. revised the manuscript; F.M.C. analyzed the data and revised the manuscript; S.S. conceived of and designed the experiments. Revised the manuscript at all the steps; V.P.A.S. conceived of and designed the experiments.

Acknowledgments: We would like to acknowledge the "Departamento de Biomateriais e Biologia Oral" from Universidade de São Paulo for all the technical support offered on polymerization stress measurements test. In addition to Coordenação de Aperfeiçoamento de Pessoal de Nível Superior—CAPES, by granting the scholarship.This work was also supported by the research grant INDI—"Programa de Consolidación de Indicadores: Fomento Plan Estatal CEU-UCH" 2017–2018 to Salvatore Sauro. We would also like to thank all those members of the group "Dental Biomaterials Science & Research (https://www.facebook.com/groups/985792791507045/?ref=bookmarks) who helped us in enriching the background and the references of our study.

Conflicts of Interest: The authors declare no conflict of interest.

References

1. Padovani, G.C.; Feitosa, V.P.; Sauro, S.; Tay, F.R.; Durán, G.; Paula, A.J.; Durán, N. Advances in dental materials through nanotechnology: Facts, perspectives and toxicological aspects. *Trends Biotechnol.* **2015**, *33*, 621–636. [CrossRef] [PubMed]
2. Wang, X.; Cai, Q.; Zhang, X.; Wei, Y.; Xu, M.; Yang, X.; Ma, Q.; Cheng, Y.; Deng, X. Improved performance of Bis-GMA/TEGDMA dental composites by net-like structures formed from SiO_2 nanofiber fillers. *Mater. Sci. Eng. C* **2016**, *59*, 464–470. [CrossRef] [PubMed]
3. Degrazia, F.W.; Leitune, V.C.B.; Takimi, A.S.; Collares, F.M.; Sauro, S. Physicochemical and bioactive properties of innovative resin-based materials containing functional halloysite-nanotubes fillers. *Dent. Mater.* **2016**, *32*, 1133–1143. [CrossRef] [PubMed]
4. Murdoch-Kinch, C.A.; McLean, M.E. Minimally invasive dentistry. *J. Am. Dent. Assoc.* **2003**, *134*, 87–95. [CrossRef] [PubMed]
5. Bernardo, M.; Luis, H.; Martin, M.D.; Leroux, B.G.; Rue, T.; Leitão, J.; DeRouen, T.A. Survival and reasons for failure of amalgam versus composite posterior restorations placed in a randomized clinical trial. *J. Am. Dent. Assoc.* **2007**, *138*, 775–783. [CrossRef] [PubMed]
6. Demarco, F.F.; Corrêa, M.B.; Cenci, M.S.; Moraes, R.R.; Opdam, N.J.M. Longevity of posterior composite restorations: Not only a matter of materials. *Dent. Mater.* **2012**, *28*, 87–101. [CrossRef] [PubMed]
7. Khvostenko, D.; Hilton, T.J.; Ferracane, J.L.; Mitchell, J.C.; Kruzic, J.J. Bioactive glass fillers reduce bacterial penetration into marginal gaps for composite restorations. *Dent. Mater.* **2016**, *32*, 73–81. [CrossRef] [PubMed]
8. Loesche, W.J. Role of streptococcus mutans in human dental decay. *Microbiol. Rev.* **1986**, *50*, 353–380. [PubMed]
9. Fugolin, A.P.P.; Pfeifer, C.S. New resins for dental composites. *J. Dent. Res.* **2017**, *96*, 1085–1091. [CrossRef] [PubMed]
10. Wu, H.-X.; Tan, L.; Tang, Z.-W.; Yang, M.-Y.; Xiao, J.-Y.; Liu, C.-J.; Zhuo, R.-X. Highly efficient antibacterial surface grafted with a triclosan-decorated poly(n-hydroxyethylacrylamide) brush. *ACS Appl. Mater. Interfaces* **2015**, *7*, 7008–7015. [CrossRef] [PubMed]
11. Degrazia, F.W.; Genari, B.; Leitune, V.C.B.; Arthur, R.A.; Luxan, S.A.; Samuel, S.M.W.; Collares, F.M.; Sauro, S. Polymerisation, antibacterial and bioactivity properties of experimental orthodontic adhesives containing triclosan-loaded halloysite nanotubes. *J. Dent.* **2017**. [CrossRef] [PubMed]
12. Kaffashi, B.; Davoodi, S.; Oliaei, E. Poly(ε-caprolactone)/triclosan loaded polylactic acid nanoparticles composite: A long-term antibacterial bionanocomposite with sustained release. *Int. J. Pharm.* **2016**, *508*, 10–21. [CrossRef] [PubMed]
13. Rathke, A.; Staude, R.; Muche, R.; Haller, B. Antibacterial activity of a triclosan-containing resin composite matrix against three common oral bacteria. *J. Mater. Sci. Mater. Med.* **2010**, *21*, 2971–2977. [CrossRef] [PubMed]
14. Wicht, M.; Haak, R.; Kneist, S.; Noack, M. A triclosan-containing compomer reduces spp. predominant in advanced carious lesions. *Dent. Mater.* **2005**, *21*, 831–836. [CrossRef] [PubMed]
15. Feitosa, S.A.; Münchow, E.A.; Al-Zain, A.O.; Kamocki, K.; Platt, J.A.; Bottino, M.C. Synthesis and characterization of novel halloysite-incorporated adhesive resins. *J. Dent.* **2015**, *43*, 1316–1322. [CrossRef] [PubMed]
16. Guimarães, L.; Enyashin, A.N.; Seifert, G.; Duarte, H.A. Structural, electronic, and mechanical properties of single-walled halloysite nanotube models. *J. Phys. Chem. C* **2010**, *114*, 11358–11363. [CrossRef]
17. Bottino, M.C.; Batarseh, G.; Palasuk, J.; Alkatheeri, M.S.; Windsor, L.J.; Platt, J.A. Nanotube-modified dentin adhesive—Physicochemical and dentin bonding characterizations. *Dent. Mater.* **2013**, *29*, 1–8. [CrossRef] [PubMed]
18. Cavallaro, G.; Lazzara, G.; Milioto, S.; Parisi, F. Halloysite nanotubes for cleaning, consolidation and protection. *Chem. Rec.* **2018**. [CrossRef] [PubMed]
19. Alkatheeri, M.S.; Palasuk, J.; Eckert, G.J.; Platt, J.A.; Bottino, M.C. Halloysite nanotube incorporation into adhesive systems—Effect on bond strength to human dentin. *Clin. Oral Investig.* **2015**, *19*, 1905–1912. [CrossRef] [PubMed]

20. Ghaderi-Ghahfarrokhi, M.; Haddadi-Asl, V.; Zargarian, S.S. Fabrication and characterization of polymer-ceramic nanocomposites containing drug loaded modified halloysite nanotubes. *J. Biomed. Mater. Res. Part A* **2018**. [CrossRef] [PubMed]

21. Palasuk, J.; Windsor, L.J.; Platt, J.A.; Lvov, Y.; Geraldeli, S.; Bottino, M.C. Doxycycline-loaded nanotube-modified adhesives inhibit MMP in a dose-dependent fashion. *Clin. Oral Investig.* 2017. [CrossRef] [PubMed]

22. Feitosa, S.A.; Palasuk, J.; Kamocki, K.; Geraldeli, S.; Gregory, R.L.; Platt, J.A.; Windsor, L.J.; Bottino, M.C. Doxycycline-encapsulated nanotube-modified dentin adhesives. *J. Dent. Res.* **2014**, *93*, 1270–1276. [CrossRef] [PubMed]

23. Liu, M.; Jia, Z.; Jia, D.; Zhou, C. Recent advance in research on halloysite nanotubes-polymer nanocomposite. *Prog. Polym. Sci.* **2014**, *39*, 1498–1525. [CrossRef]

24. Svizero Nda, R.; Silva, M.S.; Alonso, R.C.; Rodrigues, F.P.; Hipólito, V.D.; Carvalho, R.M.; D'Alpino, P.H. Effects of curing protocols on fluid kinetics and hardness of resin cements. *Dent. Mater. J.* **2013**, *32*, 32–41. [CrossRef] [PubMed]

25. Rodrigues, N.S.; de Souza, L.C.; Feitosa, V.P.; Loguercio, A.D.; D'Arcangelo, C.; Sauro, S.; Saboia, V.d.P.A. Effect of different conditioning/deproteinization protocols on the bond strength and degree of conversion of self-adhesive resin cements applied to dentin. *Int. J. Adhes. Adhes.* 2017. [CrossRef]

26. ISO 4049. *Dentistry—Polymer-Based Filling, Restorative and Luting Materials; 7.10 Depth of cure, Class 2 materials*; International Organization for Standardization: Geneva,Switzerland, 2000.

27. Gonçalves, F.; Boaro, L.C.; Ferracane, J.L.; Braga, R.R. A comparative evaluation of polymerization stress data obtained with four different mechanical testing systems. *Dent. Mater.* **2012**, *28*, 680–686. [CrossRef] [PubMed]

28. Duarte, S.; Gregoire, S.; Singh, A.P.; Vorsa, N.; Schaich, K.; Bowen, W.H.; Koo, H. Inhibitory effects of cranberry polyphenols on formation and acidogenicity of streptococcus mutans biofilms. *FEMS Microbiol. Lett.* **2006**, *257*, 50–56. [CrossRef] [PubMed]

29. Peralta, S.L.; Carvalho, P.H.A.; van de Sande, F.H.; Pereira, C.M.P.; Piva, E.; Lund, R.G. Self-etching dental adhesive containing a natural essential oil: Anti-biofouling performance and mechanical properties. *Biofouling* **2013**, *29*, 345–355. [CrossRef] [PubMed]

30. Chen, Q.; Zhao, Y.; Wu, W.; Xu, T.; Fong, H. Fabrication and evaluation of Bis-GMA/TEGDMA dental resins/composites containing halloysite nanotubes. *Dent. Mater.* **2012**, *28*, 1071–1079. [CrossRef] [PubMed]

31. Abzal, M.; Rathakrishnan, M.; Prakash, V.; Vivekanandhan, P.; Subbiya, A.; Sukumaran, V. Evaluation of surface roughness of three different composite resins with three different polishing systems. *J. Conserv. Dent.* **2016**, *19*. [CrossRef]

32. Sauro, S.; Osorio, R.; Watson, T.F.; Toledano, M. Therapeutic effects of novel resin bonding systems containing bioactive glasses on mineral-depleted areas within the bonded-dentine interface. *J. Mater. Sci. Mater. Med.* **2012**, *23*, 1521–1532. [CrossRef] [PubMed]

33. Braga, R.R.; Ballester, R.Y.; Ferracane, J.L. Factors involved in the development of polymerization shrinkage stress in resin-composites: A systematic review. *Dent. Mater.* **2005**, *21*, 962–970. [CrossRef] [PubMed]

34. Luo, R.; Sen, A. Rate enhancement in controlled radical polymerization of acrylates using recyclable heterogeneous lewis acid. *Macromolecules* **2007**, *40*, 154–156. [CrossRef]

35. Karabela, M.M.; Sideridou, I.D. Synthesis and study of properties of dental resin composites with different nanosilica particles size. *Dent. Mater.* **2011**, *27*, 825–835. [CrossRef] [PubMed]

36. Bacchi, A.; Yih, J.A.; Platta, J.; Knight, J.; Pfeifer, C.S. Shrinkage/stress reduction and mechanical properties improvement in restorative composites formulated with thio-urethane oligomers. *J. Mech. Behav. Biomed. Mater.* **2018**, *78*, 235–240. [CrossRef] [PubMed]

37. Aydınoğlu, A.; Yoruç, A.B.H. Effects of silane-modified fillers on properties of dental composite resin. *Mater. Sci. Eng. C* 2017, *79*, 382–389. [CrossRef] [PubMed]

38. Boaro, L.C.C.; Gonalves, F.; Guimarães, T.C.; Ferracane, J.L.; Versluis, A.; Braga, R.R. Polymerization stress, shrinkage and elastic modulus of current low-shrinkage restorative composites. *Dent. Mater.* **2010**, *26*, 1144–1150. [CrossRef] [PubMed]

39. Rosa, R.S.; Balbinot, C.E.; Blando, E.; Mota, E.G.; Oshima, H.M.; Hirakata, L.; Pires, L.A.; Hübler, R. Evaluation of mechanical properties on three nanofilled composites. *Stomatologija* **2012**, *14*, 126–130. [PubMed]

40. Gonçalves, F.; Boaro, L.C.C.; Miyazaki, C.L.; Kawano, Y.; Braga, R.R. Influence of polymeric matrix on the physical and chemical properties of experimental composites. *Braz. Oral Res.* **2015**, *29*, 1–7. [CrossRef] [PubMed]

41. Boaro, L.C.C.; Fróes-Salgado, N.R.; Gajewski, V.E.S.; Bicalho, A.A.; Valdivia, A.D.C.M.; Soares, C.J.; Miranda Júnior, W.G.; Braga, R.R. Correlation between polymerization stress and interfacial integrity of composites restorations assessed by different in vitro tests. *Dent. Mater.* **2014**, *30*, 984–992. [CrossRef] [PubMed]

42. Rodrigues, F.P.; Lima, R.G.; Muench, A.; Watts, D.C.; Ballester, R.Y. A method for calculating the compliance of bonded-interfaces under shrinkage: Validation for Class i cavities. *Dent. Mater.* **2014**, *30*, 936–944. [CrossRef] [PubMed]

43. Labella, R.; Lambrechts, P.; Van Meerbeek, B.V.G. Polymerization shrinkage and elasticity of flowable adhesives and filled composites. *Dent. Mater.* **1999**, *15*, 128–137. [CrossRef]

44. Rosatto, C.M.P.; Bicalho, A.A.; VerÃssimo, C.; BraganÃ§a, G.F.; Rodrigues, M.P.; Tantbirojn, D.; Versluis, A.; Soares, C.J. Mechanical properties, shrinkage stress, cuspal strain and fracture resistance of molars restored with bulk-fill composites and incremental filling technique. *J. Dent.* **2015**, *43*, 1519–1528. [CrossRef] [PubMed]

materials

MDPI

Article

Properties of Experimental Dental Composites Containing Antibacterial Silver-Releasing Filler

Robert Stencel [1], Jacek Kasperski [2], Wojciech Pakieła [3], Anna Mertas [4], Elżbieta Bobela [4], Izabela Barszczewska-Rybarek [5] and Grzegorz Chladek [3,*]

[1]   Private Practice, Center of Dentistry and Implantology, ul. Karpińskiego 3, 41-500 Chorzów, Poland; robert.stencel@op.pl

[2]   Department of Prosthetic Dentistry, School of Medicine with the Division of Dentistry in Zabrze, Medical University of Silesia, pl. Akademicki 17, 41-902 Bytom, Poland; kroczek91@interia.pl

[3]   Faculty of Mechanical Engineering, Institute of Engineering Materials and Biomaterials, Silesian University of Technology, ul. Konarskiego 18a, 44-100 Gliwice, Poland; wojciech.pakiela@polsl.pl

[4]   Chair and Department of Microbiology and Immunology, School of Medicine with the Division of Dentistry in Zabrze, Medical University of Silesia in Katowice, ul. Jordana 19, 41-808 Zabrze, Poland; amertas@sum.edu.pl (A.M.); ebobela@sum.edu.pl (E.B.)

[5]   Department of Physical Chemistry and Technology of Polymers, Silesian University of Technology, 44-100 Gliwice, Poland; Izabela.Barszczewska-Rybarek@polsl.pl

*   Correspondence: grzegorz.chladek@polsl.pl; Tel.: +48-32-237-2907

Received: 20 May 2018; Accepted: 11 June 2018; Published: 18 June 2018

Abstract: Secondary caries is one of the important issues related to using dental composite restorations. Effective prevention of cariogenic bacteria survival may reduce this problem. The aim of this study was to evaluate the antibacterial activity and physical properties of composite materials with silver sodium hydrogen zirconium phosphate (SSHZP). The antibacterial filler was introduced at concentrations of 1%, 4%, 7%, 10%, 13%, and 16% (w/w) into model composite material consisting of methacrylate monomers and silanized glass and silica fillers. The in vitro reduction in the number of viable cariogenic bacteria *Streptococcus mutans* ATCC 33535 colonies, Vickers microhardness, compressive strength, diametral tensile strength, flexural strength, flexural modulus, sorption, solubility, degree of conversion, and color stability were investigated. An increase in antimicrobial filler concentration resulted in a statistically significant reduction in bacteria. There were no statistically significant differences caused by the introduction of the filler in compressive strength, diametral tensile strength, flexural modulus, and solubility. Statistically significant changes in degree of conversion, flexural strength, hardness (decrease), solubility (increase), and in color were registered. A favorable combination of antibacterial properties and other properties was achieved at SSHZP concentrations from 4% to 13%. These composites exhibited properties similar to the control material and enhanced in vitro antimicrobial efficiency.

Keywords: dental composites; antibacterial properties; silver; mechanical properties; degree of conversion; sorption; solubility; color stability

1. Introduction

Worldwide, around 2.4 billion people (33% of the population) suffer from dental caries in permanent teeth, and the percentage of this chronic disease increased between 2005 and 2015 by 14% [1]. Moreover, in some countries like Poland, more than 90% of the adult population has experienced dental caries and use dental fillings or dentures [2]. These facts illustrate the progressive extent of the demand for dental materials and the role of constant development in this specific field of material science. Dental caries, but sometimes also dental trauma or extensive wear caused, e.g., by bruxism, may lead to the loss of hard tissues of the teeth. One of the strategies allowing reconstruction of the teeth structure is using direct restorative materials, which are shaped intraorally to create restorations directly in teeth cavities [3]. Currently, the most common of them are photopolymerizable resin-based composites, introduced few decades ago as a substitute for amalgams [4]. This type of material is also considered to be the most prospective, which has resulted in a growing number of new products on the market and numerous investigations in this area. In comparison with other direct restorative materials, composites show optimal esthetic properties, which are related to possibilities of color matching (translucency, shades), satisfying color stability and polishability [5]. Composites are also reasonably easy to use and need less invasive preparation techniques than amalgams [6], which should be considered as additional clinical advantages. As a result of many years of evolution, modern composites show good mechanical and physical properties [7], with wear rates similar to human enamel [8] as well as suitable biocompatibility [9]. Nevertheless, use of resin composites may still lead to higher failure rates in comparison to amalgams [10,11]. The two most frequent reasons for composite failures are fractures and secondary caries [12,13]. Pereira-Cenci et al. [14], in their extensive review, concluded that secondary caries is the cause of up to 55% of resin composite filling replacements. It is defined as "positively diagnosed carious lesion, which occurs at the margins of an existing restoration" [15]. However, currently, it is commonly accepted that it is a primary carious lesion of teeth at the margin of a filling, but it occurs after some time from placing the restoration [15,16], in contrast to the remaining caries, which are caused by incomplete elimination of infected tooth tissues during cavity preparation [15]. Secondary caries is often linked to the presence of microleakage caused by various factors [17–19], which may be the reason for the occurrence of liquids, chemical substances, and finally bacteria between the tooth and the restoration [20,21]. Regardless of the doubts about the etiology of caries after the placement of fillings, it is recognized as a serious and widespread clinical problem. Moreover, composites accumulate more biofilm and plaque than other direct restorative materials [22]. For this reason, it is believed that the perfect resin composite filling should not only have suitable mechanical and esthetic properties but also ought to possess antibacterial properties to avoid colonization of the tooth/restoration interface by pathogenic bacteria, such as *Streptococcus mutans* (*S. mutans*) [23,24].

Diverse research with different additives has been carried out to develop effective antibacterial composites. Numerous experiments have focused on resins containing polymerizable antibacterial additives, such as quaternary ammonium dimethacrylate (QADM) [25], 12-methacryloyloxydodecylpyridinium bromide (MDPB) [26], dimethylaminohexadecyl methacrylate (DMAHDM) [27], dimethyl-hexadecyl-methacryloxyethyl-ammonium iodide (DHMAI) [28], or dimethylaminododecyl and dimethylaminohexadecyl methacrylates [29]. Other organic materials including quaternary ammonium polyethylenimine (PEI) nanoparticles [30], chlorhexidine [31,32], triclosan [33], chitosan [34], and benzalkonium chloride and acrylic acid [35] were also tested with varying degrees of success. The use of different experimental fillers is another important strategy for developing antimicrobial composites. Tavassoli Hojati et al. [36], Kasraei et al. [37], and Aydin Sevinç et al. [38] reported the reduction of cariogenic bacteria after incorporation of zinc oxide nanoparticles, probably due to the mechanism of production of active oxygen species, such as H_2O_2 or the possible leaching of Zn^{2+} ions. Khvostenko et al. [39] used bioactive glass (65% SiO_2, 31% CaO, 4% P_2O_5) and obtained a 61% reduction of *S. mutans* penetration of the gap depth under laboratory conditions, which suggests that the release of ions from glass into the gap may help control the local chemistry by creating an antimicrobial environment that reduces biofilm propagation. Łukomska-Szymańska et al. [20] noted that composites additionally filled with calcium fluoride had shown a significant reduction of *S. mutans* and *L. acidophilus*, which was probably related with creating hydrofluoric acid that can penetrate the bacterial membrane, generate acidification of cytoplasm, and inhibit enzymes. Sodagar et al. [40] modified the commercially available orthodontic composite with titanium dioxide nanoparticles and proved inhibition of *S. mutans* and *S. sanguinis* growth. The most widely tested materials in previous years were those containing silver. Niu et al. [41] successfully applied tetrapod-like zinc oxide whiskers to increase antibacterial resistance. Chatzistavrou et al. [42] confirmed a significant reduction of *S. mutans* for Ag-doped bioactive glass and additional bioactivity of tested materials. Ai et al. [43] investigated composite resin reinforced with silver nanoparticle-laden hydroxyapatite nanowires, where nanowires were used as reinforcement and nanosilver as an antimicrobial agent. The reduction of microorganisms was noted, however, only when the experimental filler was added into the matrix and its concentration was limited to 10%, so those interesting results needs confirmation in follow-up experiments on materials with typical reinforcing fillers. Łukomska-Szymańska et al. [44], reported a viability of *S. mutans* from 48% to 87% in comparison to control samples on the surface of experimental composites with the addition of silver particles alone and combined titanium dioxide, silica dioxide, and zirconium dioxide nanoparticles or microparticles. Kasraei et al. [37] and Azarsina et al. [45] modified commercially available composites with silver nanoparticles and noted a reduction of bacterial colonies. However, amber to brown discoloration of materials with nanosilver has been noted, which is a limitation for an esthetic material [37,44,45]. Also, the inhibitory effect against *S. mutans* of resin composites with silver-containing inorganic particles like silica gel have been confirmed, which the authors linked not with silver ion release but with the presence of active oxygen, including hydroxyl radicals, created by the catalytic action of silver during photoactivation or contact with water at polar surfaces [46]. Additionally, simultaneous effects of silver nanoparticles with hydroxyapatite nanoparticles [47] or antimicrobial monomers [27,48] were also investigated.

Silver sodium hydrogen zirconium phosphate (SSHZP) is a silver-releasing ceramic. This submicron-sized antimicrobial material is white and stable, so as opposed to silver nanoparticles, it should not cause the typical initial amber or brown discoloration due to the plasmon effect [49], which is problematic in the case of dental materials. However, the question of further color changes related with silver ion release during contact with a wet environment and its oxidation remains open.

So far, SSHZP has been reported as an additive into a polymethyl methacrylate (PMMA) denture base material [50] and a polydimethylsiloxane-based soft denture lining [51]. SSHZP was also previously investigated as an antimicrobial additive into chitosan and alginate fibers [52,53]. Moreover, it is incorporated into some currently available alginate and carboxymethylcellulose wound dressings [54,55]. In this study, we report the use of SSHZP as antibacterial filler for a distinctly different material—a experimental direct restorative photopolymerizable resin-based composite, reinforced with varied filler types at high concentrations. Therefore, the aim of the presented work was to investigate the impact of the proposed filler (SSHZP), introduced into resin-based composites intended as direct restorative materials, for its antimicrobial effectiveness, mechanical properties, degree of conversion, sorption, solubility, and color changes. Our hypothesis was that composites additionally filled with SSHZP would show antimicrobial effectiveness against cariogenic bacteria and suitable properties for dental restorative materials.

2. Materials and Methods

2.1. Materials Preparation

The matrix consisted of three mixed monomers: bisphenol A glycidyl methacrylate (bis-GMA), urethane-dimethacrylate (UDMA), and triethylene glycol dimethacrylate (TEGDMA) at a weight ratio of 42:38:20, respectively (all purchased form Sigma-Aldrich, St. Louis, MO, USA). Additionally, 0.4% (*w/w*) of camphorquinone (CQ, Sigma-Aldrich, St. Louis, MO, USA) as the photosensitizer and 1% (*w/w*) of *N,N*-dimethylaminoethyl methacrylate (DMAEMA) as a photoaccelerator (both Sigma-Aldrich, St. Louis, MO, USA) were introduced. The reinforcing fillers were two silanized barium borosilicate glass fillers (Esschem, Linwood, PA, USA), with a mean particle size declared by the manufacturer of 2 μm (G1) or 0.7 μm (G2), and silanized silica nanofiller Aerosil R7200 (AR) (Evonic Industries, Essen, Germany), used at a weight ratio of 50:35:15, respectively. Silver sodium hydrogen zirconium phosphate containing approximately 10% of silver (*w/w*), with molecular formula $Ag_{0.46} Na_{0.29} H_{0.25} Zr_2 (PO_4)_3$ [56] (Milliken Chemical, Spartanburg, SC, USA) was used as an antimicrobial filler. The SSHZP was compounded at concentrations of 1%, 4%, 7%, 10%, 13%, and 16% (w/w), and the masses necessary to prepare the composites were calculated according to the equation:

$$m_{SSHZP} = \frac{c_{SSHZP} \times m_{MRF}}{1 - c_{SSHZP}} \tag{1}$$

where m_{SSHZP} was the SSZHP g; c_{SSHZP} was the SSZHP concentration, % (w/w); and m_{MRF} was the matrix with reinforcing filler mass (always constant).

The fillers were compounded into a matrix in 50 mL glass Griffin form beakers at room temperature in the following order: SSHZP, G1, G2, and AR as the last one. All composites were prepared in standardized portions based on the same masses of matrix and reinforcing fillers. The compositions of standardized portions of tested materials are listed in Table 1. The introducing process was carried out gradually in standard portions of 1 g (SSHZP, G1, G2) or 0.5 g (AR). For the lowest concentration of SSHZP, or when the last portion of particular fillers was added, they were smaller. Compounding was effected by multiple spreadings and mixings of materials with a stainless steel spatula on the wall of the beaker to apply shear forces. The subsequent doses of fillers were added when a homogeneous consistency for the previous dose was achieved. The process of compounding for one material took about 2.5–4 h; the longer time was needed for materials with higher filler concentrations due to their increasing viscosity. The obtained compositions were placed under the pressure of 80 mbar for 25 min in a modified vacuum stirrer (Twister evolution, Renfert GmbH, Hilzingen, Germany). All materials were polymerized with a DY400-4 LED lamp (Denjoy Dental, Changsha, China), power 5 W, intensity 1400–2000 mW/cm^2, optical wave length 450–470 nm.

Table 1. Compositions of investigated materials with the masses of components needed to prepare standard portions.

Code	Matrix, g	Matrix, % (w/w)	RF, g	RF, % (w/w)	SSHZP, g	SSHZP, % (w/w)	TF, % (w/w)
Control	15.00	35.00	27.86	65.00	0	0	65.00
AC 1	15.00	34.35	27.86	64.65	0.43	1	65.35
AC 4	15.00	33.60	27.86	62.40	1.76	4	66.40
AC 7	15.00	32.55	27.86	60.45	3.22	7	67.45
AC 10	15.00	31.50	27.86	58.50	4.76	10	68.50
AC 13	15.00	30.45	27.86	56.55	6.40	13	69.55
AC 16	15.00	29.40	27.86	54.60	8.16	16	70.60

AC—antibacterial composite, RF—reinforcing fillers, SSHZP—silver sodium hydrogen zirconium phosphate, TF—total concentration of compounded fillers.

2.2. Scanning Electron Microscopy (SEM) Investigations

Fillers were added to 99.8% ethanol, ultrasonically homogenized, and dropped on carbon tape. Polymerized samples for composite morphology observations measured $10 \times 2 \times 2$ mm. Two types of specimens were used. The first type was subjected to the standard procedure which involved wet-grinding and polishing using diamond pastes. The other type was immersed in liquid nitrogen and broken. Composite samples after polishing were also etched with orthophosphoric acid. All samples were sputtered with gold. Observations were performed using a Zeiss SUPRA 35 scanning electron microscope (Zeiss, Oberkochen, Germany) at accelerating voltages from 3 kV to 20 kV.

2.3. Antibacterial Test

Specimens measured 11 mm in diameter and 2 mm in thickness and were prepared in Teflon molds. The mold was placed at a microscope slide covered with 50 μm thick polyester foil. The material was placed into the mold and covered with the foil and microscope slide. Then, the upper microscope slide was manually pressed and taken away. When the sample was polymerized, the polyester foil was removed. The molds with samples were wet-ground sequentially with P800- and P1200-grit abrasive papers to remove excess of material and to standardize the surface. Next, the samples were rinsed with distilled water and pushed out of the molds.

The in vitro reduction of bacteria was examined according to the previously described method [51,57,58] with some modifications. The standard strain of bacterium *Streptococcus mutans* ATCC 33535 was used. Sterilized samples of composites were immersed individually in 2 mL of bacterial suspensions in tryptone water, which contained approximately 1.5×10^5 CFU/mL (CFU—colony forming units) of *S. mutans*. A suspension of bacteria in tryptone water was tested as a positive control. Pure tryptone water was tested as a negative control. Incubation was carried out in a shaking incubator for 17 h at 37 °C. After incubation, 20 μL of suspension was seeded onto Columbia agar (bioMerieux, Marcy l'Etoille, France) with 5% sheep blood plates. The cultured plates were finally incubated at 37 °C for 24 h, and the numbers of bacterial colonies were counted. The relative reduction in the number of viable bacteria colonies (RB) was calculated according to the equation:

$$RB = \frac{V_c - V_t}{V_c} \times 100\% \tag{2}$$

where V_c was the number of viable microorganism colonies of the positive control (BLANK) and V_t was the number of viable microorganism colonies of the test specimen.

2.4. Compressive Strength

Compressive strength was examined according to the method presented by Mota et al. [59], with some necessary specifications concerning sample preparation. Cylindrical specimens (3 mm in diameter and 6 mm in height) were prepared as described for the microbiological test. However, due to their height, polymerization was carried out at the top and at the bottom before the removal of the polyester foil. Furthermore, after removing them from the mold, the samples were cured on four lateral surfaces, according to the recommendation of Galvão et al. [60]. Ten samples were prepared from each composite. The samples were conditioned in distilled water at 37 ± 1 °C for 24 h. Tests were conducted using a universal testing machine (Zwick Z020, Zwick GmbH & Com, Ulm, Germany) at a cross-head speed of 0.5 mm/min. Compressive strength was calculated according to the equation:

$$\sigma_{cs} = \frac{F}{A} \tag{3}$$

where σ_{cs} was the compressive strength, MPa; F was force at fracture, N; and A was the initial cross-sectional area of specimen, mm^2.

2.5. Diametral Tensile Strength

The samples for the diametral tensile strength (DTS) tests (6 mm in diameter and 3 mm in height) [61] were prepared with a method similar to the microbiological test, but irradiation was carried out at the top and at the bottom before removing the polyester foil. Ten samples were prepared from each composite. The samples were conditioned in distilled water at 37 ± 1 °C for 24 h [61]. Compressive load was applied on the lateral surface of the samples at a cross-head speed of 0.5 mm/min [20] using a universal testing machine Zwick Z2.5. The DTS values were calculated according to the equation:

$$DTS = \frac{2F}{\pi dh} \tag{4}$$

where DTS was the ultimate diametral tensile strength, MPa; F was the force at fracture, N; d was the diameter, mm; and h was the thickness, mm.

2.6. Flexural Strength

Three-point bending tests were carried out using a universal testing machine Zwick Z2.5 in accordance with the ISO 4049 standard [62], with specifications concerning sample preparation. Specimens measuring $25 \times 2 \times 2$ mm were prepared using silicone (Zetalabor Platinum 85Touch,

Zhrmack SpA, Badia Polesine, Italy) molds placed in a stainless-steel frame. Materials were packed into a mold and polymerized by a method similar to the previous test, but five overlapping irradiations were carried out, starting from the center of the sample. After curing, samples were taken out of the mold, the excess of material was cut off with a scalpel, and the specimens were then wet-ground with P800- and P1200-grit abrasive papers. Ten samples were prepared from each composite. The samples were stored in distilled water at $37 \pm 1\,^\circ\text{C}$ for 24 h. The test was performed at a cross-head speed of 0.75 mm/min and the distance between the supports was 20 mm. Flexural strength and flexural modulus were calculated according to the equations:

$$\sigma_{fl} = \frac{3Pl}{2bh^2} \tag{5}$$

$$E = \frac{P_1 l^3}{4bh^3 \delta} \tag{6}$$

where σ_{fl} was flexural strength, MPa; E was flexural modulus, GPa; l was distance between the supports, mm; b and h were the specimen width and height, mm; P was maximal force, N; P_1 was the load at chosen point at the elastic region of the stress-strain plot, kN; and δ was the deflection at P_1, mm.

2.7. Vickers Hardness

Vickers microhardness was measured on specimens like for DTS, however, samples after wet-grinding were also polished with 6-μm and 3-μm diamond suspensions (Struers GmbH, Willich, Germany). Three samples were made from each composite. The samples were stored in distilled water at $37 \pm 1\,^\circ\text{C}$ for 24 h. Hardness was measured 10 times for each specimen at randomly chosen locations using the microhardness tester (Future-Tech FM-700, Future-Tech Corp, Tokyo, Japan) at a 100-g load and a loading time of 15 s [63]. Vickers hardness was calculated according to the equation:

$$E = \frac{1.8544 \times F}{d^2} \tag{7}$$

where F was the load, N, and d was the average length of the diagonal left by the indenter, mm.

2.8. Degree of Conversion

The degree of conversion (DC) was determined using the method described by Atira et al. [64] with modifications made during sample preparation. Specimens, measuring 5 mm in diameter and 2 mm in height, were prepared in Teflon molds as previously described, but irradiation was carried out only at the top. The samples were removed from the molds and dried in desiccators with freshly dried silica gel at $37 \pm 1\,^\circ\text{C}$ for 24 h. Spectra were recorded by a Fourier transform infrared spectroscopy (FTIR) spectrophotometer (Perkin Elmer Spectrum Two, Perkin Elmer, Waltham, MA, USA), equipped with an attenuated total reflectance (ATR) crystal. The absorption intensity of selected peaks was measured in the range of 1800–1500 cm^{-1} and recorded with 128 scans at a resolution of 1 cm^{-1}. The DC was calculated from the decrease of the absorption band at 1637 cm^{-1}, referring to the C=C stretching vibration ($A_{C=C}$) in relation to the peak at 1608 cm^{-1}, and assigned to the aromatic stretching vibrations (A_{Ar}) in accordance with the equation [65]:

$$DC(\%) = \left(1 - \frac{(A_{C=C}/A_{Ar})_{after\ curing}}{(A_{C=C}/A_{Ar})_{before\ curing}}\right) \times 100 \tag{8}$$

2.9. Sorption and Solubility

The specimens measuring 15 mm in diameter and 1 mm in height were prepared using Teflon molds [66] and polymerized at nine overlapping irradiation zones in accordance with the method described in the ISO standard [62]. After curing, they were ground with P1200-grit abrasive paper to remove excess material with potentially poorly polymerized layers [67] and to standardize the surface.

Then, the samples were removed from the molds. Five test samples of each material were made. The measurement of sorption and solubility was performed in accordance with ISO 4049. The samples were dried inside desiccators with freshly dried silica gel in a dryer at $37 \pm 1\,^{\circ}\text{C}$ and weighed daily (AS 110/C/2, Radwag, Radom, Poland) with an accuracy of 0.1 mg. When the changes in mass were no higher than 0.1 mg, the mass values were recorded as m_1, and the thickness and diameter were measured with a digital caliper with an accuracy of 0.1 mm. Each sample was placed in 10 mL of distilled water for 7 days at $37 \pm 1\,^{\circ}\text{C}$. After storing, the samples were removed from water with tweezers, dried from visible moisture with filter paper, kept at room temperature for 15 s, and weighed (m_2 mass values were denoted). The drying process was repeated as described above, and stable mass was denoted as m_3. Sorption and solubility were calculated using equations:

$$W_{sp} = \frac{m_2 - m_3}{V} \tag{9}$$

$$W_{sl} = \frac{m_1 - m_3}{V} \tag{10}$$

where w_{sp} was sorption, w_{sl} was solubility, m_1 was the initial mass of dried sample, µg; m_2 was the mass after storing, µg, and m_3 was the mass after the second drying, µg; and V was the volume of the sample, mm^3.

2.10. Color Change Measurement

To evaluate the color changes, the specimens measuring 7 mm in diameter and 3 mm in thickness were prepared in Teflon molds. The mold was placed on a microscope slide. The material was placed into the mold, covered with polyester foil and finally with second microscope slide. Then, the upper microscope slide was manually pressed and taken away. The form prepared in this way was inverted (the slide was on top, foil on the bottom). This was important to do because during polymerization, the elastic foil allowed the material to move due to polymerization shrinkage (typical meniscus was formed), while the working surface of the composite in contact with the slide adhered to it and remained flat. The cured sample was pushed out of the mold. Five samples were prepared from each material. After preparation, samples were stored in dry and dark conditions at $37\,^{\circ}\text{C}$ for 24 h and next were immersed in 10 mL of distilled water in darkness at $37 \pm 1\,^{\circ}\text{C}$. Distilled water was replaced after the second and fourth day. Color measurements were obtained 24 h after polymerization (baseline) and after 7 days of immersion. A spectrophotometer (CM2600d, Konica Minolta, Takyo, Japan) was used to record the CIE L*a*b* parameters with a D65 illuminant on a white ceramic tile. The CIELab system is composed of three axes: L* is the lightness from 0 (black) to 100 (white), a* represents the red (+a* value)—green (−a* value) axis, and b* represents the blue (−b* value)—yellow (+b* value) axis. The color change (ΔE*) was calculated using the equation [68]:

$$\Delta E* = \sqrt{(\Delta L*)^2 + (\Delta a*)^2 + (\Delta b*)^2} \tag{11}$$

where ΔL* = $L_{(7\,days)} - L_{(baseline)}$; Δa* = $a_{(7\,days)} - a_{(baseline)}$; and Δb* = $b_{(7\,days)} - b_{(baseline)}$.

2.11. Statistical Analysis

Statistical analysis of the results was done with the use of the Statistica software (software version 13.1, TIBCO Software Inc., Palo Alto, CA, USA). The distributions of the residuals were tested with the Shapiro–Wilk test, and the equality of variances was tested with the Levene test. When the distribution of the residuals was normal and the variances were equal, the one-way or two-way ANOVA with Tukey HSD post hoc tests were used ($\alpha = 0.05$), otherwise the nonparametric Kruskal–Wallis test ($\alpha = 0.05$) was used. Regression analysis was performed to determine the correlation between DC and hardness ($\alpha = 0.05$).

3. Results

3.1. Scanning Electron Microscopy Investigations

Figure 1 presents the morphologies of the used fillers. For both glass fillers (Figure 1a,b), numerous particles showed a much smaller (starting from 50 nm) or larger (up to 8 µm) size than the mean size declared by the manufacturer (2 µm and 0.7 µm). The shapes of the particles were irregular. Nanoparticle aggregations measuring up to 50 nm were noted for silica filler (Figure 1c). For SSHZP particles measured approximately from 100 nm to 500 nm (Figure 1d) but also larger structures, consisting of particles connected to each other, were observed (Figure 1e).

(a)

(b)

(c)

(d)

(e)

Figure 1. Scanning electron microscopy images presenting the morphologies of used fillers: glass fillers with a mean particle size of 0.7 µm (**a**); 2 µm (**b**); silica nanofiller (**c**); and silver sodium hydrogen zirconium phosphate (**d,e**).

SEM images illustrating the morphologies of composite reinforced with glass and silica fillers are presented in Figure 2a,b. The morphologies of materials with additional antibacterial filler are presented in Figure 2c–f. Good distribution of silica nanoparticles between glass submicroparticles and microparticles in the matrix was observed (Figure 2b). Large aggregations of AR were not detected. The SSHZP was also well distributed up to the highest concentrations. Single particles were clearly visible, however, clusters measuring up to 2 μm were also noted. Observations for frozen-broken but not etched samples (Figure 2e,f) showed good contact between the particles and the matrix.

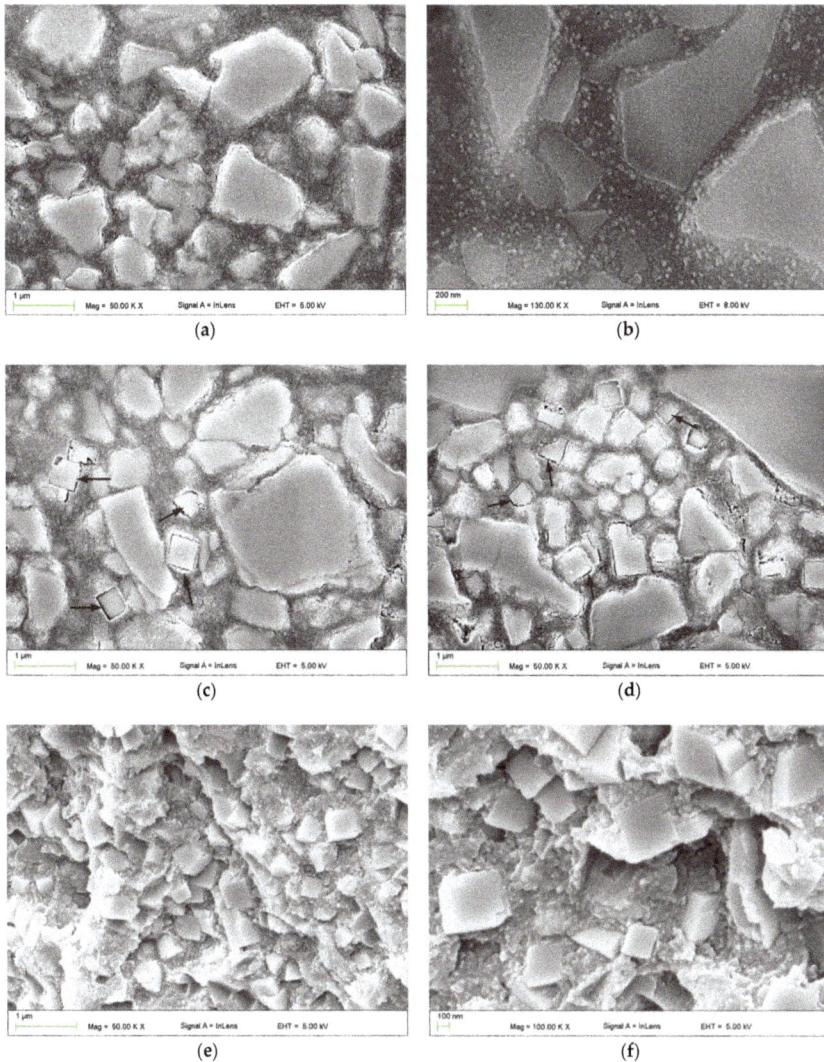

Figure 2. Representative SEM images presenting the morphologies of the cured base composite compounded with: reinforcing fillers (**a,b**); addition of 7% (**c**) and 16% (**c–f**) of silver sodium hydrogen zirconium phosphate; (**a–d**)—wet-ground, polished, etched samples (**e,f**)—frozen-broken but not etched samples, black arrows (**c,d**) indicate the gaps between SSZHP and matrix after etching.

3.2. Antibacterial Test

The achieved results of the antibacterial tests are listed in Table 2. Introducing the SSZHP into the composites had a significant effect ($p = 0.0002$) on the reduction of *S. mutans* colonies. For material without antimicrobial filler, RB values were comparable to the positive control. Composites with filler concentrations from 1% to 4% showed RB medians from 43.8% to 70.1%, and those values should be considered as different if we take into account the obtained minimal and maximal RB values. For concentrations starting from 7%, all obtained RB values were 100%.

Table 2. The reduction in the number of viable colonies (RB) of *Streptococcus mutans* ATCC 33535, after 17 h of incubation with composites samples.

c_{SSZHP}, %	CFU/mL (V_t) $\times 10^4$			RB, %		
	Med	Max	Min	Med	Max	Min
0	3.53	3.99	3.13	4.7	15.5	−7.7
1	2.08	2.89	1.87	43.8	49.6	21.9
4	0.68	0.13	0.00	70.1	93.2	65.7
7	0.00	0.00	0.00	100.0	100.0	100.0
10	0.00	0.00	0.00	100.0	100.0	100.0
13	0.00	0.00	0.00	100.0	100.0	100.0
16	0.00	0.00	0.00	100.0	100.0	100.0

c_{SSZHP}—concentration of silver sodium hydrogen zirconium phosphate; CFU—colony forming units; RB—the relative reduction in the number of viable bacteria colonies; Med—median, Min—minimal value, Max—maximal value.

3.3. Compressive Strength

The mean compressive strength values are presented in Figure 3. The SSHZP concentration did not have a significant influence on the compressive strength of the composites ($p = 0.0524$). The mean values were from 284 MPa to 307 MPa.

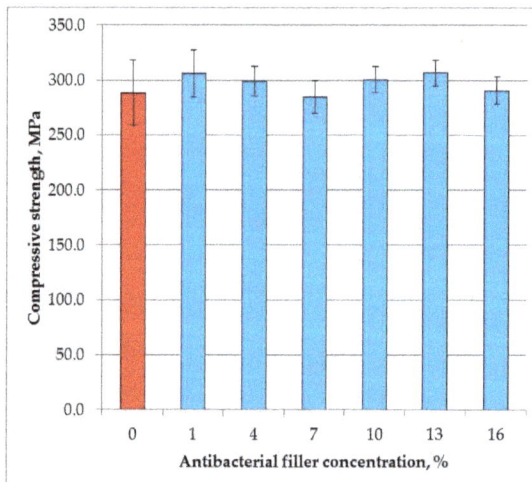

Figure 3. Mean values and standard deviations of compressive strength.

3.4. Diametral Tensile Strength

The mean diametral tensile strength values are presented in Figure 4. The SSHZP concentration did not have a significant influence on the compressive strength of the composites ($p = 0.2986$). The mean values were from 40.3 MPa to 43.1 MPa.

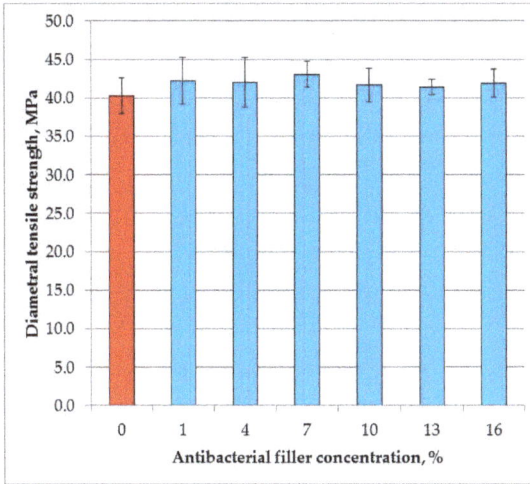

Figure 4. Mean values and standard deviations of diametral tensile strength.

3.5. Flexural Strength

The mean flexural strength values are presented in Figure 5a. The SSHZP introduction had a significant influence on flexural strength ($p = 0.0178$). The post hoc test showed a significant ($p < 0.05$) decrease in flexural strength for the composite with the antibacterial filler concentration of 16% (88 MPa). However, these values were not significantly different ($p > 0.05$) in comparison to the results obtained for other materials with SSHZP. The highest mean flexural strength value was registered for the control material (96 MPa).

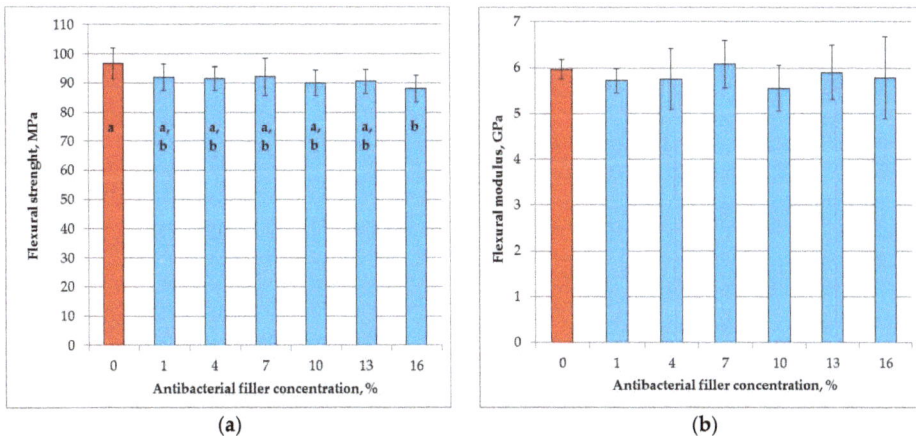

(a)   (b)

Figure 5. Mean flexural strength (**a**) and flexural modulus (**b**) values with standard deviations; different lowercase letters show significantly different results at the $p < 0.05$ level.

The mean flexural modulus values are presented in Figure 5b. The SSHZP concentration did not have a significant effect on the flexural modulus ($p = 0.5351$). The mean values were from 5.6 GPa to 6.1 GPa.

3.6. Vickers Hardness

The mean Vickers hardness values are presented in Figure 6. The SSHZP introduction had a significant influence on flexural strength ($p < 0.0001$), and the post hoc test showed a significant ($p < 0.05$) decrease in hardness starting from the antibacterial filler concentration of 4%. However, the values for SSHZP concentration from 4% to 13% and from 7% to 16% were not significantly different. The highest mean hardness value was registered for the control material (52.7 HV0.1), and the lowest value was for a composite with 16% of SSHZP (48.2 HV0.1).

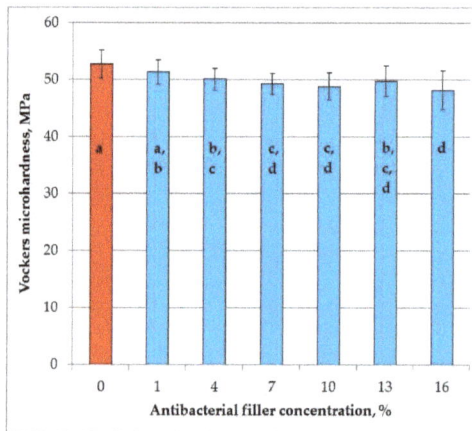

Figure 6. Mean Vickers microhardness values with standard deviations; different lowercase letters show significantly different results at the $p < 0.05$ level.

3.7. Degree of Conversion

The mean degrees of conversion values are presented in Figure 7. At the top of the samples (Figure 7a) the degree of conversion significantly decreased ($p < 0.0001$) with increasing SSHZP concentrations, from 68.7% for the control material to 58.7% for the composite with an SSHZP concentration of 16%. The post hoc test showed that the results for concentrations from 1% to 7%, from 4% to 13%, and from 7% to 16% were not significantly different. Degrees of conversion values obtained at the top were significantly lower in comparison to the values registered at the bottom of the samples ($p < 0.0001$). At the bottom of the samples (Figure 7b), the degree of conversion significantly decreased ($p < 0.0134$) with increasing SSHZP concentrations, from 53.3% for the control material to 47.6% for the composite with an SSHZP concentration of 16%. However, the post hoc test showed that the results for concentrations from 1% to 13% were not significantly different, and only the mean value for the material with the highest SSHZP concentration was significantly lower.

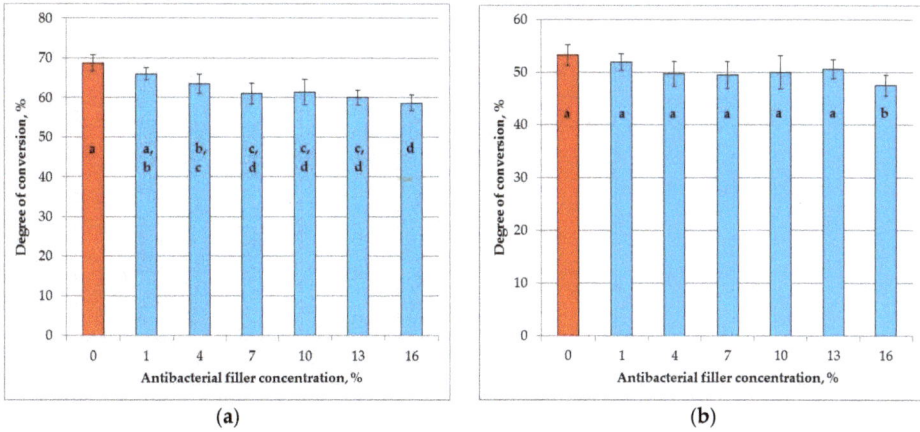

Figure 7. Mean degree of conversion values with standard deviations at the top (**a**); and at the bottom (**b**) of the samples, different lowercase letters show significantly different results at the $p < 0.05$ level.

3.8. Sorption and Solubility

The mean sorption values are presented in Figure 8a. SSHZP introduction had a significant influence on sorption values ($p < 0.0004$). The post hoc test showed a significant increase in sorption values for composites with 13% and 16% of SSHZP. The mean sorption for the composite with the highest SSHZP concentration was 44% greater than for the control material.

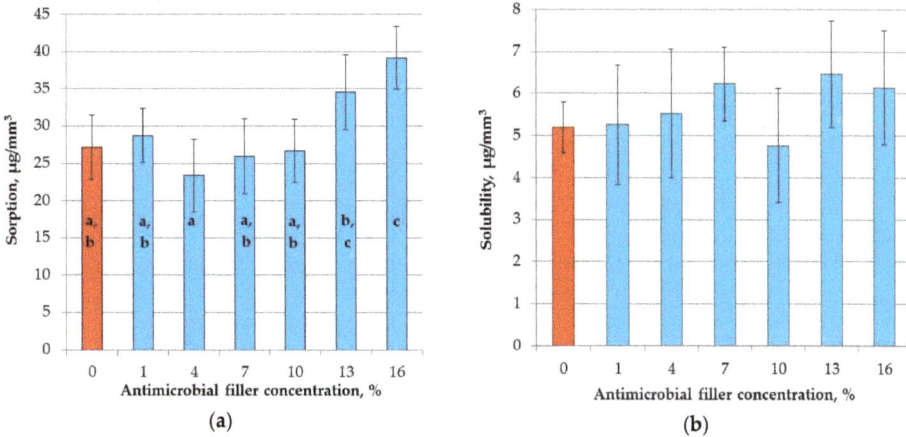

Figure 8. Mean values with standard deviations of sorption (**a**); and solubility (**b**), different lowercase letters show significantly different results at the $p < 0.05$ level.

The mean solubility values are presented in Figure 8b, and there were no statistically significant differences ($p = 0.4185$) between the results obtained for the investigated materials.

3.9. Color Measurement

The results of initial color measurements are presented in Table 3. The obtained L* axis values showed a significant increase ($p < 0.0001$) with the increasing SSHZP concentration. The a* and b* axis values showed a significant decrease with the increasing SSHZP concentration ($p < 0.0001$).

The color changes of different composites after immersion in distilled water are presented in Table 4. The ΔE values for the different composites showed a significant increase ($p < 0.0001$) with the increasing SSHZP concentration. However, the post hoc test indicated significant differences in comparison to reference materials for composites with 13% and 16% of SSHZP. A similar situation was registered for ΔL^* values. The statistically significant influence ($p < 0.0001$) of SSHZP concentration was also noted for Δa^* and Δb^* values. The post hoc test showed a significant increase for composites with 13% and 16% of antibacterial filler.

Table 3. The color of different composites before immersion.

c_{SSZHP}, %	L^*	a^*	b^*
0	51.34 ± 0.36 [a]	5.97 ± 0.08 [a]	12.83 ± 0.26 [a]
1	54.48 ± 0.47 [b]	4.53 ± 0.20 [b]	11.21 ± 0.38 [b]
4	70.42 ± 0.21 [c]	2.28 ± 0.21 [c]	10.59 ± 0.35 [c]
7	76.50 ± 0.18 [d]	1.79 ± 0.15 [d]	9.55 ± 0.22 [d]
10	81.09 ± 0.09 [e]	1.00 ± 0.04 [e]	6.82 ± 0.18 [e]
13	82.84 ± 0.16 [f]	0.96 ± 0.07 [e]	5.63 ± 0.10 [f]
16	85.84 ± 0.16 [g]	0.51 ± 0.11 [f]	5.32 ± 0.19 [f]

Groups with the same lowercase superscript letters for each column are not significantly different at the $p < 0.05$ level.

Table 4. The color changes of different composites after immersion.

c_{SSZHP}, %	ΔL^*	Δa^*	Δb^*	ΔE^*
0	-0.88 ± 0.15 [a]	0.37 ± 0.07 [a]	0.61 ± 0.04 [a]	1.14 ± 0.10 [a]
1	-0.95 ± 0.18 [a,b]	0.41 ± 0.06 [a]	0.64 ± 0.04 [a]	1.22 ± 0.13 [a]
4	-1.10 ± 0.08 [a,b]	0.47 ± 0.10 [a,b]	0.60 ± 0.05 [a]	1.34 ± 0.09 [a]
7	-1.13 ± 0.14 [a,b]	0.44 ± 0.12 [a]	0.64 ± 0.07 [a]	1.38 ± 0.10 [a]
10	-0.98 ± 0.15 [a,b]	0.43 ± 0.09 [a]	0.70 ± 0.06 [a]	1.29 ± 0.10 [a]
13	-1.21 ± 0.14 [b]	0.64 ± 0.05 [b]	1.11 ± 0.12 [b]	1.77 ± 0.14 [b]
16	-1.88 ± 0.16 [c]	1.10 ± 0.05 [c]	1.41 ± 0.08 [c]	2.60 ± 0.11 [c]

Groups with the same lowercase superscript letters for each column are not significantly different at the $p < 0.05$ level.

4. Discussion

In the current study, experimental composites based on a photopolymerizable matrix were considered as direct antibacterial restorative materials. Materials were developed by introducing a filler with confirmed antimicrobial properties: silver sodium hydrogen zirconium phosphate particles. In previous experiments, we tested SSHZP as an antimicrobial additive in two different dental materials: a PMMA denture base material [50] and silicone soft denture lining [51]. Both types of composites had shown enhanced antimicrobial properties, but for the PMMA-based materials, a significant deterioration of mechanical properties had been registered (results unpublished yet), while for polydimethylsiloxane-based composites, they were at the appropriate level [51]. In the presented work, we modified different materials in terms of the final application, polymerization, mechanical properties, and composition.

The applied filler compounding method allowed us to obtain satisfactory dispersion of the used fillers. Observations carried out on polished and etched samples clearly showed typical, irregular shapes of milled glass particles and a very good distribution of nanoparticles between them (Figure 2b). Aggregations of glass fillers or large AR aggregations were not detected. When SSHZP was additionally introduced, cubic-shaped particles were well distributed between glass particles. Due to the used filler types and the obtained morphology, all used materials may be classified as nanohybrid composites [69]. Gaps were visible between the matrix and SSHZP particles (Figure 2c,d), which was related to etching during sample preparation. For nonetched samples, slits were not detected with the used method (Figure 2e,f). However, the observed gaps may suggest the possibility of easier liquid migration between SSHZP and the matrix than between silanized glass and the matrix. For frozen-broken nonetched samples, glass particles were usually not visible, which suggests their good connection with the matrix and is related to the salinization process used by the manufacturer. Large aggregations of SSHZP were not observed, which was favored because of their potential influence on the properties of the composites [70]. However, some structures consisting of connected particles, probably coming from the used antimicrobial filler (Figure 1e), were observed. Observations have also shown some air bubbles in the polymerized composites. They were probably caused by the used procedure of manual preparation of composite and/or by the process of sample preparation. Bubbles measured from a few up to 50 μm. Those structural defects might decrease the mechanical properties because they may act as stress concentrators. In the future, the bubbles can also have a negative effect on the mechanical properties at the bonded interface.

In previous works related with dental materials, the antimicrobial effectiveness of SSHZP against *Candida albican* (*C. albicans*), *Staphylococcus aureus* (*S. aureus*), and *Escherichia coli* (*E. coli*) was confirmed [50,51]. However, only two tested microorganisms (*C. albicans*, *S. aureus*) had clinically proven relevance, which is related to using partial or complete dentures [71–74], but none of them was associated with tooth decay. The problem of caries appearing between teeth and composite restorations is widely disputed in the literature, and its mechanisms are probably multifactorial. Bacterial species associated with secondary caries and primary caries seem to be the same. However, a higher proportion of caries-related bacteria (*mutans streptococci, lactobacilli*) was found on restored surfaces than on unrestored dentin or enamel [75], which is an additional argument for the development of antibacterial materials. Despite the fact that both *mutans streptococci* and *lactobacilli* have a confirmed role in dental caries, the *S. mutans* strains are usually investigated in the context of antimicrobial composites, so this bacterium was also used in our experiment.

Due to differences in microbiological test protocols, the results obtained for the antimicrobial fillers mentioned in the introduction cannot be directly compared to one another or to the results of the present study. In our experiment, samples were stored in an *S. mutans* suspension. All specimens were finally finished with P1200-grit abrasive paper, which gives well standardized and smooth surfaces. This may be confusing in the context of the recognized fact that higher values of roughness promote bacterial adhesion and dental plaque retention [76–78]. However, in our study, adherence of bacteria and biofilm formation were not investigated. Samples were immersed in bacteria suspension in a shaking incubator, and the changes in the number of bacteria were investigated. In this experiment, silver ions released into the environment determined the reduction of bacteria, so using a smooth surface would have created stricter test conditions due to the smaller surface area responsible for the release of antibacterial ions.

After incubation, the reduction of the bacteria population in the environment was registered for all materials, but starting from a concentration of 7%, it was complete. Antimicrobial properties of the used filler are initiated in humid environments by the mechanism of silver ion release from an inorganic, insoluble carrier, which was described by Kampmann et al. [79]. With time, this mechanism may lead to the loss of antibacterial properties due to the continuous silver ion release, so further investigations in this context should be made. Additionally, restorative materials in oral cavities are subject to tribological processes. This, on the one hand, may be the reason for the selective removal of particles from the matrix, but on the other hand, it can also cause "refreshment" of antimicrobial properties by gradual abrasion of the surface with fillers. Both mentioned conceptions may be checked in future experiments.

The microbiological properties of the newly developed antibacterial composites are regularly tested, whereas their mechanical and functional properties are much less frequently reported. Compressive strength, flexural strength, flexural modulus, diametral tensile strength, and hardness are frequently tested mechanical properties for dental restorative materials.

In the present study, we used different curing protocols for each mechanical evaluation. It was justified by the varied sample dimensions, which were dictated by the requirements of the procedures.

It is known that light intensity, polymerization time, and curing depth determine whether or not dental composites are properly cured [80]. Mechanical tests as well as sorption/solubility and antibacterial tests require samples having a length or diameter much larger than the area effectively covered by the used lamp. The use of overlapping light-curing areas for flexural test samples has been subject to criticism due to the risk of preparation of nonhomogeneous specimens [81], although the effect of that method on the flexural properties has been questioned [82]. The height of the samples is equally important due to the expected decrease of the degree of conversion with the depth [83]. A thickness of 2 mm can be considered in that context as a safe value [84]. Moreover, Koran et al. [85] established that if the total dose of light intensity (interpreted as light intensity in exposure time) delivered to the photopolymerizable dental composite is high enough to achieve complete polymerization, the surface hardness, as well residual monomer concentration, tends to remain constant. This shows that the best way to standardize samples is to use overlapping areas of irradiation on both the bottom and top surface, and for samples higher than 4 mm, to use additional irradiations on the lateral surface. However, such an approach is a simplification because it does not take into account other changes occurring in the material during the polymerization and postirradiation polymerization [86–88].

Composites often replace a large bulk of the teeth structure, so the dental restorative materials are usually subjected to compressive forces generated during mastication [89]. If we consider that the compressive strength and plastic limit of tooth tissues [3,90] may be recommended as a standard for the strength of composites [60], we can accept a 230 MPa as a secure value for a composite. The obtained results were higher and were additionally comparable with values reported for numerous commercially available materials with BisGMA, TEGDMA, and UDMA matrices and similar filler content [59,91], including those releasing fluoride [92]. Compressive strength after antimicrobial filler addition was investigated in only a few previous works. The effect of the used additives was varied. An increase of compressive strength values at low concentrations and a decrease of them at larger concentrations was noted for nanosilver [93], zinc oxide [36], and tetretrapod-like zinc oxide whisker [41]. Yoshida et al. [74] have shown no effect of silver-containing ceramic microparticles. In our study, the filler addition also had no effect on compressive strength.

Stress analyses have shown that restorative composites can fracture under tension [94] and tensile strength data may have equal, if not greater, importance than compressive strength, especially in the area near the teeth–composite interface [95,96]. The diametral tensile strength test is an alternative method to evaluate the tensile strength of brittle materials, and it is the default for investigating dental restorative materials. Nevertheless, it gives correct results only if minimal or no plastic deformations occur and when deformations at fracture are small because the area of contact is still near to theoretical [61]. For this reason, this test should not be used for resins or experimental composites with a low filler concentration because of their stress-strain characteristics. Usually DTS values for different types of commercially available composite materials range from 25 MPa to 50 MPa [97–99], but for modern nanocomposites, they may reach over 80 MPa [100]. For all investigated materials, mean DTS values were above 40 MPa, which can be considered to be satisfactory values. The antibacterial filler introduction had no statistically significant effect on DTS, although the mean value for the control group was the lowest. Those findings are in opposition to the results obtained by Łukomska-Szymańska et al. [101], Diaz et al. [102], and Sokołowski et al. [103], where nonfunctionalized calcium fluoride microparticles, zinc oxide three-dimensional microstructures, nanosilver, and nanogold decreased the DTS values of modified composites.

Flexural strength and flexural modulus have been reported as indicators of clinical wear of composites in some studies [104,105]. Composite fillings are also exposed to flexural stress, especially in stress-bearing cavities for restoration classes I, II, and IV [82]. The flexural test is also indicated as a method that relates well to tensile failure [104]. Flexural strength is the only mechanical property specified by the ISO 4049 standard for composite restorative materials, which requires minimal values of 80 MPa for occlusal tooth surface restorations and 50 MPa for others [62], so all investigated materials meet these requirements. The obtained results were additionally comparable to other materials with similar matrix or filler concentrations [4,106–108]. A parallel situation was noted for flexural moduli [109], the values of which are not defined by the standard and may be diversified for different clinical situations. Cervical cavities demand composites characterized by a relatively lower modulus to flex with the teeth, but posterior composites need a high modulus to withstand the occlusal forces [110]. In the presented study, the flexural strength significantly decreased only for the highest antimicrobial filler concentration, whereas the flexural modulus values were stable. Similar trends were noted for other experimental materials compounded with antibacterial particles [36,111]. However, for composites with calcium fluoride, after 24 h storing in wet conditions, a significant deterioration of flexural properties has been registered [112]. The addition of zinc oxide whiskers [41] and nanoparticle-laden hydroxyapatite nanowires [36] caused increases in flexural strength and modulus, but for the concentration of 10%, properties decreased for both fillers. This suggests that the obtained effects are related to both filler type and filler loading. The decrease in flexural strength noted in our study for the highest SSHZP concentration should be treated with some caution, also in the context of the obtained compressive and tensile strength values. For one sample, the flexural strength was 77 MPa. If we deleted this result, we would have a mean value of 90.1 MPa, and this value is not statistically different from any other. After consideration, we decided not to remove that result because samples for flexural strength for this material were the most difficult to prepare due the increasing viscosity of composition in combination with a small working area of silicone mold (2 × 25 mm), which created an increased risk of making structural defects during the packing of the material. The registered statistical difference can be an indicator of those problems, especially if we consider that flexural strength has been noted to be more sensitive to subtle changes in material substructure than, for example, compressive strength test [113].

The Vickers microhardness test is a known method used to compare composite resins, especially in the context of their wear resistance prediction [50,114]. Direct restorative composites used in dentistry demonstrate Vickers hardness starting from 40 kgf/mm^2, but for some materials, values exceed 100 kgf/mm^2 [10,114–116]. The values obtained in this work were within this range but were rather at the lower limit. Increasing antibacterial filler concentration caused a small but systematic lowering of microhardness. This is in opposition to some findings for commercially available materials, where higher filler content was correlated with higher microhardness [117], but is in accordance with some research, where introducing antibacterial additives decreased microhardness [44,101,111].

Degree of conversion is an important property of restorative composites due to the potential risk of biological responses related to monomer release and affection of pulp tissues [118]. The obtained values were in agreement with the findings from other studies, performed on similar dimethacrylate systems, and measured with the same method [64]. The reduction of DC values at the bottom of the samples was also expected because when light moves through a material with increasing density, its intensity is reduced. The reduction of DC values with increasing SSHZP content was probably related to the effect of light scattering by particles. Moreover, some reports showed larger scattering when the particle size is circa one half or close to that of the curing light wavelength [119,120]. In the presented study, this situation took place because the optical wave length was 450–470 nm, so it was similar to the observed SSHZP particles size, which may explain unfavorable DC changes. Additionally, the DC values at the top of the samples were well correlated ($R^2 = 0.9058$, $p = 0.001$) with microhardness results. This is in accordance with other reports, where surface microhardness has been identified as a good indicator of DC changes [121,122].

The solubility and sorption properties are important from the viewpoint of biocompatibility concerns over monomer releasing and in relation to the stability of the composites due to degradation from the uptake of solvents and the wash-out of ingredients of materials [104]. Sorption values for all samples were lower than 40 µg/mm^3, so all investigated materials met the requirements of the ISO standard. Solubility values' samples were lower or, for one sample, equal to 7.5 µg/mm^3, so all materials also met the requirements of the ISO standard. The sorption values were comparable to numerous commercially available materials [123,124] but increased for the highest antimicrobial filler concentration. The solubility for all composites, including the control group, was generally higher than for most modern materials, however, for some of them, a similar value has also been registered [66,123,124]. The increasing solubility with SSHZP content was not statistically significant. The principal factor influencing sorption and solubility of dental restorative materials is the composition of polymer matrix [67]. The sorption is influenced by the polarity of the molecular structure, the presence of hydroxyl groups (which may crate hydrogen bonds), and with the degree of crosslinking of the matrix [125]. Water may penetrate into the free volume between the chains and nanopores formed during polymerization, or it can be successively attached to polymer chains via hydrogen bonds [123]. In this study, a hydrophilic monomers system (Bis-GMA, TEGDMA, UDMA) [123] was used, from which TEGDMA and Bis-GMA create networks characterized by higher sorption due to the presence of the ether linkages and hydroxyl groups, respectively [66,126]. Thus, those kinds of matrices usually show relatively large values of water sorption and solubility. Additionally, after the introduction of nonsilanized SSHZP, water would migrate in the interface between the filler particles/particle aggregations and the matrix, which may explain the enhanced sorption values for the higher concentration. Also, decreased DC may have some influence on water sorption. The uptake of water also allows diffusion out (into the storage medium) to residual monomers, fillers, degradation products, and other leachable components, so sorption is often correlated with solubility [66]. This was not found in this study, although a statistically insignificant increase in solubility was registered. This may suggest that an experiment should be prepared in future with longer storage periods to allow for more extensive changes in the materials, which will be easier to detect, or with more sensitive methods.

The introduction of SSHZP caused significant changes in composite color. The materials with increasing antimicrobial filler concentrations show a narrow whitening effect, represented by increasing L* axis values. Composites have also shown a reduction of reddish and yellow coloration related to decreasing positive values of a^* and b^*, respectively. These changes, at the starting point, might be considered as beneficial. After being stored in distilled water, color changes expressed by ΔE values can be classified as noticeable only for an experienced observer (ΔE values from 1 to 2) for all materials, excluding the highest concentration for which unexperienced observer may notice the difference (ΔE values from 2 to 2.5) [127]. However, all achieved ΔE values were comparable to the results obtained for commercial [128] and experimental [129,130] materials after a similar period of storing in distilled water. For all composites, the darkening, reddening, and yellowing effects were registered, which were also registered for other photopolymerizable resin-based composites [120,128,130]. The reasons for the noted color changes might be multifactorial. De Oliwiera et al. [120] suggest that reduced monomer conversion can lead to poorer color stability in the composites during storing due to the oxidation of the amines or monomers. In our study, the reduction of DC values was also noted, which may partially be the reason for the reddening and yellowing of the materials. However, the slight, progressive changes in DC values cannot explain much of the decreased color stability of the materials for the two highest concentrations of SSHZP. The effect of color changes can be also associated with the leaching of components [130], including the antibacterial filler. The silver ions released from the filler, their deposition on the surface, and further oxidation may be considered as the reason for the darkening, reddening, and yellowing of the composites. However, additional investigations ought to be conducted to clarify this behavior.

The findings presented here should be enhanced with further in vitro investigations. As an especially important part of future research, the cytotoxic potential of composites should be examined. The toxicological data for the used SSHZP [56] let us suppose that materials should not present unfavorable properties in that aspect, although toxic effects in dental materials with silver have been registered [70]. The ion release into the environment and the dynamics of this process should be investigated, as an indication of the persistence of the material's antibacterial activity. The SSHZP introduced into the PMMA denture base material in a previous work have shown antimicrobial properties decreasing with time during a three-month experiment [50], so it should be expected that the durability of the antibacterial activity of the tested composites will also be limited. The used antimicrobial test additionally did not reflect real conditions, especially in the context of the expected lifetime of the dental composites and the processes occurring in the interface area between tooth tissues and composite. Nevertheless, the presented results of antimicrobial tests are a promising base for further experiments. All properties studied here may also be tested with long-term experiments.

Another limitation of the present study was that the mechanical properties of the proposed composites were not evaluated after biofilm exposure. Acid production by bacteria during metabolic processes may be the cause of changes in some mechanical properties of composites. Microhardness is usually stable after exposure to *S. mutans* biofilm [77,131], but Fúcio et al. [131] have shown that it can be reduced for particular restorative composites. Moreover, the presence of biofilm influences the mechanical properties of resin–dentin bonds. Melo et al. [132], in self-designed experiments with quasi-static and fatigue performance tests, have shown that the *S. mutans* growth may be the cause of the reduction in the mechanical properties of the bonded interface. The degradation in the dentin–composite interface in the biofilm environment was confirmed by other researchers [133,134], who indicated that the studied experimental composites should be examined in the future in this respect. The problem of acid production by bacteria may also be important in the context of the observed gaps between SSZHP particles and the matrix after the etching of samples for SEM investigations. This may suggest the possibility of accelerated changes in this area due to the influence of the acidic environment. Depending on whether these changes occur as a result of the presence of a biofilm, this phenomenon could negatively affect the mechanical properties of materials due to the creation and propagation of structural defects.

In the present study, we used quasi-static tests for the mechanical properties' evaluation, which was sufficient at the planned initial stage. However, fatigue tests in the past decade have gained increased importance because dental materials under clinical conditions are subject to cyclic loading [135], caused by thousands of cycles of mastication per day. It has been proven that flexural static strength is higher than values obtained after flexural cyclic loading [136]. Cyclic loading also reduces fracture toughness [137]. Fatigue tests are also used to evaluate mechanical properties of resin–dentin bonds [138,139], also in the context of bacteria presence [133]. These results suggest the desirability of conducting research in this direction, including comparative studies with commercially available composites.

5. Conclusions

Within the limits of this study, it can be concluded that the experimental composites showed a high initial reduction of bacteria colonies for the tested *S. mutans* strain. The satisfactory combination of the reduction of bacteria colonies with physical properties was achieved for filler concentrations ranging from 4% to 13%. Those materials exhibited mechanical properties similar to the base material, as well as the degree of conversion, sorption, solubility, and color stability at acceptable levels. The cytotoxic tests and long-term investigations, including silver ion release into the environment, need to be performed in future experiments.

Author Contributions: Conceptualization, R.S. and G.C.; Formal analysis, R.S.; Investigation, R.S., W.P., A.M., E.B., I.B.-R. and G.C.; Methodology, R.S., W.P., A.M., I.B.-R. and G.C.; Resources, R.S. and G.C.; Supervision, G.C.; Visualization, R.S.; Writing—original draft, R.S. and G.C.; Writing—review & editing, R.S., J.K. and G.C.

Funding: This work was financially supported with statutory funds of Faculty of Mechanical Engineering of Silesian University of Technology in 2018 and by Robert Stencel.

Acknowledgments: The authors would like to thank the Konica Minolta Poland for help during the color measurements.

Conflicts of Interest: The authors declare no conflict of interest.

References

1. Vos, T.; Allen, C.; Arora, M.; Barber, R.M.; Brown, A.; Carter, A.; Casey, D.C.; Charlson, F.J.; Chen, A.Z.; Coggeshall, M.; et al. Global, regional, and national incidence, prevalence, and years lived with disability for 310 diseases and injuries, 1990–2015: A systematic analysis for the Global Burden of Disease Study 2015. *Lancet Lond. Engl.* **2016**, *388*, 1545–1602. [CrossRef]

2. Polish Ministry of Health. *Monitoring the Oral Health of the Polish Population in the Years 2016–2020*; Polish Ministry of Health: Warsaw, Poland, 2015. (In Polish)

3. Anusavice, K.; Shen, C.; Rawls, H.R. *Phillips' Science of Dental Materials—12th Edition*, 12th ed.; Saunders: St. Louis, MO, USA, 2013; ISBN 978-0-32-324205-9.

4. Jayanthi, N.; Vinod, V. Comparative evaluation of compressive strength and flexural strength of conventional core materials with nanohybrid composite resin core material an in vitro study. *J. Indian Prosthodont. Soc.* **2013**, *13*, 281–289. [CrossRef] [PubMed]

5. Ilie, N.; Hickel, R. Resin composite restorative materials. *Aust. Dent. J.* **2011**, *56* (Suppl. 1), 59–66. [CrossRef] [PubMed]

6. Ástvaldsdóttir, Á.; Dagerhamn, J.; van Dijken, J.W.V.; Naimi-Akbar, A.; Sandborgh-Englund, G.; Tranæus, S.; Nilsson, M. Longevity of posterior resin composite restorations in adults—A systematic review. *J. Dent.* **2015**, *43*, 934–954. [CrossRef] [PubMed]

7. Alzraikat, H.; Burrow, M.; Maghaireh, G.; Taha, N. Nanofilled Resin Composite Properties and Clinical Performance: A Review. *Oper. Dent.* **2018**, in press. [CrossRef] [PubMed]

8. Lambrechts, P.; Goovaerts, K.; Bharadwaj, D.; De Munck, J.; Bergmans, L.; Peumans, M.; Van Meerbeek, B. Degradation of tooth structure and restorative materials: A review. *Wear* **2006**, *261*, 980–986. [CrossRef]

9. Chan, K.H.S.; Mai, Y.; Kim, H.; Tong, K.C.T.; Ng, D.; Hsiao, J.C.M. Review: Resin Composite Filling. *Materials* **2010**, *3*, 1228–1243. [CrossRef]

10. Rasines Alcaraz, M.G.; Veitz-Keenan, A.; Sahrmann, P.; Schmidlin, P.R.; Davis, D.; Iheozor-Ejiofor, Z. Direct composite resin fillings versus amalgam fillings for permanent or adult posterior teeth. *Cochrane Database Syst. Rev.* **2014**, *3*, CD005620. [CrossRef] [PubMed]

11. Hurst, D. Amalgam or composite fillings—Which material lasts longer? *Evid. Based Dent.* **2014**, *15*, 50–51. [CrossRef] [PubMed]

12. Demarco, F.F.; Corrêa, M.B.; Cenci, M.S.; Moraes, R.R.; Opdam, N.J.M. Longevity of posterior composite restorations: Not only a matter of materials. *Dent. Mater. Off. Publ. Acad. Dent. Mater.* **2012**, *28*, 87–101. [CrossRef] [PubMed]

13. Opdam, N.J.M.; van de Sande, F.H.; Bronkhorst, E.; Cenci, M.S.; Bottenberg, P.; Pallesen, U.; Gaengler, P.; Lindberg, A.; Huysmans, M.C.; van Dijken, J.W. Longevity of posterior composite restorations: A systematic review and meta-analysis. *J. Dent. Res.* **2014**, *93*, 943–949. [CrossRef] [PubMed]

14. Pereira-Cenci, T.; Cenci, M.S.; Fedorowicz, Z.; Azevedo, M. Antibacterial agents in composite restorations for the prevention of dental caries. In *The Cochrane Library*; John Wiley & Sons, Ltd.: Hoboken, NJ, USA, 2013.

15. Lai, G.; Li, M. Secondary Caries. In *Contemporary Approach to Dental Caries*; InTech: Rijeka, Croatia, 2012; pp. 403–422. ISBN 978-9-53-510305-9.

16. Mjör, I.A.; Toffenetti, F. Secondary caries: A literature review with case reports. *Quintessence Int. Berl. Ger. 1985* **2000**, *31*, 165–179.

17. Shih, W.-Y. Microleakage in different primary tooth restorations. *J. Chin. Med. Assoc.* **2016**, *79*, 228–234. [CrossRef] [PubMed]

18. Gale, M.S.; Darvell, B.W. Thermal cycling procedures for laboratory testing of dental restorations. *J. Dent.* **1999**, *27*, 89–99. [CrossRef]

19. Nedeljkovic, I.; Teughels, W.; De Munck, J.; Van Meerbeek, B.; Van Landuyt, K.L. Is secondary caries with composites a material-based problem? *Dent. Mater. Off. Publ. Acad. Dent. Mater.* **2015**, *31*, e247–e277. [CrossRef] [PubMed]

20. Łukomska-Szymańska, M.; Zarzycka, B.; Grzegorczyk, J.; Sokołowski, K.; Półtorak, K.; Sokołowski, J.; Łapińska, B. Antibacterial Properties of Calcium Fluoride-Based Composite Materials: In VitroStudy. *BioMed Res. Int.* **2016**, *2016*, 1048320. [CrossRef] [PubMed]

21. Chatzistavrou, X.; Lefkelidou, A.; Papadopoulou, L.; Pavlidou, E.; Paraskevopoulos, K.M.; Fenno, J.C.; Flannagan, S.; González-Cabezas, C.; Kotsanos, N.; Papagerakis, P. Bactericidal and Bioactive Dental Composites. *Front. Physiol.* **2018**, *9*, 103. [CrossRef] [PubMed]

22. Imazato, S. Antibacterial properties of resin composites and dentin bonding systems. *Dent. Mater. Off. Publ. Acad. Dent. Mater.* **2003**, *19*, 449–457. [CrossRef]

23. Chen, L.; Shen, H.; Suh, B.I. Antibacterial dental restorative materials: A state-of-the-art review. *Am. J. Dent.* **2012**, *25*, 337–346. [PubMed]

24. Maas, M.S.; Alania, Y.; Natale, L.C.; Rodrigues, M.C.; Watts, D.C.; Braga, R.R. Trends in restorative composites research: What is in the future? *Braz. Oral Res.* **2017**, *31*, e55. [CrossRef] [PubMed]

25. Cheng, L.; Zhang, K.; Zhou, C.-C.; Weir, M.D.; Zhou, X.-D.; Xu, H.H.K. One-year water-ageing of calcium phosphate composite containing nano-silver and quaternary ammonium to inhibit biofilms. *Int. J. Oral Sci.* **2016**, *8*, 172–181. [CrossRef] [PubMed]

26. Imazato, S.; Torii, M.; Tsuchitani, Y.; McCabe, J.F.; Russell, R.R. Incorporation of bacterial inhibitor into resin composite. *J. Dent. Res.* **1994**, *73*, 1437–1443. [CrossRef] [PubMed]

27. Zhang, K.; Li, F.; Imazato, S.; Cheng, L.; Liu, H.; Arola, D.D.; Bai, Y.; Xu, H.H.K. Dual antibacterial agents of nano-silver and 12-methacryloyloxydodecylpyridinium bromide in dental adhesive to inhibit caries. *J. Biomed. Mater. Res. B Appl. Biomater.* **2013**, *101*, 929–938. [CrossRef] [PubMed]

28. Cherchali, F.Z.; Mouzali, M.; Tommasino, J.B.; Decoret, D.; Attik, N.; Aboulleil, H.; Seux, D.; Grosgogeat, B. Effectiveness of the DHMAI monomer in the development of an antibacterial dental composite. *Dent. Mater. Off. Publ. Acad. Dent. Mater.* **2017**, *33*, 1381–1391. [CrossRef] [PubMed]

29. Rego, G.F.; Vidal, M.L.; Viana, G.M.; Cabral, L.M.; Schneider, L.F.J.; Portela, M.B.; Cavalcante, L.M. Antibiofilm properties of model composites containing quaternary ammonium methacrylates after surface texture modification. *Dent. Mater. Off. Publ. Acad. Dent. Mater.* **2017**, *33*, 1149–1156. [CrossRef] [PubMed]

30. Beyth, N.; Yudovin-Farber, I.; Bahir, R.; Domb, A.J.; Weiss, E.I. Antibacterial activity of dental composites containing quaternary ammonium polyethylenimine nanoparticles against *Streptococcus mutans*. *Biomaterials* **2006**, *27*, 3995–4002. [CrossRef] [PubMed]

31. Zhang, J.F.; Wu, R.; Fan, Y.; Liao, S.; Wang, Y.; Wen, Z.T.; Xu, X. Antibacterial dental composites with chlorhexidine and mesoporous silica. *J. Dent. Res.* **2014**, *93*, 1283–1289. [CrossRef] [PubMed]

32. Leung, D.; Spratt, D.A.; Pratten, J.; Gulabivala, K.; Mordan, N.J.; Young, A.M. Chlorhexidine-releasing methacrylate dental composite materials. *Biomaterials* **2005**, *26*, 7145–7153. [CrossRef] [PubMed]

33. Rathke, A.; Staude, R.; Muche, R.; Haller, B. Antibacterial activity of a triclosan-containing resin composite matrix against three common oral bacteria. *J. Mater. Sci. Mater. Med.* **2010**, *21*, 2971–2977. [CrossRef] [PubMed]

34. Ali, S.; Sangi, L.; Kumar, N. Exploring antibacterial activity and hydrolytic stability of resin dental composite restorative materials containing chitosan. *Technol. Health Care Off. J. Eur. Soc. Eng. Med.* **2017**, *25*, 11–18. [CrossRef] [PubMed]

35. Wang, J.; Dong, X.; Yu, Q.; Baker, S.N.; Li, H.; Larm, N.E.; Baker, G.A.; Chen, L.; Tan, J.; Chen, M. Incorporation of antibacterial agent derived deep eutectic solvent into an active dental composite. *Dent. Mater. Off. Publ. Acad. Dent. Mater.* **2017**, *33*, 1445–1455. [CrossRef] [PubMed]

36. Tavassoli Hojati, S.; Alaghemand, H.; Hamze, F.; Ahmadian Babaki, F.; Rajab-Nia, R.; Rezvani, M.B.; Kaviani, M.; Atai, M. Antibacterial, physical and mechanical properties of flowable resin composites containing zinc oxide nanoparticles. *Dent. Mater. Off. Publ. Acad. Dent. Mater.* **2013**, *29*, 495–505. [CrossRef] [PubMed]

37. Kasraei, S.; Sami, L.; Hendi, S.; AliKhani, M.-Y.; Rezaei-Soufi, L.; Khamverdi, Z. Antibacterial properties of composite resins incorporating silver and zinc oxide nanoparticles on *Streptococcus mutans* and Lactobacillus. *Restor. Dent. Endod.* **2014**, *39*, 109–114. [CrossRef] [PubMed]

38. Aydin Sevinç, B.; Hanley, L. Antibacterial activity of dental composites containing zinc oxide nanoparticles. *J. Biomed. Mater. Res. B Appl. Biomater.* **2010**, *94*, 22–31. [CrossRef] [PubMed]

39. Khvostenko, D.; Hilton, T.J.; Ferracane, J.L.; Mitchell, J.C.; Kruzic, J.J. Bioactive glass fillers reduce bacterial penetration into marginal gaps for composite restorations. *Dent. Mater. Off. Publ. Acad. Dent. Mater.* **2016**, *32*, 73–81. [CrossRef] [PubMed]

40. Sodagar, A.; Akhoundi, M.S.A.; Bahador, A.; Jalali, Y.F.; Behzadi, Z.; Elhaminejad, F.; Mirhashemi, A.H. Effect of TiO$_2$ nanoparticles incorporation on antibacterial properties and shear bond strength of dental composite used in Orthodontics. *Dent. Press J. Orthod.* **2017**, *22*, 67–74. [CrossRef] [PubMed]

41. Niu, L.N.; Fang, M.; Jiao, K.; Tang, L.H.; Xiao, Y.H.; Shen, L.J.; Chen, J.H. Tetrapod-like zinc oxide whisker enhancement of resin composite. *J. Dent. Res.* **2010**, *89*, 746–750. [CrossRef] [PubMed]

42. Chatzistavrou, X.; Velamakanni, S.; DiRenzo, K.; Lefkelidou, A.; Fenno, J.C.; Kasuga, T.; Boccaccini, A.R.; Papagerakis, P. Designing dental composites with bioactive and bactericidal properties. *Mater. Sci. Eng. C Mater. Biol. Appl.* **2015**, *52*, 267–272. [CrossRef] [PubMed]

43. Ai, M.; Du, Z.; Zhu, S.; Geng, H.; Zhang, X.; Cai, Q.; Yang, X. Composite resin reinforced with silver nanoparticles-laden hydroxyapatite nanowires for dental application. *Dent. Mater. Off. Publ. Acad. Dent. Mater.* **2017**, *33*, 12–22. [CrossRef] [PubMed]

44. Łukomska-Szymańska, M.M.; Kleczewska, J.; Bieliński, D.M.; Jakubowski, W.; Sokołowski, J. Bactericidal properties of experimental dental composites based on dimethacrylate resins reinforced by nanoparticles. *Eur. J. Chem.* **2014**, *5*, 419–423. [CrossRef]

45. Azarsina, M.; Kasraei, S.; Yousef-Mashouf, R.; Dehghani, N.; Shirinzad, M. The antibacterial properties of composite resin containing nanosilver against *Streptococcus mutans* and Lactobacillus. *J. Contemp. Dent. Pract.* **2013**, *14*, 1014–1018. [CrossRef] [PubMed]

46. Yoshida, K.; Tanagawa, M.; Matsumoto, S.; Yamada, T.; Atsuta, M. Antibacterial activity of resin composites with silver-containing materials. *Eur. J. Oral Sci.* **1999**, *107*, 290–296. [CrossRef] [PubMed]

47. Sodagar, A.; Akhavan, A.; Hashemi, E.; Arab, S.; Pourhajibagher, M.; Sodagar, K.; Kharrazifard, M.J.; Bahador, A. Evaluation of the antibacterial activity of a conventional orthodontic composite containing silver/hydroxyapatite nanoparticles. *Prog. Orthod.* **2016**, *17*, 40. [CrossRef] [PubMed]

48. Cheng, L.; Weir, M.D.; Xu, H.H.K.; Antonucci, J.M.; Kraigsley, A.M.; Lin, N.J.; Lin-Gibson, S.; Zhou, X. Antibacterial amorphous calcium phosphate nanocomposites with a quaternary ammonium dimethacrylate and silver nanoparticles. *Dent. Mater.* **2012**, *28*, 561–572. [CrossRef] [PubMed]

49. Fan, C.; Chu, L.; Rawls, H.R.; Norling, B.K.; Cardenas, H.L.; Whang, K. Development of an antimicrobial resin—A pilot study. *Dent. Mater. Off. Publ. Acad. Dent. Mater.* **2011**, *27*, 322–328. [CrossRef] [PubMed]

50. Chladek, G.; Basa, K.; Mertas, A.; Pakieła, W.; Żmudzki, J.; Bobela, E.; Król, W. Effect of Storage in Distilled Water for Three Months on the Antimicrobial Properties of Poly(methyl methacrylate) Denture Base Material Doped with Inorganic Filler. *Materials* **2016**, *9*, 328. [CrossRef] [PubMed]

51. Jabłońska-Stencel, E.; Pakieła, W.; Mertas, A.; Bobela, E.; Kasperski, J.; Chladek, G. Effect of Silver-Emitting Filler on Antimicrobial and Mechanical Properties of Soft Denture Lining Material. *Materials* **2018**, *11*, 318. [CrossRef] [PubMed]

52. Qin, Y. Silver-containing alginate fibres and dressings. *Int. Wound J.* **2005**, *2*, 172–176. [CrossRef] [PubMed]

53. Qin, Y.; Zhu, C. Antimicrobial Properties of Silver-Containing Chitosan Fibers. In *Medical and Healthcare Textiles*; Woodhead Publishing Series in Textiles; Woodhead Publishing: Sawston, UK, 2010; pp. 7–13. ISBN 978-1-84-569224-7.

54. Ågren, M. *Wound Healing Biomaterials—Volume 2: Functional Biomaterials*; Woodhead Publishing: Sawston, UK, 2016; ISBN 978-0-08-100606-1.

55. Qin, Y. Antimicrobial dressings for the management of wound infection. In *Medical Textile Materials*; Woodhead Publishing Series in Textiles; Woodhead Publishing: Sawston, UK, 2016; pp. 145–160. ISBN 978-0-08-100618-4.

56. National Industrial Chemicals Notification and Assessment Scheme (Nicnas)-Silver Sodium Hydrogen Zirconium Phosphate. Available online: https://www.pharosproject.net/uploads/files/sources/1185/alphasan-std1081fr.pdf (accessed on 11 March 2004).

57. Melaiye, A.; Sun, Z.; Hindi, K.; Milsted, A.; Ely, D.; Reneker, D.H.; Tessier, C.A.; Youngs, W.J. Silver(I)-imidazole cyclophane gem-diol complexes encapsulated by electrospun tecophilic nanofibers: Formation of nanosilver particles and antimicrobial activity. *J. Am. Chem. Soc.* **2005**, *127*, 2285–2291. [CrossRef] [PubMed]

58. Xu, X.; Yang, Q.; Wang, Y.; Yu, H.; Chen, X.; Jing, X. Biodegradable electrospun poly(L-lactide) fibers containing antibacterial silver nanoparticles. *Eur. Polym. J.* **2006**, *42*, 2081–2087. [CrossRef]

59. Mota, E.G.; Weiss, A.; Spohr, A.M.; Oshima, H.M.S.; de Carvalho, L.M.N. Relationship between filler content and selected mechanical properties of six microhybrid composites. *Rev. Odonto Ciênc.* **2011**, *26*, 151–155. [CrossRef]

60. Galvão, M.R.; Caldas, S.G.F.R.; Calabrez-Filho, S.; Campos, E.A.; Bagnato, V.S.; Rastelli, A.N.S.; Andrade, M.F. Compressive strength of dental composites photo-activated with different light tips. *Laser Phys.* **2013**, *23*, 045604. [CrossRef]

61. Penn, R.W.; Craig, R.G.; Tesk, J.A. Diametral tensile strength and dental composites. *Dent. Mater.* **1987**, *3*, 46–48. [CrossRef]

62. International Organization for Standardization (ISO). *EN ISO 4049:2009 Dentistry—Polymer-Based Restorative Materials*; ISO: Geneva, Switzerland, 2009.

63. Chladek, G.; Basa, K.; Żmudzki, J.; Malara, P.; Nowak, A.J.; Kasperski, J. Influence of aging solutions on wear resistance and hardness of selected resin-based dental composites. *Acta Bioeng. Biomech.* **2016**, *18*, 43–52. [PubMed]

64. Atria, P.J.; Sampaio, C.S.; Cáceres, E.; Fernández, J.; Reis, A.F.; Giannini, M.; Coelho, P.G.; Hirata, R. Micro-computed tomography evaluation of volumetric polymerization shrinkage and degree of conversion of composites cured by various light power outputs. *Dent. Mater. J.* **2018**, *37*, 33–39. [CrossRef] [PubMed]

65. Barszczewska-Rybarek, I.; Jurczyk, S. Comparative Study of Structure-Property Relationships in Polymer Networks Based on Bis-GMA, TEGDMA and Various Urethane-Dimethacrylates. *Materials* **2015**, *8*, 1230–1248. [CrossRef] [PubMed]

66. Leal, J.P.; da Silva, D.J.; Leal, R.F.M.; Oliveira-Júnior, C.D.C.; Prado, V.L.G.; Vale, G.C. Effect of Mouthwashes on Solubility and Sorption of Restorative Composites. *Int. J. Dent.* **2017**, *2017*, 5865691. [CrossRef] [PubMed]

67. Toledano, M.; Osorio, R.; Osorio, E.; Fuentes, V.; Prati, C.; Garcia-Godoy, F. Sorption and solubility of resin-based restorative dental materials. *J. Dent.* **2003**, *31*, 43–50. [CrossRef]

68. Vichi, A.; Ferrari, M.; Davidson, C.L. Color and opacity variations in three different resin-based composite products after water aging. *Dent. Mater. Off. Publ. Acad. Dent. Mater.* **2004**, *20*, 530–534. [CrossRef] [PubMed]

69. Swift, E.J.; Swift, E.J. Nanocomposites. *J. Esthet. Restor. Dent.* **2005**, *17*, 3–4. [CrossRef]

70. Nuñez-Anita, R.E.; Acosta-Torres, L.S.; Vilar-Pineda, J.; Martínez-Espinosa, J.C.; de la Fuente-Hernández, J.; Castaño, V.M. Toxicology of antimicrobial nanoparticles for prosthetic devices. *Int. J. Nanomed.* **2014**, *9*, 3999–4006. [CrossRef]

71. Scully, C. *Oral and Maxillofacial Medicine—E-Book: The Basis of Diagnosis and Treatment*, 3rd ed.; Churchill Livingstone Elsevier: Edinburgh, UK, 2013; ISBN 978-0-70-205205-7.

72. Pinto, T.M.S.; Neves, A.C.C.; Leão, M.V.P.; Jorge, A.O.C. Vinegar as an antimicrobial agent for control of *Candida* spp. in complete denture wearers. *J. Appl. Oral Sci. Rev. FOB* **2008**, *16*, 385–390. [CrossRef]

73. Gupta Effect of Comonomer of Methacrylic Acid on Flexural Strength and Adhesion of Staphylococcus aureus to Heat Polymerized Poly (Methyl Methacrylate) Resin: An in Vitro Study. Available online: http://www.j-ips.org/article.asp?issn=0972-4052;year=2017;volume=17;issue=2;spage=149;epage=155; aulast=Gupta#ref12 (accessed on 12 January 2018).

74. Sumi, Y.; Miura, H.; Sunakawa, M.; Michiwaki, Y.; Sakagami, N. Colonization of denture plaque by respiratory pathogens in dependent elderly. *Gerodontology* **2002**, *19*, 25–29. [CrossRef] [PubMed]

75. Thomas, R.Z.; van der Mei, H.C.; van der Veen, M.H.; de Soet, J.J.; Huysmans, M.C. Bacterial composition and red fluorescence of plaque in relation to primary and secondary caries next to composite: An in situ study. *Oral Microbiol. Immunol.* **2008**, *23*, 7–13. [CrossRef] [PubMed]

76. Tanner, J.; Carlén, A.; Söderling, E.; Vallittu, P.K. Adsorption of parotid saliva proteins and adhesion of *Streptococcus mutans* ATCC 21752 to dental fiber-reinforced composites. *J. Biomed. Mater. Res. B Appl. Biomater.* **2003**, *66*, 391–398. [CrossRef] [PubMed]

77. Beyth, N.; Bahir, R.; Matalon, S.; Domb, A.J.; Weiss, E.I. *Streptococcus mutans* biofilm changes surface-topography of resin composites. *Dent. Mater. Off. Publ. Acad. Dent. Mater.* **2008**, *24*, 732–736. [CrossRef] [PubMed]

78. Cazzaniga, G.; Ottobelli, M.; Ionescu, A.; Garcia-Godoy, F.; Brambilla, E. Surface properties of resin-based composite materials and biofilm formation: A review of the current literature. *Am. J. Dent.* **2015**, *28*, 311–320. [PubMed]

79. Kampmann, Y.; De Clerck, E.; Kohn, S.; Patchala, D.K.; Langerock, R.; Kreyenschmidt, J. Study on the antimicrobial effect of silver-containing inner liners in refrigerators. *J. Appl. Microbiol.* **2008**, *104*, 1808–1814. [CrossRef] [PubMed]

80. Mahn, E. Clinical criteria for the successful curing of composite materials. *Rev. Clín. Periodoncia Implantol. Rehabil. Oral* **2013**, *6*, 148–153. [CrossRef]

81. Yap, A.U.J.; Teoh, S.H. Comparison of flexural properties of composite restoratives using the ISO and mini-flexural tests. *J. Oral Rehabil.* **2003**, *30*, 171–177. [CrossRef] [PubMed]

82. Chang, M.; Dennison, J.; Yaman, P. Physical property evaluation of four composite materials. *Oper. Dent.* **2013**, *38*, E144–E153. [CrossRef] [PubMed]

83. Gonçalves, F.; Campos, L.M.P.; Rodrigues-Júnior, E.C.; Costa, F.V.; Marques, P.A.; Francci, C.E.; Braga, R.R.; Boaro, L.C.C. A comparative study of bulk-fill composites: Degree of conversion, post-gel shrinkage and cytotoxicity. *Braz. Oral Res.* **2018**, *32*, e17. [CrossRef] [PubMed]

84. Monte Alto, R.V.; Guimarães, J.G.A.; Poskus, L.T.; da Silva, E.M. Depth of cure of dental composites submitted to different light-curing modes. *J. Appl. Oral Sci. Rev. FOB* **2006**, *14*, 71–76. [CrossRef]

85. Koran, P.; Kürschner, R. Effect of sequential versus continuous irradiation of a light-cured resin composite on shrinkage, viscosity, adhesion, and degree of polymerization. *Am. J. Dent.* **1998**, *11*, 17–22. [CrossRef] [PubMed]

86. Catelan, A.; Mainardi, M.D.C.A.J.; Soares, G.P.; de Lima, A.F.; Ambrosano, G.M.B.; Lima, D.A.N.L.; Marchi, G.M.; Aguiar, F.H.B. Effect of light curing protocol on degree of conversion of composites. *Acta Odontol. Scand.* **2014**, *72*, 898–902. [CrossRef] [PubMed]

87. Yamamoto, T.; Hanabusa, M.; Kimura, S.; Momoi, Y.; Hayakawa, T. Changes in polymerization stress and elastic modulus of bulk-fill resin composites for 24 hours after irradiation. *Dent. Mater. J.* **2018**, *37*, 87–94. [CrossRef] [PubMed]

88. Achilias, D.S. A Review of Modeling of Diffusion Controlled Polymerization Reactions. *Macromol. Theory Simul.* **2007**, *16*, 319–347. [CrossRef]

89. Korkut, E.; Torlak, E.; Altunsoy, M. Antimicrobial and mechanical properties of dental resin composite containing bioactive glass. *J. Appl. Biomater. Funct. Mater.* **2016**, *14*, e296–e301. [CrossRef] [PubMed]

90. Chun, K.; Choi, H.; Lee, J. Comparison of mechanical property and role between enamel and dentin in the human teeth. *J. Dent. Biomech.* **2014**, *5*, 1758736014520809. [CrossRef] [PubMed]

91. Moezzyzadeh, M. Evaluation of the Compressive Strength of Hybrid and Nanocomposites. *Shahid Beheshti Univ. Dent. J.* **2012**, *30*, 23–28.

92. Xu, X.; Burgess, J.O. Compressive strength, fluoride release and recharge of fluoride-releasing materials. *Biomaterials* **2003**, *24*, 2451–2461. [CrossRef]

93. Das Neves, P.B.A.; Agnelli, J.A.M.; Kurachi, C.; de Souza, C.W.O. Addition of silver nanoparticles to composite resin: Effect on physical and bactericidal properties in vitro. *Braz. Dent. J.* **2014**, *25*, 141–145. [CrossRef] [PubMed]

94. Williams, P.D.; Smith, D.C. Measurement of the tensile strength of dental restorative materials by use of a diametral compression test. *J. Dent. Res.* **1971**, *50*, 436–442. [CrossRef] [PubMed]

95. Ausiello, P.; Ciaramella, S.; Fabianelli, A.; Gloria, A.; Martorelli, M.; Lanzotti, A.; Watts, D.C. Mechanical behavior of bulk direct composite versus block composite and lithium disilicate indirect Class II restorations by CAD-FEM modeling. *Dent. Mater. Off. Publ. Acad. Dent. Mater.* **2017**, *33*, 690–701. [CrossRef] [PubMed]

96. Dejak, B.; Młotkowski, A. A comparison of stresses in molar teeth restored with inlays and direct restorations, including polymerization shrinkage of composite resin and tooth loading during mastication. *Dent. Mater. Off. Publ. Acad. Dent. Mater.* **2015**, *31*, e77–e87. [CrossRef] [PubMed]

97. Sakaguchi, R.; Powers, J. *Craig's Restorative Dental Materials*, 13th ed.; Elsevier Mosby: Philadelphia, PA, USA, 2012; ISBN 978-0-32-308108-5.

98. Casselli, D.S.M.; Worschech, C.C.; Paulillo, L.A.M.S.; Dias, C.T.D.S. Diametral tensile strength of composite resins submitted to different activation techniques. *Braz. Oral Res.* **2006**, *20*, 214–218. [CrossRef] [PubMed]

99. Della Bona, A.; Benetti, P.; Borba, M.; Cecchetti, D. Flexural and diametral tensile strength of composite resins. *Braz. Oral Res.* **2008**, *22*, 84–89. [CrossRef] [PubMed]

100. Mitra, S.B.; Wu, D.; Holmes, B.N. An application of nanotechnology in advanced dental materials. *J. Am. Dent. Assoc. 1939* **2003**, *134*, 1382–1390. [CrossRef]

101. Łukomska-Szymańska, M.; Kleczewska, J.; Nowak, J.; Pryliński, M.; Szczesio, A.; Podlewska, M.; Sokołowski, J.; Łapińska, B. Mechanical Properties of Calcium Fluoride-Based Composite Materials. *BioMed Res. Int.* **2016**, *2016*, 2752506. [CrossRef] [PubMed]

102. Dias, H.B.; Bernardi, M.I.B.; Ramos, M.A.D.S.; Trevisan, T.C.; Bauab, T.M.; Hernandes, A.C.; de Souza Rastelli, A.N. Zinc oxide 3D microstructures as an antimicrobial filler content for composite resins. *Microsc. Res. Tech.* **2017**, *80*, 634–643. [CrossRef] [PubMed]

103. Sokołowski, J.; Szynkowska, M.I.; Kleczewska, J.; Kowalski, Z.; Sobczak-Kupiec, A.; Pawlaczyk, A.; Sokołowski, K.; Łukomska-Szymańska, M. Evaluation of resin composites modified with nanogold and nanosilver. *Acta Bioeng. Biomech.* **2014**, *16*, 51–61.

104. Ferracane, J.L. Resin-based composite performance: Are there some things we can't predict? *Dent. Mater. Off. Publ. Acad. Dent. Mater.* **2013**, *29*, 51–58. [CrossRef] [PubMed]

105. Peutzfeldt, A.; Asmussen, E. Modulus of resilience as predictor for clinical wear of restorative resins. *Dent. Mater. Off. Publ. Acad. Dent. Mater.* **1992**, *8*, 146–148. [CrossRef]

106. Feiz, A.; Samanian, N.; Davoudi, A.; Badrian, H. Effect of different bleaching regimens on the flexural strength of hybrid composite resin. *J. Conserv. Dent.* **2016**, *19*, 157–160. [CrossRef] [PubMed]

107. Pala, K.; Tekçe, N.; Tuncer, S.; Demirci, M.; Öznurhan, F.; Serim, M. Flexural strength and microhardness of anterior composites after accelerated aging. *J. Clin. Exp. Dent.* **2017**, *9*, e424–e430. [CrossRef] [PubMed]

108. Kumar, G.; Shivrayan, A. Comparative study of mechanical properties of direct core build-up materials. *Contemp. Clin. Dent.* **2015**, *6*, 16–20. [CrossRef] [PubMed]

109. Randolph, L.D.; Palin, W.M.; Leloup, G.; Leprince, J.G. Filler characteristics of modern dental resin composites and their influence on physico-mechanical properties. *Dent. Mater. Off. Publ. Acad. Dent. Mater.* **2016**, *32*, 1586–1599. [CrossRef] [PubMed]

110. Rodrigues Junior, S.A.; Zanchi, C.H.; de Carvalho, R.V.; Demarco, F.F. Flexural strength and modulus of elasticity of different types of resin-based composites. *Braz. Oral Res.* **2007**, *21*, 16–21. [CrossRef] [PubMed]

111. Brandão, N.L.; Portela, M.B.; Maia, L.C.; Antônio, A.; Silva, V.L.M.E.; Silva, E.M.D. Model resin composites incorporating ZnO-NP: Activity against *S. mutans* and physicochemical properties characterization. *J. Appl. Oral Sci. Rev. FOB* **2018**, *26*, e20170270. [CrossRef] [PubMed]

112. Łapińska, B.; Łukomska-Szymańska, M.; Sokołowski, J.; Nowak, J. Experimental composite material modified with calcium fluoride—Three-point bending flexural test. *J. Achiev. Mater. Manuf. Eng.* **2016**, *74*, 72–77. [CrossRef]

113. Azillah, M.A.; Anstice, H.M.; Pearson, G.J. Long-term flexural strength of three direct aesthetic restorative materials. *J. Dent.* **1998**, *26*, 177–182. [CrossRef]

114. Poggio, C.; Lombardini, M.; Gaviati, S.; Chiesa, M. Evaluation of Vickers hardness and depth of cure of six composite resins photo-activated with different polymerization modes. *J. Conserv. Dent.* **2012**, *15*, 237–241. [CrossRef] [PubMed]

115. Abed, Y.A.; Sabry, H.A.; Alrobeigy, N.A. Degree of conversion and surface hardness of bulk-fill composite versus incremental-fill composite. *Tanta Dent. J.* **2015**, *12*, 71–80. [CrossRef]

116. Roy, K.K.; Kumar, K.P.; John, G.; Sooraparaju, S.G.; Nujella, S.K.; Sowmya, K. A comparative evaluation of effect of modern-curing lights and curing modes on conventional and novel-resin monomers. *J. Conserv. Dent.* **2018**, *21*, 68–73. [CrossRef] [PubMed]

117. Schmidt, C.; Ilie, N. The effect of aging on the mechanical properties of nanohybrid composites based on new monomer formulations. *Clin. Oral Investig.* **2013**, *17*, 251–257. [CrossRef] [PubMed]

118. Randolph, L.D.; Palin, W.M.; Bebelman, S.; Devaux, J.; Gallez, B.; Leloup, G.; Leprince, J.G. Ultra-fast light-curing resin composite with increased conversion and reduced monomer elution. *Dent. Mater. Off. Publ. Acad. Dent. Mater.* **2014**, *30*, 594–604. [CrossRef] [PubMed]

119. AlShaafi, M.M. Factors affecting polymerization of resin-based composites: A literature review. *Saudi Dent. J.* **2017**, *29*, 48–58. [CrossRef] [PubMed]

120. De Oliveira, D.C.R.S.; de Menezes, L.R.; Gatti, A.; Correr Sobrinho, L.; Ferracane, J.L.; Sinhoreti, M.A.C. Effect of Nanofiller Loading on Cure Efficiency and Potential Color Change of Model Composites. *J. Esthet. Restor. Dent.* **2016**, *28*, 171–177. [CrossRef] [PubMed]

121. AlShaafi, M.M. Effects of Different Temperatures and Storage Time on the Degree of Conversion and Microhardness of Resin-based Composites. *J. Contemp. Dent. Pract.* **2016**, *17*, 217–223. [CrossRef] [PubMed]

122. Catelan, A.; de Araújo, L.S.N.; da Silveira, B.C.M.; Kawano, Y.; Ambrosano, G.M.B.; Marchi, G.M.; Aguiar, F.H.B. Impact of the distance of light curing on the degree of conversion and microhardness of a composite resin. *Acta Odontol. Scand.* **2015**, *73*, 298–301. [CrossRef] [PubMed]

123. Bociong, K.; Szczesio, A.; Sokolowski, K.; Domarecka, M.; Sokolowski, J.; Krasowski, M.; Lukomska-Szymanska, M. The Influence of Water Sorption of Dental Light-Cured Composites on Shrinkage Stress. *Materials* **2017**, *10*, 1142. [CrossRef] [PubMed]

124. Boaro, L.C.; Gonçalves, F.; Guimarães, T.C.; Ferracane, J.L.; Pfeifer, C.S.; Braga, R.R. Sorption, solubility, shrinkage and mechanical properties of "low-shrinkage" commercial resin composites. *Dent. Mater. Off. Publ. Acad. Dent. Mater.* **2013**, *29*, 398–404. [CrossRef] [PubMed]

125. Mortier, E.; Gerdolle, D.A.; Jacquot, B.; Panighi, M.M. Importance of water sorption and solubility studies for couple bonding agent—Resin-based filling material. *Oper. Dent.* **2004**, *29*, 669–676. [PubMed]

126. Sideridou, I.; Tserki, V.; Papanastasiou, G. Study of water sorption, solubility and modulus of elasticity of light-cured dimethacrylate-based dental resins. *Biomaterials* **2003**, *24*, 655–665. [CrossRef]

127. Mokrzycki, W.S.; Tatol, M. Color difference ΔE: A survey. *Mach. Graph. Vis.* **2011**, *20*, 383–411.

128. De Oliveira, D.C.R.S.; Ayres, A.P.A.; Rocha, M.G.; Giannini, M.; Puppin Rontani, R.M.; Ferracane, J.L.; Sinhoreti, M.A.C. Effect of Different In Vitro Aging Methods on Color Stability of a Dental Resin-Based Composite Using CIELAB and CIEDE2000 Color-Difference Formulas. *J. Esthet. Restor. Dent.* **2015**, *27*, 322–330. [CrossRef] [PubMed]

129. Jeong, T.-S.; Kang, H.-S.; Kim, S.-K.; Kim, S.; Kim, H.-I.; Kwon, Y.H. The effect of resin shades on microhardness, polymerization shrinkage, and color change of dental composite resins. *Dent. Mater. J.* **2009**, *28*, 438–445. [CrossRef] [PubMed]

130. Albuquerque, P.P.A.C.; Bertolo, M.L.; Cavalcante, L.M.A.; Pfeifer, C.; Schneider, L.F.S. Degree of conversion, depth of cure, and color stability of experimental dental composite formulated with camphorquinone and phenanthrenequinone photoinitiators. *J. Esthet. Restor. Dent.* **2015**, *27* (Suppl. 1), S49–S57. [CrossRef] [PubMed]

131. Fúcio, S.B.P.; Carvalho, F.G.; Sobrinho, L.C.; Sinhoreti, M.A.C.; Puppin-Rontani, R.M. The influence of 30-day-old *Streptococcus mutans* biofilm on the surface of esthetic restorative materials—An in vitro study. *J. Dent.* **2008**, *36*, 833–839. [CrossRef] [PubMed]

132. Melo, M.A.; Orrego, S.; Weir, M.D.; Xu, H.H.K.; Arola, D.D. Designing Multiagent Dental Materials for Enhanced Resistance to Biofilm Damage at the Bonded Interface. *ACS Appl. Mater. Interfaces* **2016**, *8*, 11779–11787. [CrossRef] [PubMed]

133. Mutluay, M.M.; Zhang, K.; Ryou, H.; Yahyazadehfar, M.; Majd, H.; Xu, H.H.K.; Arola, D. On the fatigue behavior of resin-dentin bonds after degradation by biofilm. *J. Mech. Behav. Biomed. Mater.* **2013**, *18*, 219–231. [CrossRef] [PubMed]

134. Li, Y.; Carrera, C.; Chen, R.; Li, J.; Lenton, P.; Rudney, J.D.; Jones, R.S.; Aparicio, C.; Fok, A. Degradation in the dentin-composite interface subjected to multi-species biofilm challenges. *Acta Biomater.* **2014**, *10*, 375–383. [CrossRef] [PubMed]

135. Drummond, J.L. Degradation, fatigue and failure of resin dental composite materials. *J. Dent. Res.* **2008**, *87*, 710–719. [CrossRef] [PubMed]

136. Lohbauer, U.; Belli, R.; Ferracane, J.L. Factors involved in mechanical fatigue degradation of dental resin composites. *J. Dent. Res.* **2013**, *92*, 584–591. [CrossRef] [PubMed]

137. Ravindranath, V.; Gosz, M.; De Santiago, E.; Drummond, J.L.; Mostovoy, S. Effect of cyclic loading and environmental aging on the fracture toughness of dental resin composite. *J. Biomed. Mater. Res. B Appl. Biomater.* **2007**, *80*, 226–235. [CrossRef] [PubMed]

138. Mutluay, M.M.; Yahyazadefar, M.; Ryou, H.; Majd, H.; Do, D.; Arola, D. Fatigue of the Resin-Dentin Interface: A New Approach for Evaluating the Durability of Dentin Bonds. *Dent. Mater. Off. Publ. Acad. Dent. Mater.* **2013**, *29*, 437–449. [CrossRef] [PubMed]

139. Yahyazadehfar, M.; Mutluay, M.M.; Majd, H.; Ryou, H.; Arola, D. Fatigue of the resin-enamel bonded interface and the mechanisms of failure. *J. Mech. Behav. Biomed. Mater.* **2013**, *21*, 121–132. [CrossRef] [PubMed]

materials

MDPI

Article

Modified Polymeric Nanoparticles Exert In Vitro Antimicrobial Activity Against Oral Bacteria

Manuel Toledano-Osorio [1], **Jegdish P. Babu** [2], **Raquel Osorio** [1,*], **Antonio L. Medina-Castillo** [3], **Franklin García-Godoy** [2] and **Manuel Toledano** [1]

[1] Dental School, University of Granada, Campus de Cartuja s/n, 18071 Granada, Spain; toledano@correo.ugr.es (M.T.-O.); toledano@ugr.es (M.T.)
[2] College of Dentistry, University of Tennessee Health Science Center, 875 Union Avenue, Memphis, TN 381632110, USA; jbabu@uthsc.edu (J.P.B.); fgarciagodoy@gmail.com (F.G.-G.)
[3] NanoMyP, Spin-Off Enterprise from University of Granada, Edificio BIC-Granada, Av. Innovación 1, Armilla, 18016 Granada, Spain; amedina@nanomyp.com
*   Correspondence: rosorio@ugr.es; Tel.: +34-958-243-789; Fax: +34-958-240-908

Received: 16 May 2018; Accepted: 13 June 2018; Published: 14 June 2018

Abstract: Polymeric nanoparticles were modified to exert antimicrobial activity against oral bacteria. Nanoparticles were loaded with calcium, zinc and doxycycline. Ions and doxycycline release were measured by inductively coupled plasma optical emission spectrometer and high performance liquid chromatography. *Porphyromonas gingivalis, Lactobacillus lactis, Streptoccocus mutans, gordonii* and *sobrinus* were grown and the number of bacteria was determined by optical density. Nanoparticles were suspended in phosphate-buffered saline (PBS) at 10, 1 and 0.1 mg/mL and incubated with 1.0 mL of each bacterial suspension for 3, 12, and 24 h. The bacterial viability was assessed by determining their ability to cleave the tetrazolium salt to a formazan dye. Data were analyzed by ANOVA and Scheffe's F ($p < 0.05$). Doxycycline doping efficacy was 70%. A burst liberation effect was produced during the first 7 days. After 21 days, a sustained release above 6 µg/mL, was observed. Calcium and zinc liberation were about 1 and 0.02 µg/mL respectively. The most effective antibacterial material was found to be the Dox-Nanoparticles (60% to 99% reduction) followed by Ca-Nanoparticles or Zn-Nanoparticles (30% to 70% reduction) and finally the non-doped nanoparticles (7% to 35% reduction). *P. gingivalis, S. mutans* and *L. lactis* were the most susceptible bacteria, being *S. gordonii* and *S. sobrinus* the most resistant to the tested nanoparticles.

Keywords: antibacterial; calcium; doxycycline; nanoparticles; zinc

---

1. Introduction

Bacteria are the main cause of prevalent oral diseases as caries and periodontitis. Oral administration of antibacterial agents presents an important limitation, as it is accessing the dentin interface, the radicular canal or the subgingival pockets where these bacteria grow. In these cases, benefits of local versus systemic delivery routes are clear [1]. This work explores the design of nanoparticles (NPs) that locally administered will exert therapeutic antibacterial properties against oral bacteria.

Resin-based restorative materials are commonly employed in clinical treatments to seal the interfaces and as a result, bonding to dentin is challenging [2]. Gaps at these bonded interfaces lead to microleakage, which also facilitate the invasion of cariogenic pathogens to cause secondary caries infections [2,3]. Therefore, in order to minimize the incidence of secondary caries, it would be desirable the existence of an antibacterial agent able to inhibit cariogenic pathogens at the dentin interface [4]. For this purpose, studying antibacterial effects against *Streptococcus mutans* (Sm), *Streptococcus gordonii* (Sg), *Streptococcus sobrinus* (Ss), and *Lactobacillus lactis* (Ll) have been recommended [3].

In endodontic treatment, elimination of bacteria in the root canal system, is a major challenge. Microorganisms remained after canal treatment will impair periapical healing and will facilitate developing apical lesions [5]. *Porphyromonas gingivalis* (*Pg*) is a major etiologic agent not only in the recurrent infections after endodontic treatment [6], but also in the development and progression of periapical lesions and periodontitis [7]. Biomaterials design leading to minimize the incidence of persistent or recurrent infections of the root canal system and apical periodontitis would also be desirable [8]. It is important to note that the biosafety of sodium hypochlorite during root canal treatment has been recently questioned [9].

Endogenous matrix metalloproteinases (MMPs) are interstitial collagenases present in radicular dentin, periodontal tissue and periapical bone [10–14]. MMPs have been related to chronic inflammation processes and abscesses at apical level [10,12]. Then MMPs inhibition will improve the prognosis of endodontic treatments [11]. Moreover, if resin-based materials are used for dentin bonding and tooth restoring, collagen degradation by MMPs will occur at the dentin interface jeopardizing restorations longevity [15]. When bonding to dentin in restorative dentistry, if dentin is infiltrated by MMP inhibitors, crystallite-sparse collagen fibrils of the scaffold could be protected from degradation facilitating further remineralization [2,16,17]. Metal nanoparticles (i.e., silver, gold, or zinc oxide . . .) have been previously introduced in restorative dentistry, mainly due to their antibacterial or MMPs inhibition properties [18–20].

Novel polymeric nanoparticles (NPs), about 100 nm in diameter, have been synthetized and previously tested at the resin-dentin bonded interface [17,21]. NPs have been shown to inhibit dentin MMPs collagen degradation [22], and to facilitate mineral growth at the interface without impairing bond strength [21,22]. Sequences of anionic carboxylate (i.e., COO^-) are along the backbone of the polymeric NPs. These functional groups permit the possibility of calcium and zinc quelation (1 µg Ca/mg NPs and 2.2 µg Zn/mg NPs) [23]. Cationic metals, loaded onto particles surfaces, if released, may provide for antimicrobial activity. Both metal cations have been demonstrated to have significant antibacterial effects [24,25]. Moreover, when NPs are larger than 10 nm, they do not penetrate bacteria membranes and are thought to exert further antimicrobial effects through accumulation on cell membranes [26]. At this stage, the bacterial membrane permeability may become compromised rendering the cell unable to regulate transport through it, and eventually causing cell death [27]. Doxycycline hyclate is also an antibacterial [28] and potent MMPs inhibitor [29] that is proposed to be immobilized on presented NPs.

Proposed NPs may be employed in three different dental clinical applications: (1) during the dentin bonding process, to exert antibacterial activity at the resin–dentin bonded interface [17,21]; (2) inside the radicular canal, during the endodontic treatment, to facilitate bacterial elimination; and (3) at the periodontal pocket, onto the cementum surface, to directly exert antibacterial activity [23].

Thus, the purpose of this *in vitro* study was to design and synthetize NPs doped with calcium/zinc ions or with immobilized doxycycline able to exert antibacterial activity against *Sm*, *Sg*, *Ss*, *Ll* and *Pg*. The null hypotheses to be tested are that: (1) Calcium, zinc and doxycycline are not liberated from NPs, and (2) NPs, calcium, zinc and doxycycline doped NPs do not affect bacterial viability.

2. Results

1. Loading efficacy and release of doxycycline hyclate from NPs: The amount of doxycycline in the aqueous solution before NPs immersion was 1333 µg/mL (per mg of NPs). In the supernatant, after NPs immersion, doxycycline concentration was 399.5 µg/mL (per mg of NPs). Loading efficacy was around 70%. Mean and standard deviation of doxycycline liberation (µg/mL) and cumulative liberation (%), per 10 mg of NPs at each time point are presented in Table 1. Doxycycline liberation was 106 µg/mL (per mg of NPs) at 12 h. A burst effect with rapid doxycycline liberation was observed from 12 h until the first week of storage. After 7 days, the antibiotic release was above 20 µg/mL (per mg of NPs). After 21 days, doxycycline liberation was stably sustained, being 8 and 6 µg/mL (per mg of NPs) at 21 and 28 days, respectively. After 24 h, a 57% of the immobilized doxycycline was liberated, and after 7 days and 28 days, 72% and 80% of loaded antibiotic was respectively released (Table 1).

Table 1. Mean and standard deviation (SD) of Ca^{2+}, Zn^{2+} and doxycycline liberation in µg. Cumulative liberation (CL) was expressed in percentages. Values are obtained per 10 mg of NPs, at each time point.

Time	Ca^{2+} (µg)	Ca^{2+} CL (%)	Zn^{2+} (µg)	Zn^{2+} CL (%)	Doxycycline (µg)	Doxycycline CL (%)
12 h	1.006 (0.002)	11	0.025 (0.001)	0.1	1211.29 (166.32)	30
24 h	1.007 (0.001)	21	0.025 (0.001)	0.2	1065.98 (146.15)	57
48 h	0.909 (0.003)	30	0.023 (0.002)	0.3	458.08 (63.5)	68
7 days	0.856 (0.001)	39	0.021 (0.001)	0.4	210.81 (28.33)	74
21 days	2.082 (0.05)	61	0.024 (0.002)	0.5	81.85 (10.97)	78
28 days	2.031 (0.02)	82	0.044 (0.005)	0.8	63.23 (9.01)	80

2. Calcium and zinc liberation from NPs: Mean and standard deviation of Ca^{2+} and Zn^{2+} liberation (µg) and cumulative liberation (%), per 10 mg of NPs at each time point are presented in Table 1. Calcium liberation ranged from 0.856 to 1.007 µg (per 10 mg of NPs) during the first week. This calcium release was doubled after 21 d, being around 2 µg (per 10 mg of NPs). Zn-NPs maintained a sustained zinc liberation that ranged from 0.021 to 0.025 µg (per 10 mg of NPs) between 12 to 21 days. A double fold increase was observed at day 28, when 0.044 µg were released per each 10 mg of NPs.

3. MTT assay: Mean and standard deviations of the different bacteria survival values expressed as number of viable cells after 3, 12 and 24 h of exposure to the distinct NPs and control PBS are shown in Figures 1–5.

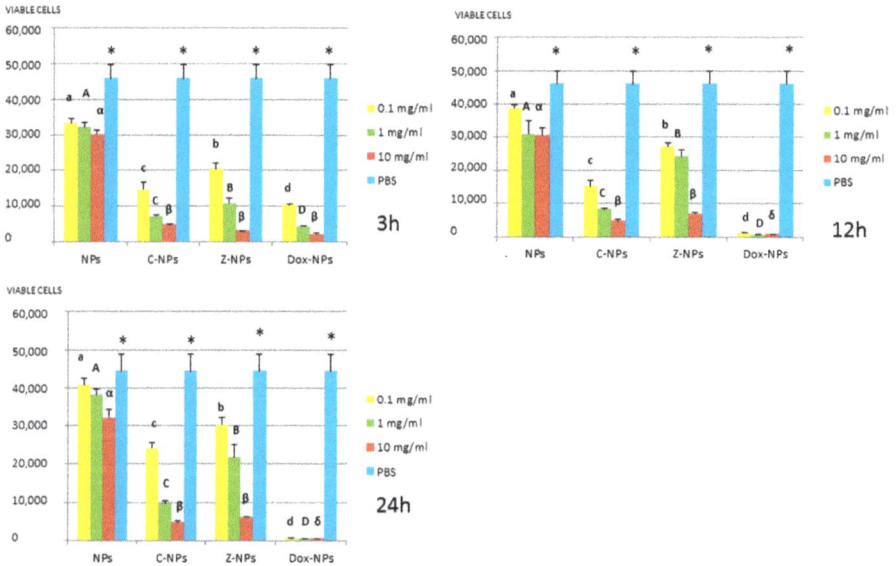

Figure 1. *P. gingivalis* survival (number of viable cells) after 3 h, 12 h, and 24 h of different concentration NPs exposure. Same letter or symbol indicates no significant difference of viable bacteria between different NPs concentrations ($p < 0.05$).

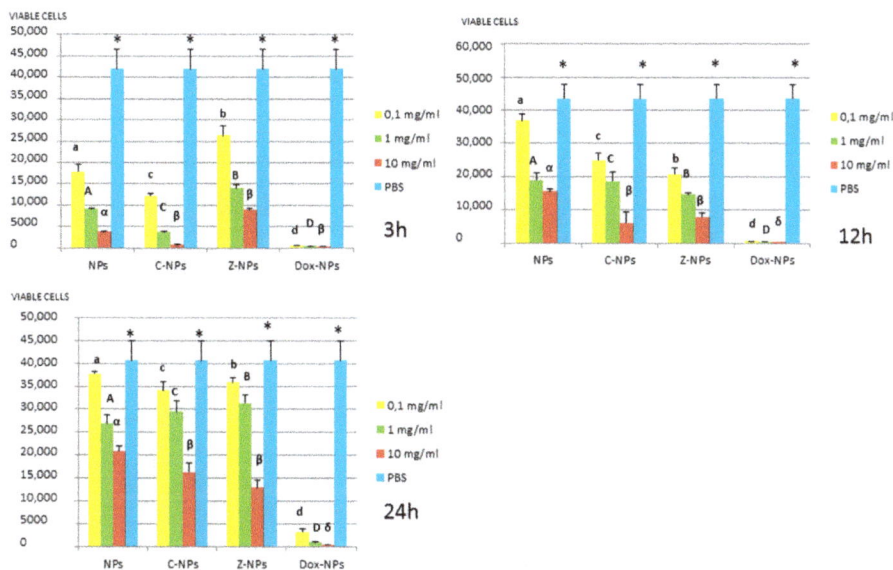

Figure 2. *S. mutans* survival (number of viable cells) after 3 h, 12 h, and 24 h of different concentration NPs exposure. Same letter or symbol indicates no significant difference of viable bacteria between different NPs concentrations (*p* < 0.05).

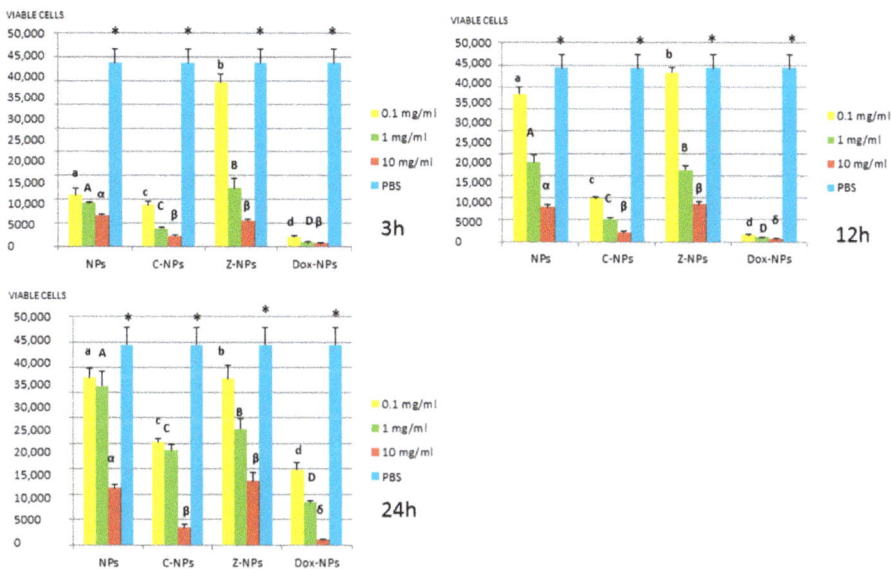

Figure 3. *L. lactis* survival (number of viable cells) after 3 h, 12 h, and 24 h of different concentration NPs exposure. Same letter or symbol indicates no significant difference of viable bacteria between different NPs concentrations (*p* < 0.05).

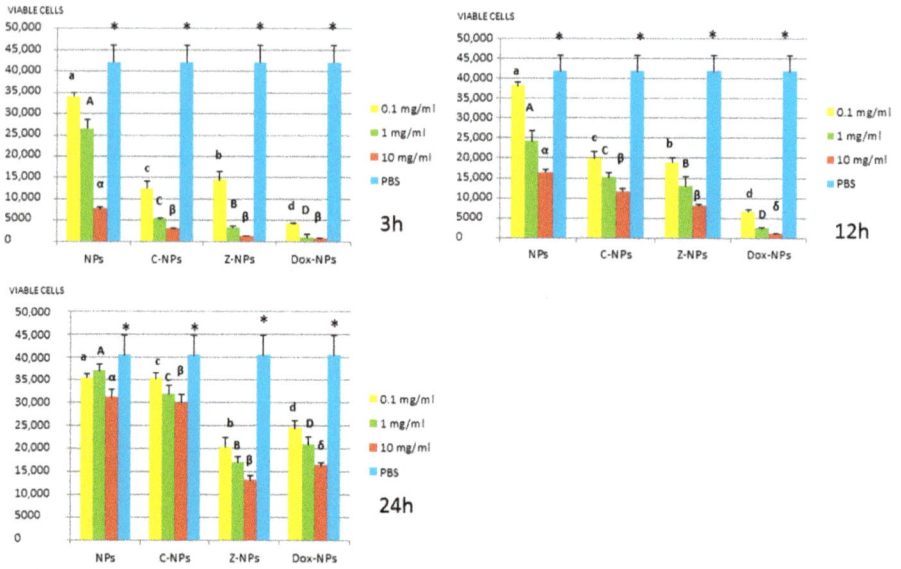

Figure 4. *S. gordonii* survival (number of viable cells) after 3 h, 12 h, and 24 h of different concentration NPs exposure. Same letter or symbol indicates no significant difference of viable bacteria between different NPs concentrations ($p < 0.05$).

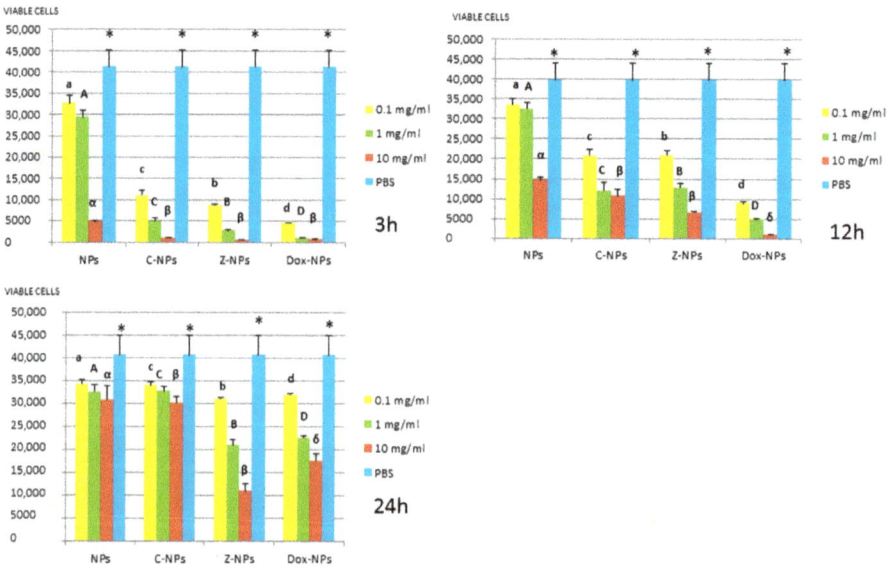

Figure 5. *S. sobrinus* survival (number of viable cells) after 3 h, 12 h, and 24 h of different concentration NPs exposure. Same letter or symbol indicates no significant difference of viable bacteria between different NPs concentrations ($p < 0.05$).

In general, all tested NPs affected the viability of bacterial suspension. The most effective were the Dox-NPs followed by Ca-NPs or Zn-NPs and finally non-doped NPs that attained the most variable and least reduction in bacterial survival (8% to 70% after 24 h).

The viability of tested bacteria following the incubation with NPs depends on the type of NPs. The two bacteria, *S. gordonii* and *S. sobrinus* were found to be the most resistant to the tested NPs. After 24 h, they were only affected by Zn-NPs (70% reduction in bacterial viability) and by Dox-NPs (60% reduction). For *P. gingivalis*, *S. mutans* and *L. lactis*, Dox-NPs reduced the bacterial viability by 60% to 99%, after 24 h depending on the concentration of doxycycline. Meanwhile the reduction in bacterial viability were from 20% to 60% for *S. gordonii* and *S. sobrinus*. The *P. gingivalis*, *L. lactis* and *S. mutans* Dox-NPs effect was not variable during the time of the study. Only in the cases of *S. gordonii* and *S. sobrinus* cultures a drop in Dox-NPs efficacy was observed after 24 h. At 24 h, for *P. gingivalis* all tested concentrations of Dox-NPs attained above 98% bacterial death. In general, most effective dosage of Dox-NPs was found to be 10 mg/mL.

When testing Ca, Zn-doped or even undoped-NPs for *S. mutans* and *L. lactis*, bacterial viability was significantly affected in doses and time dependent manners. After 24 h, only those NPs contained 10 mg/mL were effective. Both bacteria were equally susceptible to Zn-NPs (68% cells reduction). When considering Ca-NPs or undoped-NPs, *L. lactis* was more susceptible (reduction values for Ca-NPs: 90%, for unloaded NPs: 70%) than *S. mutans* (reduction values for Ca-NPs: 60%, for unloaded NPs: 50%).

Testing of NPs doped with Ca and Zn-doped NPs at 0.1 mg/mL against *P. gingivalis*, bacterial viability was significantly affected and bacterial death ranges between 55% to 27%. However, at the most effective concentration −10 mg/mL, bacterial reduction ranges were from 80% to 93%, without significant differences between both ion-doped NPs. *P. gingivalis* incubated with unloaded NPs attained low but dose and time-dependent percentages of bacterial survival reduced, from 34.3% to 7.2%.

3. Discussion

There are several in vitro testing models for the efficacy of antibacterial agents, which may involve single or multispecies bacteria. This microcosm model is the most clinically relevant, but attained results are often difficult to interpret as there is no way to control for the behavior of individual bacterial species. It is also difficult to decide which species are appropriate in each experiment and their relative amounts [3]. Using biofilm models is also challenging as the results may also be different on various materials surfaces with different chemistry and/or topography [30]. Therefore, when analyzing novel antibacterial agents planktonic monoculture tests are necessary to facilitate results interpretation. Further studies need to be conducted on clinical isolates and multi-species biofilms on different material surfaces or interfaces, which may express resistance trait against tested antibacterial effect.

P. gingivalis was selected for the present study as it is one of the most frequently detected anaerobic microorganisms in subgingival plaque samples from periodontal-endodontic combined lesions and necrotic pulp [6]. *S. mutans*, *S. gordonii*, *S. sobrinus* and *L. lactis* were used as are the most frequently detected microorganisms in cariogenic plaque [3]. *P. gingivalis* is a Gram-negative bacteria, *S. mutans*, *S. gordonii*, *S. sobrinus* and *L. lactis* are Gram positive. *P. gingivalis* has an asymmetric distribution of lipids at their cell walls, the outer face contains lipopolysaccharide (LPS), and the inner face has phospholipids [8]. *S. gordonii*, *S. sobrinus* *S. mutans* and *L. lactis* also have LPS at their membranes, which exhibits anionic charge, as a result it may facilitate cationic groups to bond and exert antimicrobial activity [8]. This may be a reason for observing low antibacterial activity of tested non-loaded NPs, as they are also anionic (potential zeta is −41 ± 5 mV measured in water at pH = 7) [20], and will not be attracted to tested bacteria which posse a zeta potential of approximately −25 mV at pH = 7 [31].

Ca-NPs and Zn-NPs exerted antibacterial activity, at 10 mg/mL 80% to 93% bacterial reduction after 48 h, was encountered (Figure 1) as a possible result of liberated calcium and zinc from NPs. After 48 h, 0.9 and 0.02 µg per 10 mg of NPs of calcium and zinc are respectively released

(Table 1). Cationic metals as calcium or zinc have been shown to be potent antimicrobials [24,25,27]. Calcium release from NPs was estimated to be 0.08 and 0.1 µg/mL (per mg of NPs) from 12 h up to 7 days, while zinc release was around 0.02 µg/mL (per mg of NPs) at the same time-points. Cummulative liberation of both ions is 30% for calcium and 0.3% for zinc after 48 h. It has been shown that lipopolysaccharides at the outer membrane of Gram-negative bacteria possess magnesium and calcium ions that bridge to negatively-charged phosphor-sugars [8]. Therefore, cationic elements may also displace these metal ions damaging the outer membrane, leading to cell death [8,26]. It has also been previously shown that zinc ions markedly enhanced the adhesion and accumulation of salivary and serum proteins on cells of *P. gingivalis* and inhibited their coaggregation when growing on biofilms [27].

Zinc has a known inhibitory effect on glycolysis and proteinase activity in many oral bacteria [27]. Zinc may affect *S. mutans* viability by inhibiting glycolysis [32]. Kinetic studies of the glucosyltransferases of *S. sobrinus* by Devulapalle and Mooser [33] showed that the Zn ion acts as a reversible, competitive inhibitor at the fructose subsite within the active site of the glucosyltransferase. This observation may well explain the reported dose-dependent effects of zinc on the tested bacteria. Even when the exact antibacterial mechanism of zinc has not been clearly identified yet, covalently or oxidatively induced damage has been claimed [32]. Zinc ions are considered useful for limiting the settlement/colonization of *P. gingivalis* in the gingival sulcus with the goal of preventing periodontal disease [27] and in the case of *S. mutans* preventing carious disease [32]. Zinc has long been known as a plaque-inhibiting compound and also can influence acid production by different microbes [32]. In addition, zinc is able to depolarize the membrane potential, it does not always cause the bacterial cell membrane to rupture and leak, but alters permeability that is closely related to the sensitivity of bacteria to ionic environment [30]. Ion homeostasis affects the proliferation, communication and metabolism of bacteria; then, zinc may sometimes produce an inhibitive instead of destructive effect against bacteria [30].

Dox-NPs exerted the highest antibacterial activity to all the tested concentrations (80% to 97% bacterial reduction after 24 h). Following our results, doxycycline was found to be released at sustained levels for over 28 days, with a significant burst effect at 24 h. It is liberated at concentrations high above to those considered effective against bacteria at any time point of the present study. For each mg of NPs 121, 106 and 46 µg/mL of doxycycline will be liberated at 12, 24 and 48 h, respectively. A burst effect with rapid doxycycline liberation was observed from 12 h until the first week of storage. After 7 days time-point, antibiotic release was maintained above 20 µg/mL (per mg of NPs). As bacterial susceptibility to doxycycline is obtained around 0.1 to 0.2 µg/mL [28], doxycycline is then liberated from NPs at concentrations high above to those considered effective against most of the tested bacteria. It was shown that doxycycline at a concentration between 0.5 and 1 µg/mL is bactericidal against different *Pg* strains [34], and between 0.1 and 6.0 µg/mL is effective against *Pg* and other putative periodontal pathogens [35,36]. It should be stressed that tested Dox-NPs after 28 days are able to liberate doxycycline concentrations above 6 µg/mL.

Doxycycline is a polar and amphoteric compound. Doxycycline as a salt (hyclate) is water soluble. Doxycycline is known to act against most bacteria by inhibiting the microbial protein synthesis that requires access into the cell wall and lipid solubility [37]. Doxycycline binds the ribosome to prevent ribonucleic acid synthesis by avoiding addition of more amino acid to the polypeptide [37]. Doxycycline is also known to provoke a potent and long-lasting inhibition of dentin matrix metalloproteinases [29] that are related to chronic inflammation processes and abscesses at apical level [10]. It may explain how long-term administration of a sub-antimicrobial dose of doxycycline, to dogs with periodontitis, is regarded as an effective treatment for periodontal inflammation, even when it does not induce antimicrobial effects [38]. It is also important to note that MMPs inhibition may also prevent collagen degradation at the resin bonded dentin interface [15]. It will also probably reduce secondary caries formation, as MMPs activity is augmented at caries affected dentin [39].

The reported doxycycline liberation data are high and sustained, if compared to the release profile of other previously proposed compounds as a cellulose-acetate-loaded doxycycline formulation studied by Tonetti et al. [40]. Kim et al. [28] introduced a biodegradable doxycycline gel and reported a mean local concentration of 20 mg/mL, after 15 min; values that were lowered to 577 μg/mL after 3 days and to 16 μg/mL after 12 days. Deasy et al. [41] used tetracycline hydrochloride in poly(hydroxybutyric acid) as a biodegradable polymer matrix and showed sustained release just over 4 to 5 days, with a significant burst effect at 24 h. Previously introduced materials are then able to liberate doxycycline at higher concentrations, but in shorter periods of time, denoting accentuated burst effects.

In general, tested NPs had little effect on the growth of (*S. sobrinus* and *S. gordonii*) and specially Dox-NPs after 24 h, which greatly affected *P. gingivalis*, *S. mutans* and *L. lactis* survival rates (at least at the evaluated time points and concentrations). Recent results on advanced caries lesions in young human teeth, using bacterial sequence analysis methods are consistent and indicate that *S. gordonii* diminishes greatly in caries-associated plaque biofilm, while *S. mutans* persists [42]. It means that NPs may selectively inhibit cariogenic and periodontal bacteria, while leaving commensal microbes. However, it should be assayed in properly designed multibacteria biofilms models in future studies. It is to be noted that the tested NPs are biocompatible against human fibroblasts [23], and the application of antibacterials is crucial if regenerative/revascularization processes are performed for the endodontic treatment [43], these NPs may be an interesting tool.

Two important limitations are recognized for the present Dox-NPs: (1) antibiotics may produce bacterial strain resistance, which is a current global concern; therefore, further research is needed. (2) The bacteria grow in biofilms, and are known to be more resistant to antimicrobial treatment than the planktonic cultures used for the present antimicrobial susceptibility testing [44]. Then, it is imperative to include the biofilm mode of growth of bacteria when testing treatments for bonded dentin interfaces, endodontic and periapical diseases. But these tests are difficult to control, in terms of knowing specifically how bacteria are involved in the process [3]. It will not be possible to ascertain if a specific toxicity of NPs against individual bacteria is being produced [44], or just a biofilm disruption interfering with first colonizers bacteria attachment to dentin.

It may be concluded after this in vitro study that experimental NPs loaded with zinc, calcium or doxycycline are effective to eradicate tested oral bacteria. For clinical applications, using these NPs at the resin-bonded interface as cavity liners may be recommended. As it was shown before, that NPs do not affect bond efficacy and improve dentin remineralization [17,21,22]. The same beneficial effect may be found if NPs are used in endodontics, before resin sealant cement application. However, as recognized in the study limitations, further investigations into antibacterial effects through biofilm models of multiple bacterial species should be implemented.

4. Material and Methods

1. Preparation of Nanoparticles (NPs): PolymP-*n* Active NPs were acquired (NanoMyP, Granada, Spain). Particles are fabricated trough polymerization precipitation. Main components of NPs are 2-hydroxyethyl methacrylate, ethylene glycol dimethacrylate and methacrylic acid; these compounds are the backbone monomer, the cross-linker and the functional monomer respectively.

Calcium-loaded NPs (Ca-NPs) and Zinc-loaded NPs (Zn-NPs) were produced. Zinc and calcium complexation was obtained immersing 30 mg of NPs during 3 days, under continuous shaking in aqueous solutions of $ZnCl_2$ or $CaCl_2$, at room temperature. Fifteen mL of the solutions containing zinc or calcium at 40 ppm were employed (pH 6.5). Then, the adsorption equilibrium of metal ions was reached [23]. To separate the NPs from the supernatant, the suspensions were centrifuged. 0.96 ± 0.04 μg Ca/mg NPs and 2.15 ± 0.05 μg Zn/mg NPs were the attained ion complexation values [23]. NPs loaded with doxycycline hyclate were also produced. An 18-mL aqueous solution, with 40 mg/mL of doxycycline hyclate (Sigma-Aldrich, Darmstadt, Germany) was prepared, and 30 mg of NPs were immersed in the solution for 4 h, under continuous shaking. Then, to separate NPs from the supernatant the suspensions were centrifuged. Following groups were tested: (1) NPs (NPs), (2) NPs loaded with Ca (Ca-NPs), (3) NPs loaded with Zn (Zn-NPs), and (4) NPs with immobilized doxycycline hyclate (Dox-NPs).

2. Loading efficacy and release of doxycycline from NPs: For loading efficacy 18 mL of 40 mg/mL aqueous solution of doxycycline hyclate was prepared and the amount of doxycycline in the initial aqueous solution was assessed in triplicate samples of 100 µL and recorded as initial doxycycline concentration (1333 µg Dox/mL). Three different samples containing 1 mg of NPs and 0.6 mL of the doxycycline solution were incubated for 4 h, under continuous shaking. Then, the suspensions were centrifuged and the particles were separated from the supernatant, 100 µL of each supernatant was analyzed for final doxycycline concentration. Final doxycycline concentration was subtracted from initial values to calculate loading efficacy [45]. To ascertain for doxycycline liberation, 30 mg of doxycycline loaded-NPs were suspended in 3 mL of phosphate buffer saline (PBS, pH 7.4, Fisher Scientific SL, Madrid, Spain), three different eppendorf tubes containing 1 mL of the Dox-NPs suspension were stored at 37 °C. After 12 h, suspensions were centrifuged and the particles were separated from the supernatant. An aliquot (0.1 mL) of each supernatant was analyzed for doxycycline concentration. NPs were washed and 1 mL of fresh PBS was used to resuspend the NPs at 10 mg/mL until the next supernatant collection. Seven different time-points were tested: 12, 24, 48 h, 7, 14, 21 and 28 days. Supernatans were stored at -20 °C until doxycycline concentration measuring [45]. The amount of doxycycline was assayed by high performance liquid chromatography (HPLC) (Waters Alliance 2690, Waters Corporation, Milford, MA, USA) equipped with a UV-Vis detector. A binary mobile phase consisting of solvent systems A and B was used in an isocratic elution with 80:20 A:B. Mobile phase A was 50 mM $KHPO_4$ in distilled H_2O and mobile phase B was 100% acetonitrile. The HPLC flow rate was 1.0 mL/min and the total run time was 10 min. The retention time was 4.85 min. The concentration of doxycycline was calculated based on a standard curve of known levels of doxycycline at 273 nm [45].

3. Calcium and zinc liberation from NPs: 150 mg of zinc and 150 mg of calcium loaded-NPs were suspended in 15 mL of deionized water. Three different eppendorf tubes containing 5 mL of the Ca-NPs suspensions and other 3 with Zn-NPs were stored at 37 °C. After 12 h, suspensions were centrifuged and the particles separated from the supernatant; 5 mL of each supernatant was analyzed for calcium and zinc concentration. NPs were washed and 5 mL of fresh deionized water was used to resuspend the NPs at 10 mg/mL until the next supernatant collection. Seven different time points were tested: 12 h, 24 h, 48 h, 7 d, 14 d, 21 d and 28 d. Supernatans were stored at -20 °C until testing. Calcium and zinc concentrations were analyzed through an inductively coupled plasma (ICP) optical emission spectrometer (ICP-OES Optima 8300, Perkin-Elmer, MA, USA) [23].

4. Bacteria: *P. gingivalis* 33,277, *S. mutans* 700,610, *S. sobrinus* 33,478, *S. gordonii* 10,558 and *L. lactis* 12,315 were obtained from ATCC (Bethesda, MD, USA). The anaerobic organism, *Pg* was grown in Tryptic Soy broth (TSB) supplemented with yeast extract (5 g/L), Hemin (5 mg/L), Menadione (1 mg/L), for 72 h. Strict anaerobic conditions were employed, *Thermo Scientific* Oxoid *AnaeroGen* (Thermo Fisher Scientific, Waltham, MA, USA) was used in an anaerobic jar, which provides 7–15% CO_2 and <0.1% O_2. The remaining test bacteria were grown in TSB for 24 h at 37 °C. The bacterial cells were harvested by centrifugation and re-suspended in the same growth media. The number of bacteria per mL was determined by measuring the optical density at 600 nm and adjusting it to a standard bacterial suspension of 1×10^7 CFU/mL [46].

5. MTT assay: The NPs were suspended in PBS at three different concentrations (10 mg/mL, 1 mg/mL and 0.1 mg/mL). NPs were placed into Eppendorf tubes with bacterial broths (1×10^7 CFU/mL for each 0.45 mL of NPs suspensions) and incubated for 3, 12 and 24 h at 37 °C. Sterile pipetting was used throughout the study. Susceptibility testing of *P. gingivalis* was conducted in an anaerobic jar as described above. At the end of each incubation period, the effect of the NPs on bacteria was evaluated by the ability of viable bacteria to cleave the tetrazolium salt (3-[4,5-dimethylthiazol-2-yl]-2,5-diphenyl tetrazolium bromide) (MTT) to a formazan dye (Sigma-Aldrich, Darmstadt, Germany). 96-well flat-bottom microtiter plates were used to place the suspensions in. The plates were incubated for 4 h at 37 °C, after the MTT labeling agent addition to each culture well. Then, the solubilizing agent that was provided by the manufacturer was added and an overnight incubation at room temperature was performed. An enzyme-linked immunosorbent assay (ELISA) reader (Spectrostar Nano, BMG Labtech, Cary, NC, USA) was employed, the purple formazan color that was produced from the MTT by viable cells, was read (560 nm) [46]. Assays were performed with three determinants, and experiments were performed in triplicate. Data expressed as mean ± standard deviation were analyzed by one-way analysis of variance (ANOVA) and the post hoc comparisons Scheffe's F tests, at $p < 0.05$, using SPSS Statistic 20.

Author Contributions: Conceptualization, M.T.-O., J.B., F.G., M.T. and R.O.; Methodology, M.T.-O., J.B., F.G., A.M., M.T. and R.O.; Software, M.T.-O., J.B.; Formal Analysis, M.T.-O., J.B., F.G., A.M., M.T. and R.O.; Investigation, M.T.-O., J.B., F.G., A.M., M.T. and R.O.; Resources, M.T.-O., J.B., A.M.; Data Curation, M.T.-O., J.B., F.G., A.M., M.T. and R.O.; Writing—Original Draft Preparation, M.T.-O., J.B., F.G., A.M., M.T. and R.O.; Writing—Review & Editing, M.T.-O., J.B., F.G., M.T. and R.O.; Supervision, J.B., F.G., M.T. and R.O; Project Administration, R.O. and M.T.; Funding Acquisition, M.T. and R.O.

Funding: This research was funded by the Spanish Ministry of Economy and Competitiveness (MINECO) and European Regional Development Fund (FEDER) grant number MAT2017-85999-P.

Acknowledgments: The research project MAT2017-85999-P was funded by the Spanish Ministry of Economy and Competitiveness (MINECO) and European Regional Development Fund (FEDER). No funds were received to cover publication costs.

Conflicts of Interest: The authors declare no conflicts of interest. The founding sponsors had no role in the design of the study; in the collection, analyses, or interpretation of data; in the writing of the manuscript, and in the decision to publish the results.

References

1. Chaves, P.; Oliveira, J.; Haas, A.; Beck, R.C.R. Applications of polymeric nanoparticles in oral diseases: A review of recent findings. *Curr. Pharm. Des.* **2018**. [CrossRef] [PubMed]
2. Spencer, P.; Ye, Q.; Park, J.; Topp, E.M.; Misra, A.; Marangos, O.; Wang, Y.; Bohaty, B.S.; Singh, V.; Sene, F.; et al. Adhesive/Dentin Interface: The Weak Link in the Composite Restoration. *Ann. Biomed. Eng.* **2010**, *38*, 1989–2003. [CrossRef] [PubMed]
3. Ferracane, J.L. Models of Caries Formation around Dental Composite Restorations. *J. Dent. Res.* **2017**, *96*, 364–371. [CrossRef] [PubMed]
4. Hirose, N.; Kitagawa, R.; Kitagawa, H.; Maezono, H.; Mine, A.; Hayashi, M.; Haapasalo, M.; Imazato, S. Development of a Cavity Disinfectant Containing Antibacterial Monomer MDPB. *J. Dent. Res.* **2016**, *95*, 1487–1493. [CrossRef] [PubMed]
5. Albuquerque, M.T.P.; Evans, J.D.; Gregory, R.L.; Valera, M.C.; Bottino, M.C. Antibacterial TAP-mimic Electrospun Polymer Scaffold—Effects on *P. gingivalis*-Infected Dentin Biofilm. *Clin. Oral Investig.* **2016**, *20*, 387–393. [CrossRef] [PubMed]
6. Kim, R.J.; Kim, M.O.; Lee, K.S.; Lee, D.Y.; Shin, J.H. An in vitro evaluation of the antibacterial properties of three mineral trioxide aggregate (MTA) against five oral bacteria. *Arch. Oral Biol.* **2015**, *60*, 1497–1502. [CrossRef] [PubMed]
7. Rôças, I.N.; Siqueira, J.F.; Santos, K.R.; Coelho, A.M. "Red complex" (*Bacteroides forsythus, Porphyromonas gingivalis*, and *Treponema denticola*) in endodontic infections: A molecular approach. *Oral Surg. Oral Med. Oral Pathol. Oral Radiol. Endod.* **2001**, *91*, 468–471. [CrossRef] [PubMed]
8. Wang, L.; Xie, X.; Weir, M.D.; Fouad, A.F.; Zhao, L.; Xu, H.H. Effect of bioactive dental adhesive on periodontal and endodontic pathogens. *J. Mater. Sci. Mater. Med.* **2016**, *27*, 168. [CrossRef] [PubMed]
9. Gu, L.; Huang, X.; Griffin, B.; Bergeron, B.R.; Pashley, D.H.; Niu, L.; Tay, F.R. Primum non nocere—The effects of sodium hypochlorite on dentin as used in endodontics. *Acta Biomater.* **2017**, *61*, 144–156. [CrossRef] [PubMed]
10. Letra, A.; Ghaneh, G.; Zhao, M. MMP-7 and TIMP-1, New Targets in Predicting Poor Wound Healing in Apical Periodontitis. *J. Endod.* **2013**, *39*, 1141–1146. [CrossRef] [PubMed]

11. Menezes-Silva, R.; Khaliq, S.; Deeley, K.; Letra, A.; Vieira, A.R. Genetic Susceptibility to Periapical Disease: Conditional Contribution of MMP2 and MMP3 Genes to the Development of Periapical Lesions and Healing Response. *J. Endod.* **2012**, *38*, 604–607. [CrossRef] [PubMed]

12. Paula-Silva, G.F.W.; Bezerra da Silva, L.A.; Kapila, L.Y. Matrix Metalloproteinase Expression in Teeth with Apical Periodontitis Is Differentially Modulated by the Modality of Root Canal Treatment. *J. Endod.* **2010**, *36*, 231–237. [CrossRef] [PubMed]

13. Ahmed, G.M.; El-Baz, A.A.; Hashem, A.A.; Shalaan, A.K. Expression levels of matrix metalloproteinase-9 and gram-negative bacteria in symptomatic and asymptomatic periapical lesions. *J. Endod.* **2013**, *39*, 444–448. [CrossRef] [PubMed]

14. Accorsi-Mendonca, T.; Silva, E.J.; Marcaccini, A.M.; Gerlach, R.F.; Duarte, K.M.; Pardo, A.P.; Line, S.R.; Zaia, A.A. Evaluation of gelatinases, tissue inhibitor of matrix metalloproteinase-2, and myeloperoxidase protein in healthy and inflamed human dental pulp tissue. *J. Endod.* **2013**, *39*, 879–882. [CrossRef] [PubMed]

15. Osorio, R.; Yamauti, M.; Osorio, E.; Ruiz-Requena, M.E.; Pashley, D.; Tay, F.R.; Toledano, M. Effect of dentin etching and chlorhexidine application on metalloproteinase-mediated collagen degradation. *Eur. J. Oral Sci.* **2011**, *119*, 79–85. [CrossRef] [PubMed]

16. Liu, Y.; Mai, S.; Li, N. Differences between top-down and bottom-up approaches in mineralizing thick, partially-demineralized collagen scaffolds. *Acta Biomater.* **2011**, *7*, 1742–1751. [CrossRef] [PubMed]

17. Toledano, M.; Osorio, R.; Osorio, E.; Medina-Castillo, A.L.; Toledano-Osorio, M.; Aguilera, F.S. Ions-modified nanoparticles affect functional remineralization and energy dissipation through the resin-dentin interface. *J. Mech. Behav. Biomed. Mater.* **2017**, *68*, 62–79. [CrossRef] [PubMed]

18. Zhu, J.; Liang, R.; Sun, C.; Xie, L.; Wang, J.; Leng, D.; Wu, D.; Liu, W. Effects of nanosilver and nanozinc incorporated mesoporous calcium-silicate nanoparticles on the mechanical properties of dentin. *PLoS ONE* **2017**, *12*, e0182583. [CrossRef] [PubMed]

19. Zhang, K.; Cheng, L.; Imazato, S.; Antonucci, J.M.; Lin, N.J.; Lin-Gibson, S.; Bai, Y.; Xu, H.H.K. Effects of dual antibacterial agents MDPB and nano-silver in primer on microcosm biofilm, cytotoxicity and dentin bond properties. *J. Dent.* **2013**, *41*, 464–474. [CrossRef] [PubMed]

20. Hashimoto, M.; Sasaki, J.I.; Yamaguchi, S.; Kawai, K.; Kawakami, H.; Iwasaki, Y.; Imazato, S. Gold Nanoparticles Inhibit Matrix Metalloproteases without Cytotoxicity. *J. Dent. Res.* **2015**, *94*, 1085–1091. [CrossRef] [PubMed]

21. Osorio, R.; Cabello, I.; Medina-Castillo, A.L.; Osorio, E.; Toledano, M. Zinc-modified nanopolymers improve the quality of resin-dentin bonded interfaces. *Clin. Oral Investig.* **2016**, *20*, 2411–2420. [CrossRef] [PubMed]

22. Osorio, R.; Osorio, E.; Medina-Castillo, A.L.; Toledano, M. Polymer nanocarriers for dentin adhesion. *J. Dent. Res.* **2014**, *93*, 1258–1263. [CrossRef] [PubMed]

23. Osorio, R.; Alfonso-Rodríguez, C.A.; Medina-Castillo, A.L.; Alaminos, M.; Toledano, M. Bioactive Polymeric Nanoparticles for Periodontal Therapy. *PLoS ONE* **2016**, *7*, e0166217. [CrossRef] [PubMed]

24. Munchow, E.A.; Albuquerque, M.T.; Zero, B.; Kamocki, K.; Piva, E.; Gregory, R.L.; Bottino, M.C. Development and characterization of novel ZnO-loaded electrospun membranes for periodontal regeneration. *Dent. Mater.* **2015**, *31*, 1038–1051. [CrossRef] [PubMed]

25. Munchow, E.A.; Pankajakshan, D.; Albuquerque, M.T.; Kamocki, K.; Piva, E.; Gregory, R.L.; Bottino, M.C. Synthesis and characterization of CaO-loaded electrospun matrices for bone tissue engineering. *Clin. Oral Investig.* **2016**, *27*, 1921–1933. [CrossRef] [PubMed]

26. Vargas-Reus, M.A.; Memarzadeh, K.; Huang, J.; Ren, G.G.; Allaker, R.P. Antimicrobial activity of nanoparticulate metal oxides against peri-implantitis pathogens. *Int. J. Antimicrob. Agents* **2012**, *40*, 135–139. [CrossRef] [PubMed]

27. Tamura, M.; Ochiai, K. Zinc and copper play a role in coaggregation inhibiting action of *Porphyromonas gingivalis*. *Mol. Oral Microbiol.* **2009**, *24*, 56–63. [CrossRef]

28. Kim, T.S.; Bürklin, T.; Schacher, B.; Ratka-Krüger, P.; Schaecken, M.T.; Renggli, H.H.; Fiehn, W.; Eickholz, P. Pharmacokinetic profile of a locally administered doxycycline gel in crevicular fluid, blood, and saliva. *J. Periodontol.* **2002**, *73*, 1285–1291. [CrossRef] [PubMed]

29. Osorio, R.; Yamauti, M.; Osorio, E.; Ruiz-Requena, M.E.; Pashley, D.H.; Tay, F.R.; Toledano, M. Zinc reduces collagen degradation in demineralized human dentin explants. *J. Dent.* **2011**, *39*, 148–153. [CrossRef] [PubMed]

30. Fan, W.; Sun, Q.; Li, Y.; Tay, F.R.; Fan, B. Synergistic mechanism of Ag^+–Zn^{2+} in anti-bacterial activity against *Enterococcus faecalis* and its application against dentin infection. *J. Nanobiotechnol.* **2018**, *16*, 10. [CrossRef] [PubMed]

31. Cowan, M.M.; Van der Mei, H.C.; Stokroos, I.; Busscher, H.J. Heterogeneity of surfaces of subgingival bacteria as detected by zeta potential measurements. *J. Dent. Res.* **1992**, *71*, 1803–1806. [CrossRef] [PubMed]

32. Wunder, D.; Bowen, W.H. Action of agents on glucosyltransferases from Streptococcus mutans in solution and adsorbed to experimental pellicle. *Arch. Oral Biol.* **1999**, *44*, 203–214. [CrossRef]

33. Devulapalle, K.S.; Mooser, G. Subsite specificity of the active site of glucosyltransferases from *Streptococcus sobrinus*. *J. Biol. Chem.* **1994**, *269*, 11967–11971. [PubMed]

34. Larsen, T. Susceptibility of *Porphyromonas gingivalis* in biofilms to amoxicillin, doxycycline and metronidazole. *Oral Microbiol. Immunol.* **2002**, *17*, 267–271. [CrossRef] [PubMed]

35. Slots, J.; Rams, T.E. Antibiotics in periodontal therapy: Advantages and disadvantages. *J. Clin. Periodontol.* **1990**, *17*, 479–493. [CrossRef] [PubMed]

36. Larsen, T. In vitro release of doxycycline from bioabsorbable materials and acrylic strips. *J. Periodontol.* **1990**, *61*, 30–34. [CrossRef] [PubMed]

37. Gamal, A.Y.; Kumper, R.M.A. Novel Approach to the Use of Doxycycline-Loaded Biodegradable Membrane and EDTA Root Surface Etching in Chronic Periodontitis: A Randomized Clinical Trial. *J. Periodontol.* **2012**, *83*, 1086–1094. [CrossRef] [PubMed]

38. Kim, S.E.; Hwang, S.Y.; Jeong, M. Clinical and microbiological effects of a subantimicrobial dose of oral doxycycline on periodontitis in dogs. *Vet. J.* **2016**, *208*, 55–59. [CrossRef] [PubMed]

39. Toledano, M.; Nieto-Aguilar, R.; Osorio, R.; Campos, A.; Osorio, E.; Tay, F.R.; Alaminos, M. Differential expression of matrix metalloproteinase-2 in human coronal and radicular sound and carious dentine. *J. Dent.* **2010**, *38*, 635–640. [CrossRef] [PubMed]

40. Tonetti, M.S.; Lang, N.P.; Cortellini, P.; Suvan, J.E.; Eickholz, P.; Fourmousis, I.; Topoll, H.; Vangsted, T.; Wallkamm, B. Effects of a single topical doxycycline administration adjunctive to mechanical debridement in patients with persistent/recurrent periodontitis but acceptable oral hygiene during supportive periodontal therapy. *J. Clin. Periodontol.* **2012**, *39*, 475–482. [CrossRef] [PubMed]

41. Deasy, P.B.; Collins, A.E.; MacCarthy, D.J.; Russell, R.J. Use of strips containing tetracycline hydrochloride or metronidazole for the treatment of advanced periodontal disease. *J. Pharm. Pharmacol.* **1989**, *41*, 694–699. [CrossRef] [PubMed]

42. Gross, E.L.; Leys, E.J.; Gasparovich, S.R.; Firestone, N.D.; Schwartzbaum, J.A.; Janies, D.A.; Asnani, K.; Griffen, A.L. Bacterial 16S sequence analysis of severe caries in young permanent teeth. *J. Clin. Microbiol.* **2010**, *48*, 4121–4128. [CrossRef] [PubMed]

43. Yassen, G.H.; Vail, M.M.; Chu, T.G.; Platt, J.A. The effect of medicaments used in endodontic regeneration on root fracture and microhardness of radicular dentine. *Int. Endod. J.* **2013**, *46*, 688–695. [CrossRef] [PubMed]

44. Garrett, T.R.; Bhakoo, M.; Zhang, Z.B. Bacterial adhesion and biofilms on surfaces. *Prog. Nat. Sci.* **2008**, *18*, 1049–1056. [CrossRef]

45. Palasuk, J.; Windsor, L.J.; Platt, J.A.; Lvov, Y.; Geraldeli, S.; Bottino, M.C. Doxycycline-loaded nanotube-modified adhesives inhibit MMP in a dose-dependent fashion. *Clin. Oral Investig.* **2018**, *22*, 1243–1252. [CrossRef] [PubMed]

46. Banzi, E.C.; Costa, A.R.; Puppin-Rontani, R.M.; Babu, J.P.; García-Godoy, F. Inhibitory effects of a cured antibacterial bonding system on viability and metabolic activity of oral bacteria. *Dent. Mater.* **2014**, *30*, e238–e244. [CrossRef] [PubMed]

![materials logo] *materials*

MDPI

Article

Dental Resin Cements—The Influence of Water Sorption on Contraction Stress Changes and Hydroscopic Expansion

Grzegorz Sokolowski [1], Agata Szczesio [2], Kinga Bociong [2], Karolina Kaluzinska [2],
Barbara Lapinska [3], Jerzy Sokolowski [3], Monika Domarecka [3] and
Monika Lukomska-Szymanska [3,*]

[1] Department of Prosthetic Dentistry, Medical University of Lodz, 251 Pomorska St., 92-213 Lodz, Poland;
    grzegorz.sokolowski@umed.lodz.pl
[2] University Laboratory of Materials Research, Medical University of Lodz, 251 Pomorska St., 92-213 Lodz,
    Poland; agata.szczesio@umed.lodz.pl (A.S.); kinga.bociong@umed.lodz.pl (K.B.);
    karolina.kaluzinska@umed.lodz.pl (K.K.)
[3] Department of General Dentistry, Medical University of Lodz, 251 Pomorska St., 92-213 Lodz, Poland;
    barbara.lapinska@umed.lodz.pl (B.L.); jerzy.sokolowski@umed.lodz.pl (J.S.);
    monika.domarecka@umed.lodz.pl (M.D.)
*   Correspondence: monika.lukomska-szymanska@umed.lodz.pl; Tel.: +48-42-675-74-64

Received: 14 May 2018; Accepted: 6 June 2018; Published: 8 June 2018

Abstract: Resin matrix dental materials undergo contraction and expansion changes due to polymerization and water absorption. Both phenomena deform resin-dentin bonding and influence the stress state in restored tooth structure in two opposite directions. The study tested three composite resin cements (Cement-It, NX3, Variolink Esthetic DC), three adhesive resin cements (Estecem, Multilink Automix, Panavia 2.0), and seven self-adhesive resin cements (Breeze, Calibra Universal, MaxCem Elite Chroma, Panavia SA Cement Plus, RelyX U200, SmartCem 2, and SpeedCEM Plus). The stress generated at the restoration-tooth interface during water immersion was evaluated. The shrinkage stress was measured immediately after curing and after 0.5 h, 24 h, 72 h, 96 h, 168 h, 240 h, 336 h, 504 h, 672 h, and 1344 h by means of photoelastic study. Water sorption and solubility were also studied. All tested materials during polymerization generated shrinkage stress ranging from 4.8 MPa up to 15.1 MPa. The decrease in shrinkage strain (not less than 57%) was observed after water storage (56 days). Self-adhesive cements, i.e., MaxCem Elite Chroma, SpeedCem Plus, Panavia SA Plus, and Breeze exhibited high values of water expansion stress (from 0 up to almost 7 MPa). Among other tested materials only composite resin cement Cement It and adhesive resin cement Panavia 2.0 showed water expansion stress (1.6 and 4.8, respectively). The changes in stress value (decrease in contraction stress or built up of hydroscopic expansion) in time were material-dependent.

Keywords: resin cements; shrinkage stress; water sorption; hydroscopic expansion; photoelastic investigation

---

1. Introduction

Resin composite cements have been widely used with ceramic, resin, or metal alloy-based prosthodontic restorations [1]. The cementation technique used in adhesive dentistry is one of the major factors, which exerts influence on the clinical success of indirect restorative procedures. Cement is used to bond tooth and restoration simultaneously creating a barrier against microbial leakage [2]. The universal cement that can be applied in all indirect restorative procedures has not been introduced into the market yet. Therefore, clinicians should understand the influence of applied material properties and preparation design on the clinical performance of the restoration [3].

The composition of resin composite cements is almost the same as of resin composites [4]. Resin cements mainly consist of various methacrylate resins and inorganic fillers which are often coated with organic silanes to provide adhesion between the filler and the matrix. These materials often include bonding agents to promote the adhesion between resin cement and tooth structure. Main monomers are, i.e., hydroxyethyl methacrylate (HEMA), 4-methacryloyloxyethy trimellitate anhydride (4-MET), carboxylic acid, and organophospate 10-methacryloxydecyl dihydrogen phosphate (10-MDP) (Figure 1). The acidic group bonds calcium ions in the tooth structure [5,6].

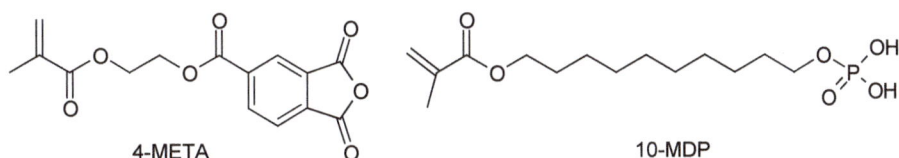

4-META   **10-MDP**

Figure 1. Monomers used as bonding agents in resin cements.

Resin composite cements are used in combination with adhesive systems. This procedure aims at creating micro-mechanical retention to both enamel and dentin. The material may also form a strong adhesion to an adequately-treated surface of the composite, ceramic, and metallic restorations [7]. Taking into account the surface preparation before the cementation process, resin cements can be divided into: (1) composite resin cement (used with total-etch adhesive systems); (2) adhesive resin cement (used with separate self-etching adhesive systems); and (3) self-adhesive resin cement (containing a self-adhesive system) [1].

The application of resin matrix-based cements is time-consuming and susceptible to manipulation errors [8]. The self-adhesive resin cements are proposed to simplify the restoration procedure. These materials bond dentin in one step without any surface conditioning or pre-treatment (priming) [9,10].

All currently available resin-based materials exhibit polymerization shrinkage. Moreover, resin cements are generally applied as a thin layer, particularly when used to lute posts, inlays, and crowns. In the aforementioned clinical cases, the cavity design has a high C-factor (high number of bonded surfaces and a low number of un-bonded surfaces) [11]. Additionally, low-viscosity composites exhibit a relatively high shrinkage amounting up to 6% (comparable to resin cements) [12,13]. These factors may generate sufficient stress resulting in debonding of the luting material, thereby increasing microleakage [14]. Nevertheless, there is little data on the stress generated by these materials. The sorption characteristic of resin-based dental cements has been extensively evaluated [15–17]. However, the analysis of the influence of water sorption on the change in contraction stress is inadequate. The purpose of this study was to evaluate the development of the stress state, i.e., the contraction stress generated during photopolymerization and hydroscopic expansion within different types of resin cements which undergo water ageing by means of photoelastic analysis.

2. Materials and Methods

The composition of investigated resin cements and bonding systems is presented in Tables 1 and 2.

Table 1. The composition of resin cements.

Material	Type	Composition	Curing Time (s)	Manufacturer
Cement-It	Composite resin cement	bis-GMA, UDMA, HDDMA, PEGDMA, barium-boro-silicate glass (65 wt %)	20	Jeneric Pentron (Wallingford, CT, USA)
NX3	Composite resin cement	TEGDMA, bis-GMA, fluoro-aluminosilicate glass (67.5 wt %/47 vol %), activators, stabilizers, radiopaque agent	20	Kerr (Orange, CA, USA)
Variolink Esthetic DC	Composite resin cement	UMDA and further methacrylate monomers, ytterbium trifluoride, spheroid mixed oxide (67 wt %/38 vol %), initiators, stabilizers and pigments	10	Ivoclar Vivadent (Ellwangen, Germany)
Estecem	Adhesive resin cement	bis-GMA, TEGDMA, bis-MPEPP; silica-zirconia filler (74 wt %), camphorquinone	20	Tokuyama Dental (Taitou, Japan)
Multilink Automix	Adhesive resin cement	dimethacrylate and HEMA, barium glass and silica filler, ytterbiumtrifluoride (68 wt %), catalysts, stabilizers, pigments	10	Ivoclar Vivadent (Ellwangen, Germany)
Panavia 2.0	Adhesive resin cement	10-MDP, BPEDMA, hydrophobic aliphatic metahrylates, hydrophilic aliphatic metahrylate, silanated silica filler, silanated barium glass filler, sodium fluoride (70.8 wt %)	20	Kuraray (Osaka, Japan)
Breeze	Self-adhesive resin cement	bis-GMA, UDMA, TEGDMA, HEMA, 4-MET, silane treated barium glass, silica, BiOCl, Ca-Al-F-silicate, curing system	20	Jeneric Pentron (Wallingford, CT, USA)
Calibra Universal	Self-adhesive resin cement	UDMA, trimethylolpropane trimethacrylate TMPTMA, bis-EMA—Bisphenol A ethoxylate dimethacrylate, TEGDMA, HEMA, 3-(acryloyloxy)-2-hydroxypropyl methacrylate, urethane modified bis-GMA, PENTA, silanated barium glass, fumed silica (48 vol %)	10	Dentsply Sirona (York, PA, USA)
MaxCem Elite Chroma	Self-adhesive resin cement	HEMA, GDM, UDMA, 1,1,3,3-tetramethylbutyl hydroperoxide TEGDMA, fluoroaluminosilicate glass, GPDM, barium glass filler, fumed silica (69 wt %)	10	Kerr (Orange, CA, USA)
Panavia SA Cement Plus	Self-adhesive resin cement	bis-GMA, TEGDMA, HEMA, 10-MDP, hydrophobic aromatic dimethacrylate, hydrophobic aliphatic dimethacrylate, sodium fluoride, silanated barium glass filler, silanated colloidal silica (70 wt %/40 vol %)	10	Kuraray (Osaka, Japan)
RelyX U200	Self-adhesive resin cement	methacrylate monomers containing phosphoric acid groups, methacrylate monomers, silanated fillers (70 wt %/43 vol %), initiator components, stabilizers, rheological additives, alkaline(basic) initiator components, stabilizers, pigments	20	3M ESPE (St. Paul, MN, USA)
SmartCem 2	Self-adhesive resin cement	UDMA, urethane modified bis-GMA, TEGDMA, PENTA, dimethacrylate resins, barium boron fluoroaluminosilicate glass amorphous silica (69 wt %/46 vol %)	10	Dentsply Sirona (York, PA, USA)
SpeedCEM Plus	Self-adhesive resin cement	UDMA, TEGDMA, PEGDMA, methacrylated phosphoric acid ester, 1,10-decandiol dimethacrylate, copolymers, dibenzoyl peroxide, ytterbium trifluoride, barium glass, silicon dioxide (75 wt %/45 vol %)	20	Ivoclar Vivadent (Ellwangen, Germany)

bis-GMA—bisphenol A glycol dimethacrylate, UDMA—urethane dimethacrylate, TEGDMA—triethylene glycol dimethacrylate, GDM—glycerol 1,3-dimethacrylate, GPDM—glycerol phosphate dimethacrylate, bis-MPEPP—bisphenol A polyethoxy methacrylate, HEMA—hydroxyethyl methacrylate, PEGDMA—polyethylene glycol dimethacrylate, NPGDMA—neopentyldimethacrylate, 10-MDP—10-methacryloxydecyl dihydrogen phosphate, BPEDMA—bisphenol-A-polyethoxy dimethacrylate, PENTA—dipentaerythritol penttacrylate monophosphate, HDDMA—1,6-hexanediol dimethacrylate, 4-MET—4-methacryloyloxyethy trimellitate anhydride, MAC-10—11-methacryloyloxy-1,1-undecanedicarboxylic acid, TMPTMA—trimethylolpropane trimethacrylate.

Table 2. The curing time of bonding systems.

Bonding system	Manufacturer	Curing Time (s)	Bonding System Dedicated to
Bond-1 C&B Primer/Adhesive	Jeneric Pentrton (Wallingford, CT, USA)	10	Cement It, Breeze
Clearfil SE bond	Kuraray (Osaka, Japan)	10	Panavia 2.0, Panavia SA Cement Plus
Easy Bond	3M ESPE (St. Paul, MN, USA)	10	RelyX U200
Estelink	Tokuyam Dental (Taitou, Japan)	10	Estecem
Monobond Plus	Ivoclar Vivadent (Ellwangen, Germany)	10	Variolink Esthetic DC, Multilink Automix, SpeedCEM Plus
OptiBond XRT	Kerr (Orange, CA, USA)	10	NX3, MaxCem Elite Chroma
Prime&Bond Elect Universal	Dentsply Sirona (York, PA, USA)	10	SmartCem 2, Calibra Universal

2.1. Absorbency Dynamic Study

Absorbency dynamic was determined by means of procedure as described by Bociong et al. [18]. The samples were prepared according to ISO 4049 [19]. Curing time was consistent with the manufacturer's instructions (Table 2).

In order to characterize absorbency dynamic, the cylindrical samples with dimensions of 15 mm in diameter and of 1 mm in width were prepared. The tested materials were applied in one layer and cured with LED light lamp (Mini L.E.D., Acteon, Mérignac Cedex, France) in nine zones partially overlapping each other with direct contact of optical fiber with the material surface. Exposure time was applied according to the manufacturer's instructions (Table 1).

Five samples were prepared for each tested cement. The samples' weight was determined (RADWAG AS 160/C/2, Poland) immediately after preparation and then for 30 consecutive days, and after 1344 h (56 days) and 2016 h (84 days). The absorbency was calculated according to the Equation (1) [20]:

$$A = \frac{m_i - m_0}{m_0} \cdot 100\% \tag{1}$$

where A is the absorbency of water, m_0 is the mass of the sample in dry condition, and m_i is the mass of the sample after storage in water for a specified (i) period of time.

2.2. Water Sorption and Solubility

Water sorption and solubility were investigated according to ISO 4049 [19]. The detailed procedure of tests has been described extensively in the previously published literature [18,21]. Curing time was consistent with the manufacturer's instructions (Table 1).

Water sorption and solubility were calculated according to following equations:

$$W_{sp} = \frac{m_2 - m_3}{V} \tag{2}$$

$$W_{sl} = \frac{m_1 - m_3}{V} \tag{3}$$

where: W_{sp} is the water sorption, W_{sl} is the water solubility, m_1 is the initial constant mass (µg), m_2 is the mass after seven days of water immersion (µg), m_3 is the final constant mass (µg), V is the specimen volume (mm³).

2.3. Photoelastic Study

Photoelastic analysis allows for quantitative measurement and visualization of stress concentration that develops during photopolymerization or water sorption of resin composites [22,23]. The modified method enables analysis of the relationship between water sorption and the change of stress state (contraction or expansion) of resin materials. This test was described extensively in our previous articles [18,24]. Photoelastically-sensitive plates of epoxy resin (Epidian 53, Organika-Sarzyna SA, Nowa Sarzyna, Poland) were used in this study. Calibrated orifices of 3 mm in diameter and of 4 mm in depth were prepared in resin plates in order to mimic an average tooth cavity. The generated strains in the plates were visualized in circular transmission polariscope FL200 (Gunt, Barsbüttel, Germany) and photoelastic strain calculations were based on the Timoshenko equation [25].

3. Results

3.1. Absorbency Dynamic Study

Water absorbency and contraction stress mean values were presented in Figures 2–14. The water immersion resulted in an increase in weight of all tested materials. The water sorption (wt%) increased for Breeze up to three times. The lowest value of absorbency after 2016 h (84 days) was observed for Variolink Esthetic DC.

3.2. Water Sorption and Solubility

Mean values of water sorption and solubility were presented in Table 3. Maxcem Elite Chroma and Breeze exhibited the highest, while Estecem showed the lowest values of water sorption.

Table 3. Stress state before and after 2016 h (84 days) of water immersion, contraction stress drop, absorbency, and solubility of tested materials.

Material	Stress State (MPa)		Contraction Stress Drop (%)	Sorption ($\mu g/mm^3$)	Solubility ($\mu g/mm^3$)
	0.5 h	2016 h			
Cement It	10.9 ± 2.2	-1.6 ± 0.4	115 *	27.8 ± 0.8	1.9 ± 0.4
NX3	6.3 ± 0.1	1.6 ± 0.1	79	23.8 ± 0.6	3.7 ± 1.2
Variolink Esthetic	10.9 ± 0.4	4.7 ± 0.1	57	22.4 ± 0.8	10.0 ± 2.0
Estecem	6.8 ± 0.9	1.6 ± 0.2	76	12.5 ± 2.2	4.6 ± 1.9
Multilink Automix	12.5 ± 0.4	2.1 ± 0.9	83	25.3 ± 1.5	2.2 ± 0.8
Panavia 2.0	5.3 ± 1.8	-4.8 ± 0.4	191 *	33.9 ± 1.7	11.1 ± 1.0
Breeze	7.8 ± 1.6	-6.3 ± 1.6	180 *	47.7 ± 3.1	3.1 ± 0.5
Calibra Universal	11.1 ± 0.7	0.0 ± 0.8	100	30.9 ± 1.5	5.0 ± 2.6
MaxCem Elite Chroma	10.4 ± 0.9	-6.3 ± 0.3	160 *	50.4 ± 1.3	8.5 ± 1.3
Panavia SA Plus	4.8 ± 0.4	-1.6 ± 0.2	133 *	26.4 ± 1.3	1.7 ± 0.4
RelyX U200	13.5 ± 0.8	2.6 ± 0.9	81	29.6 ± 1.3	0.4 ± 0.2
SmartCem 2	15.1 ± 0.9	1.6 ± 0.9	89	33.0 ± 0.9	4.9 ± 1.2
SpeedCEM Plus	11.9 ± 1.1	-1.6 ± 0.4	113 *	28.2 ± 0.5	2.5 ± 0.4

* represents materials with over-compensated polymerization stress due to water expansion.

3.3. Photoelastic Study

All materials exhibited shrinkage and the associated contraction stress during hardening process. The significant reduction in contraction stress was observed due to hygroscopic expansion of cements (Figures 2–16). Water ageing of six cements resulted in additional stress characterized by the opposite direction of forces. The investigated materials exhibited various contraction stress values.

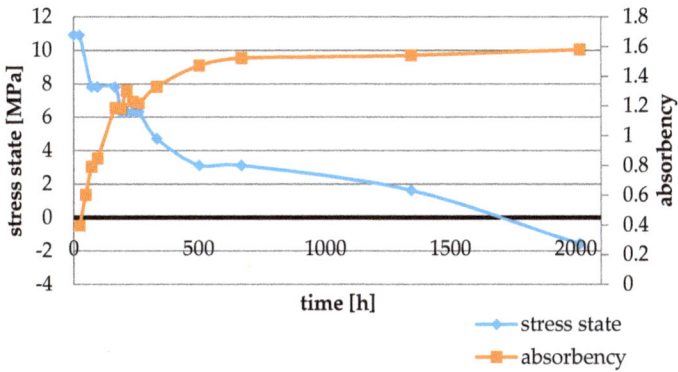

Figure 2. The influence of water sorption (2016 h water ageing) on the absorbency and contraction stress generated during the photopolymerization of Cement It.

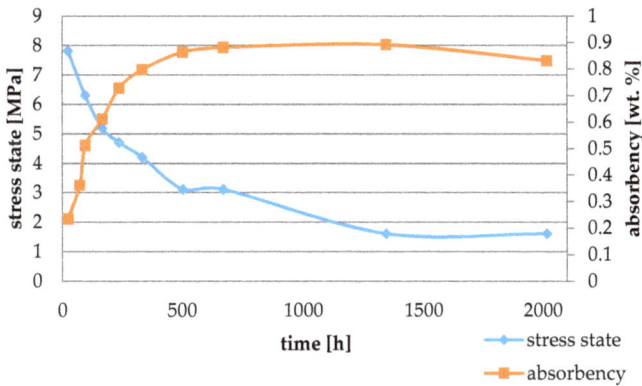

Figure 3. The influence of water sorption (2016 h water ageing) on the absorbency and contraction stress generated during the photopolymerization of NX3.

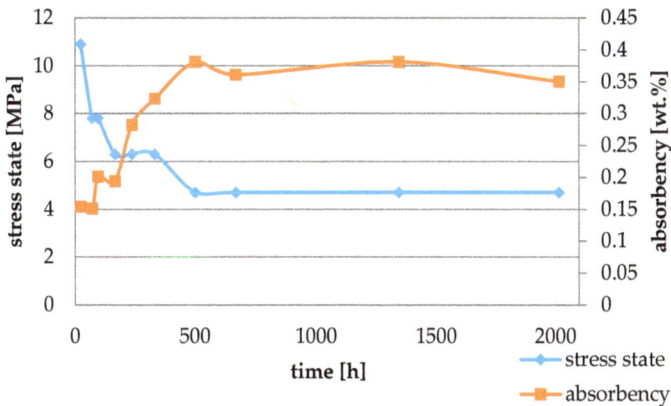

Figure 4. The influence of water sorption (2016 h water ageing) on the absorbency and contraction stress generated during the photopolymerization of Variolink Esthetic DC.

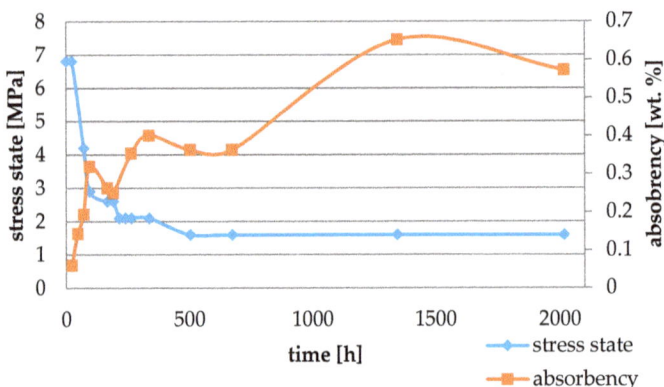

Figure 5. The influence of water sorption (2016 h water ageing) on the absorbency and contraction stress generated during the photopolymerization of Estecem.

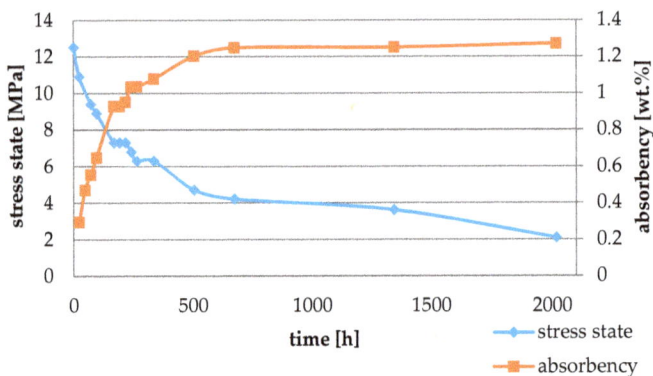

Figure 6. The influence of water sorption (2016 h water ageing) on the absorbency and contraction stress generated during the photopolymerization of Multilink Automix.

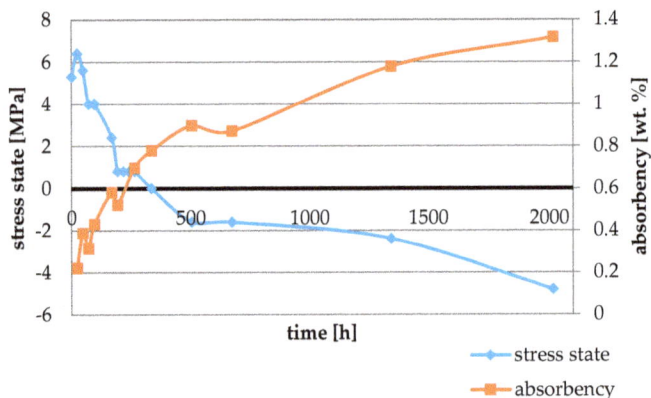

Figure 7. The influence of water sorption (2016 h water ageing) on the absorbency and contraction stress generated during the photopolymerization of Panavia 2.0.

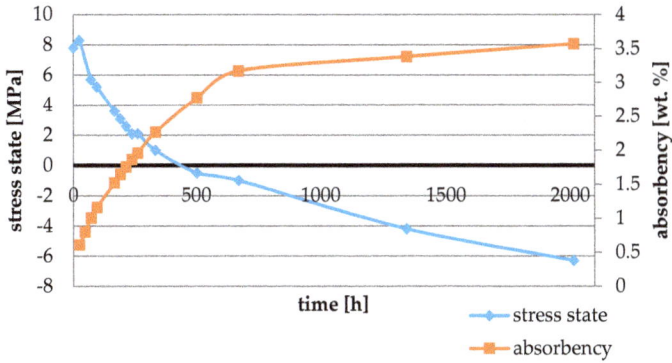

Figure 8. The influence of water sorption (2016 h water ageing) on the absorbency and contraction stress generated during the photopolymerization of Breeze.

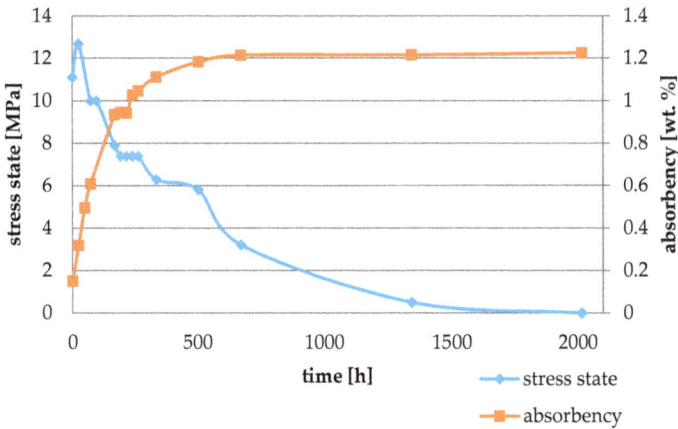

Figure 9. The influence of water sorption (2016 h water ageing) on the absorbency and contraction stress generated during the photopolymerization of Calibra Universal.

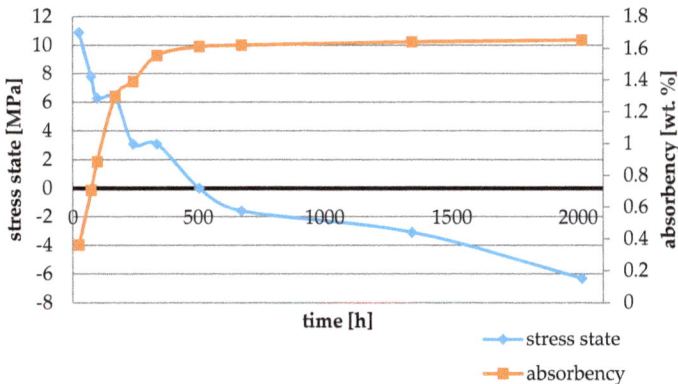

Figure 10. The influence of water sorption (2016 h water ageing) on the absorbency and contraction stress generated during the photopolymerization of Maxcem Elite Chroma.

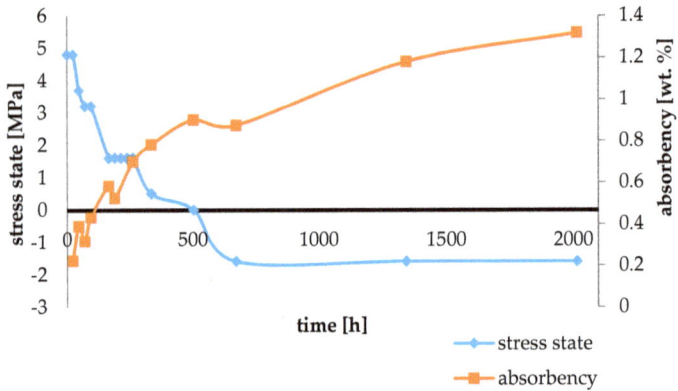

Figure 11. The influence of water sorption (2016 h water ageing) on the absorbency and contraction stress generated during the photopolymerization of Panavia SA Plus.

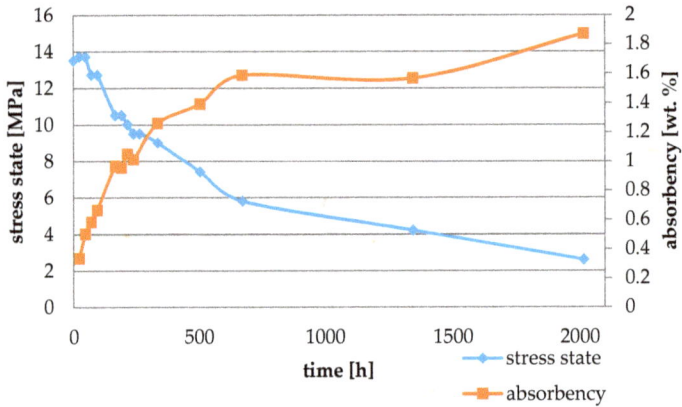

Figure 12. The influence of water sorption (2016 h water ageing) on the absorbency and contraction stress generated during the photopolymerization of Rely U200.

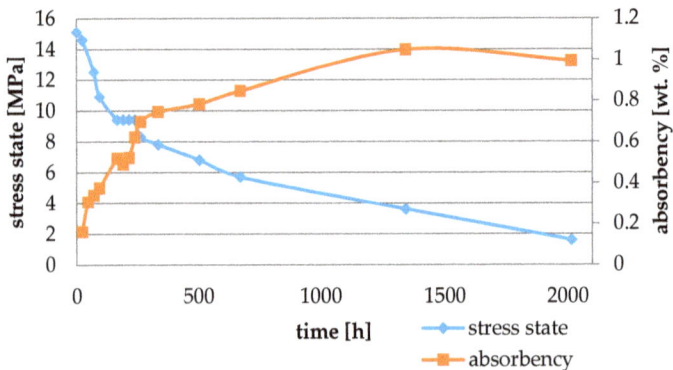

Figure 13. The influence of water sorption (2016 h water ageing) on the absorbency and contraction stress generated during the photopolymerization of SmartCem 2.

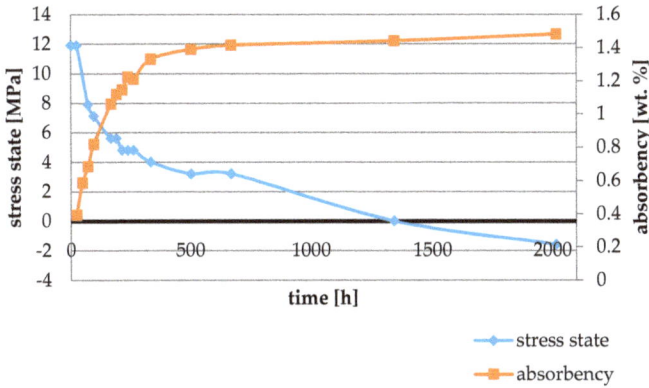

Figure 14. The influence of water sorption (2016 h water ageing) on the absorbency and contraction stress generated during the photopolymerization of SpeedCEM Plus.

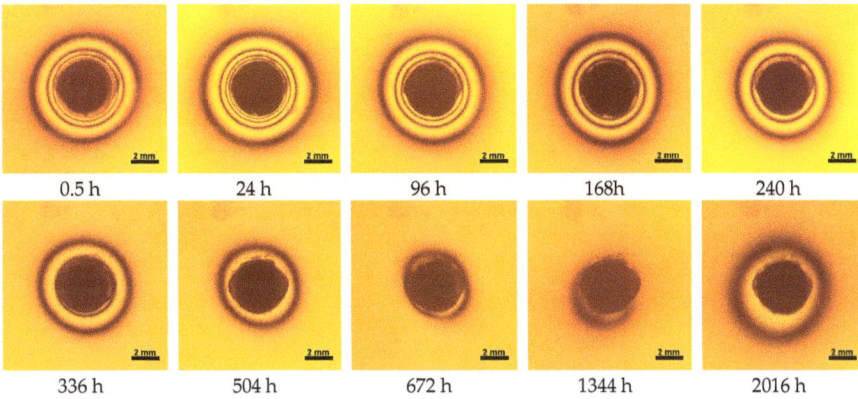

Figure 15. Isochromes in an epoxy plate around Maxcem Elite Chroma restoration before and after water storage; 0.5–2016 h.

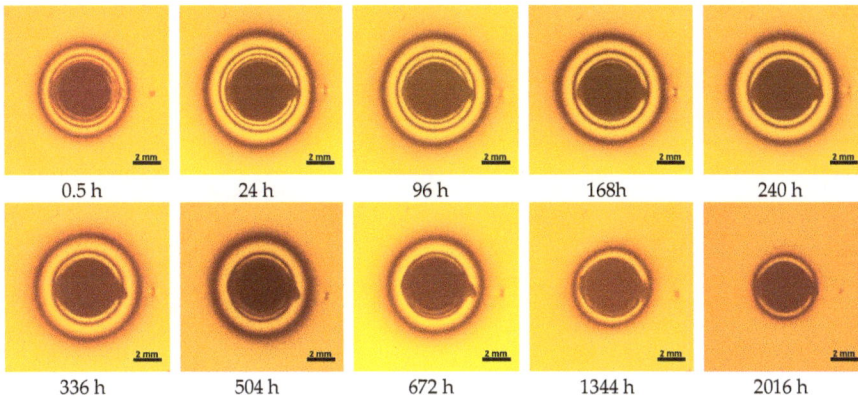

Figure 16. Isochromes in an epoxy plate around NX3 restoration before and after water storage; 0.5–2016 h.

The lowest contraction stress from tested materials exhibited Panavia SA Plus. The contraction stress decreased from 4.8 up to 0.0 MPa after 504 h (21 days) of water conditioning (Figure 11). Further water ageing resulted in additional stress: after 2016 h (84 days) stress level increased up to −1.6 MPa (Figure 11).

The highest contraction stress was observed for SmartCem 2 amounting up to 15.1 MPa. The contraction stress of SmartCem 2 after 2016 h (84 days) of water storage reduced up to 1.6 MPa (Figure 13).

4. Discussion

High configuration factor (C-factor) and the low viscosity of resin cements may generate relatively high contraction stress. This stress may cause debonding of the luting material, thereby increasing microleakage [14]. Our previous study (using an epoxy resin plate) [18,24], demonstrated that the photoelastic method can be used to evaluate the effect of water sorption on stress reduction at the tooth-restoration interface. This method shows that the contraction stress of dental resins may be partially relieved by the water uptake [18]. However, the in-depth analysis of the shrinkage stress values in various resin cement materials after water ageing is also highly demanded.

The overall results showed the development of the initial stress in the compressive direction during photopolymerization. The composition of resin cements affected the sorption and solubility processes which, in turn, exerted influence on the hygroscopic expansion and plasticization. Thus, the compensatory effect was composition-dependent [26]. This study confirmed the lower water sorption for composite resin cements as compared with self-adhesive resin cements that underwent water ageing. The presence of hydroxyl, carboxyl, and phosphate groups in monomers made them more hydrophilic and, supposedly, more prone to water sorption [27].

Variolink Esthetic DC and NX3 do not contain adhesive monomers. In the present study they exhibited sorption similar to composite materials [18]. Variolink Esthetic DC showed a high solubility value and the lowest decrease in contraction stress (of about 70%). The water immersion of resin materials might result in dissolving and leaching of some components (unreacted monomers or fillers) out of the material [28]. Variolink Esthetic DC contains a modified polymer matrix and nanofiller. The lowest contraction stress might result from a small hydrolysis and plasticization effect of the modified resin matrix.

Cement It is a composite resin cement which, in comparison with NX3 and Variolink Esthetic DC, showed higher values of water sorption and its total value of stress changes was 12.5 MPa. This could be explained by the composition of the polymer matrix containing bis-GMA and HDDMA. These monomers showed comparable characteristics: high water sorption values [29,30], similar polymer networks, and susceptibility to hydrolysis [31].

The four tested materials did not meet the requirements of ISO 4049, as they showed sorption values above 40 µg/mm^3 (Breeze and MaxCem Elite Chroma) or the solubility value above 7.5 µg/mm^3 (Variolink Esthetic DC and Panavia 2.0). The differences in water absorption of the polymer network depending on monomer type were reported. The highest water sorption was observed for MaxCem Elite Chroma (Figure 15). This material consists of HEMA and GDM that have one of the highest hydrophilicities among dental resins. HEMA was shown to induce water sorption, leading to the expansion of the polymer matrix [32]. Resin-modified glass-ionomer cements (RMGICs) absorbed more water due to hydroxyethyl methacrylate content, present in the hydrogel form in the polymerized matrices [33]. HEMA might be present either as a separate component or a grafted component into the structure of the polyacrylic acid backbone. The polymerized matrices of these materials were very hydrophilic and might include an interpenetrating network of poly(HEMA), copolymers of grafted HEMA, and polyacid salts that were more prone to water uptake [34]. Park et al. [35] showed that GDM exhibited the highest water sorption in comparison to bis-GMA, HEMA, EGDM, DEGDM, TEGDMA, GDM, and GTM. In the present study, the decrease in contraction stress after 2016 h (84 days) of water immersion varied significantly between tested materials. Figures 2–16 showed that the expansion dynamics also differed substantially. All studied materials exhibited contraction stress relief during water immersion. The value of stress decreased up to 0 MPa in different times depending on the material and its composition.

To sum up, the phenomenon of hydroscopic expansion after compensation of contraction stress should be emphasized more and evaluated. Such over-compensation could lead to internal expansion stress [36]. Hygroscopic stress could result in micro-cracks or even cusp fractures in the restored tooth [37], poorer mechanical properties [38,39], hydrolytic degradation of bonds particularly at the resin–filler interface [40], polymer plasticization leading to hardness reduction and glass transition temperature [40], and impaired wear resistance [41]. Excessive water sorption is not desired as it causes an outward movement of residual monomers and ions due to material solubility. Furthermore, water sorption might generate peeling stress in bonded layers of polymers that may cause serious clinical consequences [42], which may occur especially when prosthetic restorations are adhesively cemented [42].

The present study demonstrated that self-adhesive cements, i.e., Maxcem Elite Chroma, Speed Cem Plus, Cement It, Panavia SA Plus, Breeze, and Panavia 2.0 exhibited high stress values due to water expansion (from 0 up to almost 7 MPa). Water expansion stress of Maxcem Elite Chroma and Breeze amounted to up to ~6–7 MPa which could be associated with their composition, particularly with acidic monomer 4-MET in Breeze and HEMA, and GDM and tetramethylbutyl hydroperoxide in Maxcem Elite Chroma. The monomers, mentioned above, were responsible for water uptake and stress built-up associated with hydroscopic expansion [35]. According to the literature such high stress values were not desirable. Huang et al. [33] found that a giomer material exhibited extensive hygroscopic expansion (due to osmotic effect) enabling enclosing glass cylinders to crack after two weeks of immersion in water. Cusp fracture in endodontically-treated teeth was attributed to hygroscopic expansion of a temporary filling material [33], while cracks in all-ceramic crowns were associated with hygroscopic expansion of compomer and resin-modified glassionomer materials used as core build-up and/or luting cements [33]. Three-year clinical performance study also suggested hygroscopic expansion as a possible cause of cusp fracture in 19% of teeth restored with an ion-releasing composite [43].

Thus, the positive influence of water sorption on contraction stress relief in the case of luting cements, particularly self-adhesive materials, should be considered carefully. Shrinkage occurred within seconds, but water sorption took days and weeks. The rate of hygroscopic shear stress relief depended on the resin volume and its accessibility to water [11]. The contraction stress relief rates observed in luting resin cements could be much lower than in composite resin restorative materials. The composite restorations usually have a relatively large surface exposed to water in comparison to the overall surface of the luting material. As far as luting cements are concerned, the surface exposed to oral fluids is extremely small, while the pathway is extremely long. The consequences (slower compensation of contraction stress) might be less severe if water sorption is also possible from the dentin (dentin exposed to oral environment) [26].

The precise effect of water absorption depends on many factors including not only the material characteristics, the rate and amount of water absorbed, but also the mechanism of absorption [44]. Absorption leads to dimensional changes and has potentially important clinical implications. The positive effect of water absorption on composite restorative materials can be described as the mechanism for the compensation of polymerization shrinkage and the relaxation of stress. In clinical conditions, water absorption may help in the closure of contraction gaps around composite filling materials. It is worth emphasizing that the absorption can, in some cases, result in significant hydroscopic expansion and, thus, be damaging to the resin material and bonded tooth structure.

5. Conclusions

Among all studied resin cements, self-adhesive cements exhibited the highest water sorption due to acid monomer content, which affected the formation of hydroscopic expansion stress. The presence of this type of stress might pose a threat to prosthetic restorations. Therefore, there is still a need for research that would precisely illustrate the generated stress in clinical conditions.

Tested resin cements generated differentiated contraction stress during photopolymerization. The dynamic of hydroscopic compensation (resulting from water sorption) or over-compensation of the contraction stress is also material characteristic-dependent.

Author Contributions: M.-L.S., K.B., and J.S. conceived and designed the experiments; A.S., B.L., and K.K. performed the experiments; M.D. and G.S. analyzed the data; and G.S., A.S., and K.B. wrote the paper.

Funding: The financial support of this work by Medical University of Lodz, Poland (grant no. 502-03/2-148-03/502-24-075) is gratefully acknowledged.

Conflicts of Interest: The authors declare no conflict of interest. The founding sponsors had no role in the design of the study; in the collection, analyses, or interpretation of data; in the writing of the manuscript; or in the decision to publish the results.

References

1. Christensen, G.J. Why use resin cements? *J. Am. Dent. Assoc.* **2010**, *141*, 204–206. [CrossRef] [PubMed]
2. Diaz-Arnold, A.M.; Vargas, M.A.; Haselton, D.R. Current status of luting agents for fixed prosthodontics. *J. Prosthet. Dent.* **1999**, *81*, 135–141. [CrossRef]
3. Radovic, I.; Monticelli, F.; Goracci, C.; Vulicevic, Z.R.; Ferrari, M. Self-adhesive resin cements: A literature review. *J. Adhes. Dent.* **2008**, *10*, 251–258. [CrossRef] [PubMed]
4. Sunico-Segarra, M.; Segarra, A. *A Practical Clinical Guide to Resin Cements*; Springer: Berlin/Heidelberg, Germany, 2015; pp. 9–23.
5. Rawls, H.R.; Shen, C.; Anusavice, K.J. Dental Cements. In *Phillips' Science of Dental Materials*; Saunders: Philadelphia, PA, USA, 2013; pp. 307–339.
6. Hill, E.E.; Lott, J. A clinically focused discussion of luting materials. *Aust. Dent. J.* **2011**, *56* (Suppl. S1), 67–76. [CrossRef] [PubMed]
7. El-Mowafy, O. The use of resin cements in restorative dentistry to overcome retention problems. *J. Can. Dent. Assoc.* **2001**, *67*, 97–102. [PubMed]
8. Vrochari, A.D.; Eliades, G.; Hellwig, E.; Wrbas, K.T. Curing efficiency of four self-etching, self-adhesive resin cements. *Dent. Mater.* **2009**, *25*, 1104–1108. [CrossRef] [PubMed]
9. Turkistani, A.; Sadr, A.; Shimada, Y.; Nikaido, T.; Sumi, Y.; Tagami, J. Sealing performance of resin cements before and after thermal cycling: Evaluation by optical coherence tomography. *Dent. Mater.* **2014**, *30*, 993–1004. [CrossRef] [PubMed]
10. Hitz, T.; Stawarczyk, B.; Fischer, J.; Hämmerle, C.H.F.; Sailer, I. Are self-adhesive resin cements a valid alternative to conventional resin cements? A laboratory study of the long-term bond strength. *Dent. Mater.* **2012**, *28*, 1183–1190. [CrossRef] [PubMed]
11. Feilzer, A.J.; De Gee, A.J.; Davidson, C.L. Increased Wall-to-Wall Curing Contraction in Thin Bonded Resin Layers. *J. Dent. Res.* **1989**, *68*, 48–50. [CrossRef] [PubMed]
12. Al Sunbul, H.; Silikas, N.; Watts, D.C. Polymerization shrinkage kinetics and shrinkage-stress in dental resin-composites. *Dent. Mater.* **2016**, *32*, 998–1006. [CrossRef] [PubMed]
13. Spinell, T.; Schedle, A.; Watts, D.C. Polymerization shrinkage kinetics of dimethacrylate resin-cements. *Dent. Mater.* **2009**, *25*, 1058–1066. [CrossRef] [PubMed]
14. Frassetto, A.; Navarra, C.O.; Marchesi, G.; Turco, G.; Di Lenarda, R.; Breschi, L.; Ferracane, J.L.; Cadenaro, M. Kinetics of polymerization and contraction stress development in self-adhesive resin cements. *Dent. Mater.* **2012**, *28*, 1032–1039. [CrossRef] [PubMed]
15. Shiozawa, M.; Takahashi, H.; Asakawa, Y.; Iwasaki, N. Color stability of adhesive resin cements after immersion in coffee. *Clin. Oral Investig.* **2015**, *19*, 309–317. [CrossRef] [PubMed]
16. Petropoulou, A.; Vrochari, A.D.; Hellwig, E.; Stampf, S.; Polydorou, O. Water sorption and water solubility of self-etching and self-adhesive resin cements. *J. Prosthet. Dent.* **2015**, *114*, 674–679. [CrossRef] [PubMed]
17. Marghalani, H.Y. Sorption and solubility characteristics of self-adhesive resin cements. *Dent. Mater.* **2012**, *28*, e187–e198. [CrossRef] [PubMed]

18. Bociong, K.; Szczesio, A.; Sokolowski, K.; Domarecka, M.; Sokolowski, J.; Krasowski, M.; Lukomska-szymanska, M. The Influence of Water Sorption of Dental Light-Cured Composites on Shrinkage Stress. *Materials* **2017**, *10*, 1142. [CrossRef] [PubMed]

19. *Stomatologia—Materiały Polimerowe do Odbudowy*; PN-EN ISO 4049:2010; Polski Komitet Normalizacyjny: Warszawa, Poland, 2010.

20. Li, L.; Chen, M.; Zhou, X.; Lu, L.; Li, Y.; Gong, C.; Cheng, X. A case of water absorption and water/fertilizer retention performance of super absorbent polymer modified sulphoaluminate cementitious materials. *Constr. Build. Mater.* **2017**, *150*, 538–546. [CrossRef]

21. Müller, J.A.; Rohr, N.; Fischer, J. Evaluation of ISO 4049: Water sorption and water solubility of resin cements. *Eur. J. Oral Sci.* **2017**, *125*, 141–150. [CrossRef] [PubMed]

22. Bociong, K. Naprężenia skurczowe generowane podczas fotoutwardzania eksperymentalnego kompozytu stomatologicznego. Cz. II. *Przem. Chem.* **2017**, *1*, 72–74. [CrossRef]

23. Bociong, K.; Krasowski, M.; Domarecka, M.; Sokołowski, J. Wpływ metody fotopolimeryzacji kompozytów stomatologicznych na bazie żywic dimetakrylanowych na naprężenia skurczowe oraz wybrane właściwości utwardzonego materiału. *Polimery* **2016**, *61*, 499–508. [CrossRef]

24. Domarecka, M.; Sokołowski, K.; Krasowski, M.; Szczesio, A.; Bociong, K.; Sokołowski, J.; Łukomska-Szymańska, M. Influence of water sorption on the shrinkage stresses of dental composites. *J. Stomatol.* **2016**, *69*, 412–419.

25. Timoshenko, S.; Goodier, J.N. *Theory of Elasticity*, 2nd ed.; McGraw-Hill: New York, NY, USA, 1951.

26. Feilzer, A.J.; de Gee, A.J.; Davidson, C.L. Relaxation of polymerization contraction shear stress by hygroscopic expansion. *J. Dent. Res.* **1990**, *69*, 36–39. [CrossRef] [PubMed]

27. Santerre, J.P.; Shajii, L.; Leung, B.W. Relation of dental composite formulations to their degradation and the release of hydrolyzed polymeric-resin-derived products. *Crit. Rev. Oral Biol. Med.* **2001**, *12*, 136–151. [CrossRef] [PubMed]

28. Mahajan, R.P.; Shenoy, V.U.; MV, S.; Walzade, P.S. Comparative Evaluation of Solubilities of Two Nanohybrid Composite Resins in Saliva Substitute and Distilled Water: An in vitro Study. *J. Contemp. Dent.* **2017**, *7*, 82–85.

29. Siswomihardjo, W.; Sunarintyas, S.; Matinlinna, J.P. The influence of resin matrix on the water sorption of fiber-reinforced composites for DENTAL use. *J. Eng. Appl. Sci.* **2016**, *11*, 2678–2682.

30. Sideridou, I.; Tserki, V.; Papanastasiou, G. Study of water sorption, solubility and modulus of elasticity of light-cured dimethacrylate-based dental resins. *Biomaterials* **2003**, *24*, 655–665. [CrossRef]

31. Ling, L.; Xu, X.; Choi, G.Y.; Billodeaux, D.; Guo, G.; Diwan, R.M. Novel F-releasing composite with improved mechanical properties. *J. Dent. Res.* **2009**, *88*, 83–88. [CrossRef] [PubMed]

32. Malacarne, J.; Carvalho, R.M.; de Goes, M.F.; Svizero, N.; Pashley, D.H.; Tay, F.R.; Yiu, C.K.; de Oliveira Carrilho, M.R. Water sorption/solubility of dental adhesive resins. *Dent. Mater.* **2006**, *22*, 973–980. [CrossRef] [PubMed]

33. Huang, C.; Kei, L.; Wei, S.H.Y.; Cheung, G.S.P.; Tay, F.R.; Pashley, D.H. The influence of hygroscopic expansion of resin-based restorative materials on artificial gap reduction. *J. Adhes. Dent.* **2002**, *4*, 61–71. [PubMed]

34. Wilson, A.D. Resin-modified glass-ionomer cements. *Int. J. Prosthodont.* **1990**, *3*, 425–429. [PubMed]

35. Park, J.; Eslick, J.; Ye, Q.; Misra, A.; Spencer, P. The influence of chemical structure on the properties in methacrylate-based dentin adhesives. *Dent. Mater.* **2013**, *31*, 1713–1723. [CrossRef] [PubMed]

36. Wei, Y.J.; Silikas, N.; Zhang, Z.T.; Watts, D.C. Diffusion and concurrent solubility of self-adhering and new resin–matrix composites during water sorption/desorption cycles. *Dent. Mater.* **2011**, *21*, 97–205. [CrossRef] [PubMed]

37. Rüttermann, S.; Krüger, S.; Raab, W.H.M.; Janda, R. Polymerization shrinkage and hygroscopic expansion of contemporary posterior resin-based filling materials—A comparative study. *J. Dent.* **2007**, *35*, 806–813. [CrossRef] [PubMed]

38. Musanje, L.; Shu, M.; Darvell, B.W. Water sorption and mechanical behaviour of cosmetic direct restorative materials in artificial saliva. *Dent. Mater.* **2001**, *17*, 394–401. [CrossRef]

39. Bastioli, C.; Romano, G.; Migliaresi, C. Water sorption and mechanical properties of dental composites. *Biomaterials* **1990**, *11*, 219–223. [CrossRef]

40. Ferracane, J.L. Hygroscopic and hydrolytic effects in dental polymer networks. *Dent. Mater.* **2006**, *22*, 211–222. [CrossRef] [PubMed]

41. Göhring, T.N.; Besek, M.J.; Schmidlin, P.R. Attritional wear and abrasive surface alterations of composite resin materials in vitro. *J. Dent.* **2002**, *30*, 119–127. [CrossRef]

42. Kalachandra, S.; Turner, D.T. Water sorption of polymethacrylate networks: Bis-GMA/TEGDM copolymers. *J. Biomed. Mater. Res.* **1987**, *21*, 329–338. [CrossRef] [PubMed]

43. Versluis, A.; Tantbirojn, D.; Lee, M.S.; Tu, L.S.; Delong, R. Can hygroscopic expansion compensate polymerization shrinkage? Part I. Deformation of restored teeth. *Dent. Mater.* **2011**, *27*, 126–133. [CrossRef] [PubMed]

44. McCabe, J.F.; Rusby, S. Water absorption, dimensional change and radial pressure in resin matrix dental restorative materials. *Biomaterials* **2004**, *25*, 4001–4007. [CrossRef] [PubMed]

materials

MDPI

Article

Effects of Three Calcium Silicate Cements on Inflammatory Response and Mineralization-Inducing Potentials in a Dog Pulpotomy Model

Chung-Min Kang [1,2,†], Jiwon Hwang [1,†], Je Seon Song [1,3], Jae-Ho Lee [1,3], Hyung-Jun Choi [1,3] and Yooseok Shin [3,4,*]

1 Department of Pediatric Dentistry, College of Dentistry, Yonsei University, Seoul, 03722, Korea; zezu7@yuhs.ac (C.-M.K.); kntdent@naver.com (J.H.); songjs@yuhs.ac (J.S.S.); leejh@yuhs.ac (J.-H.L.); choihj88@yuhs.ac (H.-J.C.)
2 Department of Pharmacology, College of Medicine, Yonsei University, Seoul 03722, Korea
3 Oral Science Research Center, College of Dentistry, Yonsei University, Seoul 03722, Korea
4 Department of Conservative Dentistry, College of Dentistry, Yonsei University, 50-1 Yonseiro, Seodaemun-Gu, Seoul 03722, Korea
* Correspondence: densys@yuhs.ac; Tel.: +82-2-2228-3149; Fax: +82-2-313-7575
† These authors contributed equally to this work.

Received: 27 April 2018; Accepted: 25 May 2018; Published: 27 May 2018

Abstract: This beagle pulpotomy study compared the inflammatory response and mineralization-inducing potential of three calcium silicate cements: ProRoot mineral trioxide aggregate (MTA) (Dentsply, Tulsa, OK, USA), OrthoMTA (BioMTA, Seoul, Korea), and Endocem MTA (Maruchi, Wonju, Korea). Exposed pulp tissues were capped with ProRoot MTA, OrthoMTA, or Endocem MTA. After 8 weeks, we extracted the teeth, then performed hematoxylin-eosin and immunohistochemical staining with osteocalcin and dentin sialoprotein. Histological evaluation comprised a scoring system with eight broad categories and analysis of calcific barrier areas. We evaluated 44 teeth capped with ProRoot MTA (n = 15), OrthoMTA (n = 18), or Endocem MTA (n = 11). Most ProRoot MTA specimens formed continuous calcific barriers; these pulps contained inflammation-free palisading patterns in the odontoblastic layer. Areas of the newly formed calcific barrier were greater with ProRoot MTA than with Endocem MTA (p = 0.006). Although dentin sialoprotein was highly expressed in all three groups, the osteocalcin expression was reduced in the OrthoMTA and Endocem MTA groups. ProRoot MTA was superior to OrthoMTA and Endocem MTA in all histological analyses. ProRoot MTA and OrthoMTA resulted in reduced pulpal inflammation and more complete calcific barrier formation, whereas Endocem MTA caused a lower level of calcific barrier continuity with tunnel defects.

Keywords: calcium silicate cements; pulpal response; mineralization; calcific barrier; inflammation; odontoblastic layer

---

1. Introduction

Vital pulp therapy consists of apexogenesis, pulpotomy, pulpal debridement, indirect pulp capping, and direct pulp capping [1]. The aim of these treatments includes maintenance of vitality and preservation of the remaining pulp to enable adequate healing of the pulp-dentin complex [2,3]. These procedures offer a more conservative approach to root canal therapy for traumatic and cariously exposed pulps [4,5]. ProRoot mineral trioxide aggregate (PMTA) (Dentsply, Tulsa, OK, USA) has been recognized as an appropriate material for vital pulp therapy, and has provided higher clinical success rates in a long-term study [6].

Although PMTA possesses bioactive and antibacterial activities with high sealing properties [7–10], it has some drawbacks such as a prolonged setting time, discoloration potential, and difficult handling properties [11]. Recently, calcium silicate cement development has helped to overcome these disadvantages in the search for bioactive dental materials. These include OrthoMTA (OMTA; BioMTA) and Endocem MTA (EMTA; Maruchi, Wonju, Korea).

OMTA contains lower concentrations of heavy metals for greater biocompatibility [12] and has a low expansion rate and a good sealing ability [13]. According to the manufacturer, OMTA prevents microleakage by forming a layer of hydroxyapatite that acts as an interface between the material and the canal wall [14]. Moreover, it has been reported to create a favorable environment for bacterial entombment through intratubular mineralization [15]. Clinical studies reported that both PMTA and OMTA had high success rates in pulpotomy and partial pulpotomy when used in primary and permanent teeth [16,17]. However, OMTA showed lower biocompatibility compared with PMTA in an in vitro study [18]. EMTA has been introduced as an MTA-derived pozzolan cement. EMTA has a short setting time (around 4 min) without the addition of a chemical accelerator because it uses a pozzolanic reaction [19]. It is convenient to handle based on its adequate consistency and washout resistance [20]. Indeed, EMTA demonstrated good biocompatibility, osteogenicity, and odontogenic effects similar to PMTA in an in vitro study [19].

Despite the increased use of various calcium silicate cements in vital pulp therapy, their effects after pulpotomy in an in vivo animal model have not yet been investigated. In addition, most studies have had short durations of approximately 4 weeks. Accordingly, the present study aimed to evaluate and compare levels of calcific barrier formation, inflammation reaction, and hard tissue barrier formation histologically following the application of PMTA, OMTA, and EMTA for 8 weeks in a beagle pulpotomy model. The null hypothesis is that there is no significant difference in inflammatory response and mineralization-inducing potentials among the three materials; both OMTA and EMTA have similar biologic effects to PMTA.

2. Materials and Methods

2.1. Animal Model

The present study used a beagle dog pulpotomy model. The inclusion criteria were non-damaged dentition and healthy periodontium. The teeth of two dogs were numbered and randomly divided into three groups with a table of random sampling numbers; the three calcium silicate material groups are shown in Table 1. We performed animal selection, control, the surgical operation, and preparation as per the procedures approved by the Yonsei University Health System's Institutional Animal Care and Use Committee (certification #2013-0317-4). These experiments were performed in accordance with the National Institutes of Health Guide for the Care and Use of Laboratory Animals.

Table 1. The chemical compositions of the calcium silicate cements tested in this study.

Materials	Composition	Setting Time
PMTA	Tricalcium silicate Tricalcium aluminate Dicalcium silicate Tetracalcium aluminoferrite Gypsum Free calcium oxide Bismuth oxide	Initial setting time: 78 min (±5 min) Final setting time: 261 min (±21 min)
OMTA	Tricalcium silicate Dicalcium silicate Tricalcium aluminate Tetracalcium aluminoferrite Free calcium oxide Bismuth oxide	Initial setting time: 180 min Final setting time: 360 min (±21 min)

Table 1. *Cont.*

Materials	Composition	Setting Time
EMTA	Calcium oxide Silicon dioxide Bismuth oxide Aluminum oxide H_2O/CO_2 Magnesium oxide Sulfur trioxide Ferrous oxide Titanium dioxide	Initial setting time: 2 min ($\pm$30 s) Final setting time: 4 min ($\pm$30 s)

PMTA, ProRoot MTA® (Dentsply Tulsa, OK, USA); OMTA, Ortho MTA® (BioMTA, Seoul, Korea); EMTA, Endocem MTA® (Maruchi, Wonju, Korea).

2.2. Surgical Procedure

We performed all operations in a clean, sterilized room. We administered intravascular injections of tiletamine/zolazepam (Zoletile® 5 mg/kg; Virbac Korea, Seoul, Korea) and xylazine (Rompun® 0.2 mg/kg; Bayer Korea, Seoul, Korea), and used the inhalational anesthetic isoflurane (Gerolan®, Choongwae Pharmaceutical, Seoul, Korea) for general anesthesia induction. To prevent infection, we injected enfloxacin (5 mg/kg) subcutaneously immediately before and after the operation. We administered amoxicillin clavulanate (12.5 mg/kg) orally for 5 to 7 days after the operation.

2.3. Pulpotomy Procedure

We used lidocaine hydrochloride (2%) with 1:100,000 epinephrine (Kwangmyung Pharmaceutical, Seoul, Korea) for local anesthesia. After forming a cavity on the occlusal surface using a high-speed carbide bur #330 (H7 314 008, Brasseler, Lemgo, Germany), we mechanically exposed the pulp. We removed the crown of the pulp at the level of the cementoenamel junction and stopped bleeding by injecting sterile saline and applying slight pressure with sterile cotton pellets. We applied the MTA to the top of the cut pulp as per the manufacturer's guidelines in each group. When we applied the MTA to the pulp area, we used cotton balls soaked in saline. We performed the final cavity restoration using the self-curing glass ionomer cement Ketac-Molar (3M ESPE; St. Paul, MN, USA). Eight weeks after the operation, we euthanized the dogs by over-sedation.

2.4. Histological Analysis

We extracted the teeth with forceps and removed one-third of the root using a high-speed bur. We fixed the specimens in 10% neutral-buffered formalin (Sigma-Aldrich, St. Louis, MO, USA) for 48 h, and demineralized them in ethylenediaminetetraacetic acid (EDTA; pH 7.4; Fisher Scientific, Houston, TX, USA) for 6 weeks before embedding them in paraffin. For each specimen, we created 3-μm continuous sections in the buccolingual direction and stained them subsequently with hematoxylin and eosin (HE). We used InnerView 2.0 software (InnerView Co., Seongnam-si, Gyeonggi-do, Korea) for image analysis. Five observers blinded to the group allocations examined the specimens. We evaluated the specimens for calcific barrier formation, dental pulp inflammation, and odontoblastic cell layer formation with a scoring system reported in a previous study [21–23] (Table 2). We adopted the score agreed upon by at least three of the five observers who were blinded to specimen grouping. In addition, we measured the area of the newly formed hard tissue using ImageJ version 1.48 (National Institute of Health, Bethesda, MD, USA).

2.5. Immunohistochemistry

For immunohistochemistry (IHC), we deparaffinized 3-μm cross-sections with xylene, rehydrated them, and rinsed them with distilled water. For antigen retrieval, we used protease K (Dako North America Inc., Carpinteria, CA, USA) for osteocalcin (OC) and dentin sialoprotein (DSP) staining.

To activate endogenous peroxidase, we added 3% hydrogen peroxide, while preventing non-specific binding by incubating sections in 5% bovine serum albumin (Sigma-Aldrich). Subsequently, we incubated sections overnight with the following primary antibodies: anti-OC antibody (rabbit polyclonal, Ab109112, 1:10,000; Abcam, Cambridge, UK) or anti-DSP antibody (rabbit polyclonal, sc-33586, 1:500; Santa Cruz Biotechnology, Santa Cruz, CA, USA). Subsequently, we applied EnVision + System-Horseradish peroxidase (HRP)-Labeled Polymer anti-rabbit (K4003, Dako) for 20 min. After developing color using the labeled streptavidin biotin kit (Dako) as per the manufacturer's guidelines, we counterstained the sections with Gill's hematoxylin (Sigma-Aldrich).

Table 2. Scores used during the histological analysis of calcific barriers and dental pulp.

Scores	Calcific Barrier Continuity
1	Complete calcific barrier formation
2	Partial/incomplete calcific barrier formation extending to more than one-half of the exposure site but not completely closing the exposure site
3	Initial calcific barrier formation extending to no more than one-half of the exposure site
4	No calcific barrier formation
Scores	**Calcific barrier morphology**
1	Dentin or dentin-associated with irregular hard tissue
2	Only irregular hard tissue deposition
3	Only a thin layer of hard tissue deposition
4	No hard tissue deposition
Scores	**Tubules in calcific barrier**
1	No tubules present
2	Mild (tubules present in less than 30% of the calcific barrier)
3	Moderate to severe (tubules present in more than 30% of the calcific barrier)
4	No hard tissue deposition
Scores	**Inflammation intensity**
1	Absent or very few inflammatory cells
2	Mild (an average of <10 inflammatory cells)
3	Moderate (an average of 10–25 inflammatory cells)
4	Severe (an average >25 inflammatory cells)
Scores	**Inflammation extensity**
1	Absent
2	Mild (inflammatory cells next to the dentin bridge or area of pulp exposure only)
3	Moderate (inflammatory cells observed in one-third or more of the coronal pulp or in the mid pulp)
4	Severe (all of the coronal pulp is infiltrated or necrotic)
Scores	**Inflammation type**
1	No inflammation
2	Chronic inflammation
3	Acute and chronic inflammation
4	Acute inflammation
Scores	**Dental pulp congestion**
1	No congestion
2	Mild (enlarged blood vessels next to the dentin bridge or area of pulp exposure only)
3	Moderate (enlarged blood vessels observed in one-third or more of the coronal pulp or in the mid pulp)
4	Severe (all of the coronal pulp is infiltrated with blood cells)
Scores	**Odontoblastic cell layer**
1	Complete palisading cell pattern
2	Partial/incomplete palisading cell pattern
3	Presence of odontoblast-like cells only
4	Absent

2.6. Statistical Analysis

We performed statistical analyses using SPSS version 23 software (SPSS, Chicago, IL, USA). To analyze the area of the newly formed calcific barrier, we applied a one-way analysis of variance (ANOVA) (significance at $p < 0.05$) and the post hoc Scheffé test (Bonferroni correction; $p < 0.017$).

3. Results

The present study used two male beagle dogs aged 18–24 months that weighed approximately 12 kg. We selected 60 teeth from the two dogs. Among the 60 teeth, we excluded five teeth from the PMTA group, two from the OMTA group, and nine from the EMTA group that failed during tooth removal or specimen production; we evaluated only 44 specimens in the final analysis. Eventually, we analyzed the PMTA ($n = 15$), OMTA ($n = 18$), and EMTA ($n = 11$) specimens histologically.

3.1. Calcific Barrier Formation

Most specimens in the PMTA group formed a complete calcific barrier, while some in the OMTA and EMTA groups produced a partially discontinuous calcific barrier (Figure 1 and Table 3). We observed no calcific barrier in only one specimen from the EMTA group (Figure 2B). In addition, the PMTA group produced hard tissue most similar to the dentin, while we observed a partially irregular or thinly formed calcific layer in the OMTA and EMTA groups. Although three groups contained calcific barriers with low tubule formation (Figure 2D), moderate to severe tubule formation in calcific barrier was found in an OMTA specimen (Figure 2C). We compared the areas of the formed calcific barriers; the calcific barrier in the PMTA group was the widest, followed by those in the OMTA and EMTA groups. There was a statistically significant difference between the PMTA and EMTA groups ($p = 0.006$; Figure 3E). When the calcific barriers were standardized with coronal pulpal width according to tooth type, PMTA also had a significantly higher area than EMTA ($p = 0.0114$; Figure 3F).

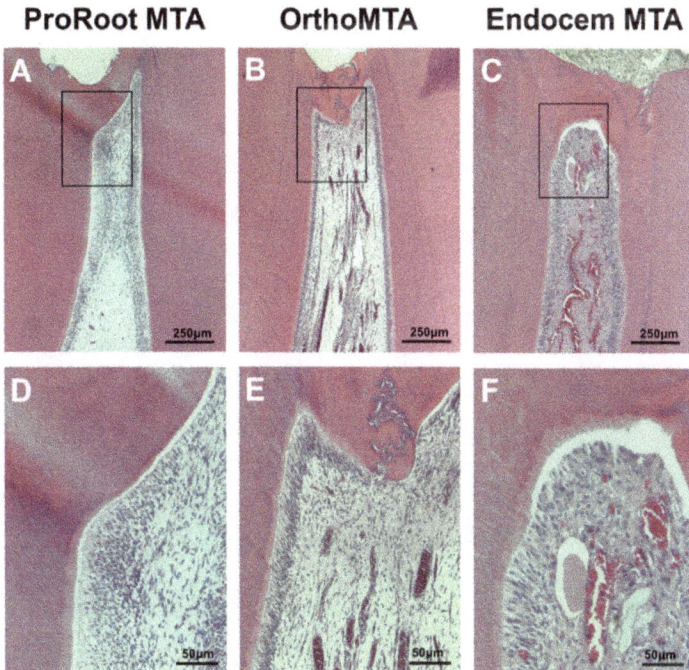

Figure 1. Hematoxylin-eosin staining for the evaluation of the histomorphologic characteristics of the newly formed calcific barrier (CB) after 8 weeks. Most PMTA and OMTA specimens formed continuous CBs and the pulps contained palisading patterns in the odontoblastic layer that were free from inflammation. However, EMTA specimens showed less favorable odontoblastic layer formation ((**A–C**): scale bars = 250 μm, (**D–F**): scale bars = 50 μm).

Table 3. Score percentages for calcific barriers.

Groups	Calcific Barrier Continuity (%)			
	1	2	3	4
PMTA	100 (15/15) *	-	-	-
OMTA	66.67 (12/18)	16.67 (3/18)	16.67 (3/18)	-
EMTA	45.45 (5/11)	18.18 (2/11)	27.27 (3/11)	9.09 (1/11)
Groups	Calcific Barrier Morphology (%)			
	1	2	3	4
PMTA	86.67 (13/15)	13.33 (2/15)	-	-
OMTA	38.89 (7/18)	27.78 (5/18)	33.33 (6/18)	-
EMTA	45.45 (5/11)	18.18 (2/11)	27.27 (3/11)	9.09 (1/11)
Groups	Tubules in Calcific Barrier (%)			
	1	2	3	4
PMTA	60 (9/15)	33.33 (5/15)	6.67 (1/15)	-
OMTA	61.11 (11/18)	27.78 (5/18)	11.11 (2/18)	-
EMTA	63.64 (7/11)	18.18 (2/11)	9.09 (1/11)	9.09 (1/11)

PMTA, ProRoot MTA®; OMTA, Ortho MTA®; EMTA, Endocem MTA®. * (number of teeth receiving the score/total number of teeth evaluated).

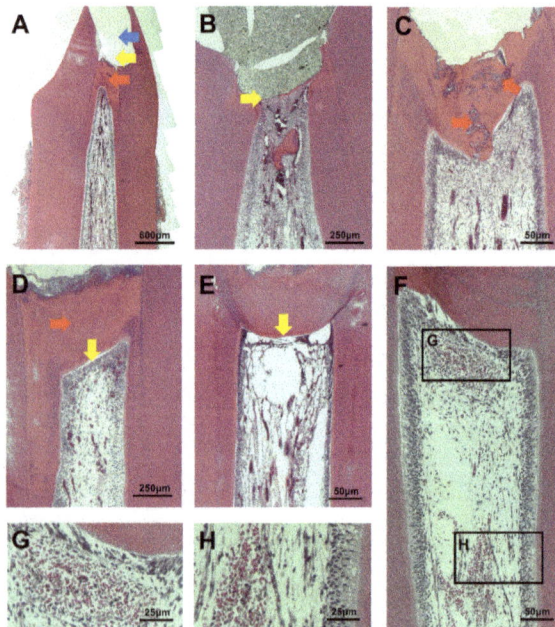

Figure 2. (**A**) An example of a pulpotomy procedure. The site of pulpotomy (yellow arrow) was an upper part of the calcific barrier (red arrow). MTA was filled in the pulp exposure area, but the material is not visible due to partial loss during specimen processing (blue arrow); (**B**) An EMTA specimen without a calcific barrier (yellow arrow). This corresponds to score 4 in the calcific barrier continuity category; (**C**) Moderate to severe tubule formation in calcific barrier of an OMTA specimen (red arrow); (**D**) A PMTA specimen with complete calcific barrier formation and no tubule in barrier (red arrow). It also has complete palisading cell pattern in odontoblastic cell layer (yellow arrow); (**E**) An EMTA specimen with absent odontoblastic cell layer under calcific barrier. This corresponds to score 4 in the odontoblastic cell layer category; (**F–H**) Inflammatory cells that are observed in coronal and middle pulps.

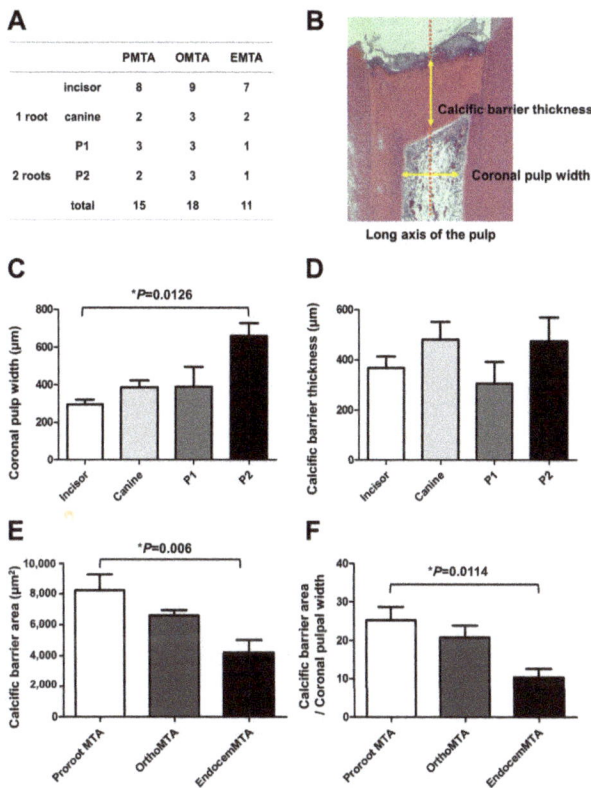

Figure 3. The area of the newly formed calcific barrier for each material after 8 weeks. (**A**) The distributions of tooth types among the three test groups; (**B**) An example measuring coronal pulpal width and calcific barrier thickness; Although the thickness of the calcific barriers did not vary according to tooth type (**C**); the width of the coronal pulp differed by tooth type (*p* = 0.0126) (**D**). Thus, the area of the calcific barrier was calculated by dividing by the horizontal width of the coronal pulp. The bars represent the mean ± the standard deviation; (**E**) In the PMTA group, the calcific barrier is widest, followed by those in the OMTA and EMTA groups. There is a statistically significant difference between the PMTA and EMTA groups (*p* = 0.006); (**F**) When the calcific barriers were standardized by coronal pulpal width, PMTA also had a significantly higher area than EMTA (*p* = 0.0114). We performed statistical analyses with a one-way ANOVA and the post hoc Scheffé test. The number of specimens is *n* = 15 in the PMTA group, *n* = 18 in the OMTA group, and *n* = 11 in the EMTA group.

3.2. Pulpal Reaction

We observed mononuclear inflammatory cells in 26.67% of the specimens from the PMTA group, 44.44% from the OMTA group, and 63.64% from the EMTA group (Table 4). When we compared the extent of inflammation among the groups, inflammation was mild or almost absent in the PMTA and OMTA groups. In 9% of the EMTA specimens, we observed inflammatory cells in one-third or more of the coronal pulp or middle pulp. While there was almost no acute inflammation in any group, a few cases demonstrated chronic inflammation exclusively. The pulpal congestion reaction was mild in the PMTA group and most severe in the EMTA group. We observed no pulpal congestion above the moderate level in any of the three groups.

Table 4. Score percentages for inflammatory responses.

Groups	Inflammation Intensity (%)			
	1	2	3	4
PMTA	73.33 (11/15) *	26.67 (4/15)	-	-
OMTA	55.56 (10/18)	44.44 (8/18)	-	-
EMTA	36.36 (4/11)	63.64 (7/11)	-	-
Groups	**Inflammation Extensity (%)**			
	1	2	3	4
PMTA	73.33 (11/15)	26.67 (4/15)	-	-
OMTA	55.56 (10/18)	44.44 (8/18)	-	-
EMTA	36.36 (4/11)	54.55 (6/11)	9.09 (1/11)	-
Groups	**Inflammation Type (%)**			
	1	2	3	4
PMTA	73.33 (11/15)	26.67 (4/15)	-	-
OMTA	55.56 (10/18)	44.44 (8/18)	-	-
EMTA	36.36 (4/11)	63.64 (7/11)	-	-
Groups	**Dental Pulp Congestion (%)**			
	1	2	3	4
PMTA	40 (6/15)	53.33 (8/15)	6.67 (1/15)	-
OMTA	27.78 (5/18)	61.11 (11/18)	11.11 (2/18)	-
EMTA	18.18 (2/11)	63.64 (7/11)	18.18 (2/11)	-

* (number of teeth receiving the score/total number of teeth evaluated).

3.3. Odontoblastic Cell Layer

In the PMTA group, a complete palisading cell pattern was visible in 60% of the specimens, with 26.67% showing Partial/incomplete palisading cell pattern (Table 5). The OMTA and EMTA groups showed less favorable results compared with the PMTA group: all OMTA specimens showed an odontoblastic cell layer. In contrast, approximately 9.09% of the EMTA group specimens showed no odontoblastic cell layer.

Table 5. Score percentages for the odontoblastic cell layer.

Groups	Odontoblastic Cell Layer (%)			
	1	2	3	4
PMTA	60 (9/15) *	26.67 (4/15)	13.33 (4/15)	-
OMTA	33.33 (6/18)	50 (9/18)	16.67 (3/18)	-
EMTA	45.45 (5/11)	18.18 (2/11)	27.27 (3/11)	9.09 (1/11)

* (number of teeth receiving the score/total number of teeth evaluated).

3.4. Immunohistochemistry

DSP and OC staining indicated hard tissue formation in all three groups. The DSP was highly expressed in all three groups, which indicated odontogenic differential potential (Figure 4A–F). Although OC was also expressed in all three groups, its expression was relatively less in the EMTA group (Figure 4G–L).

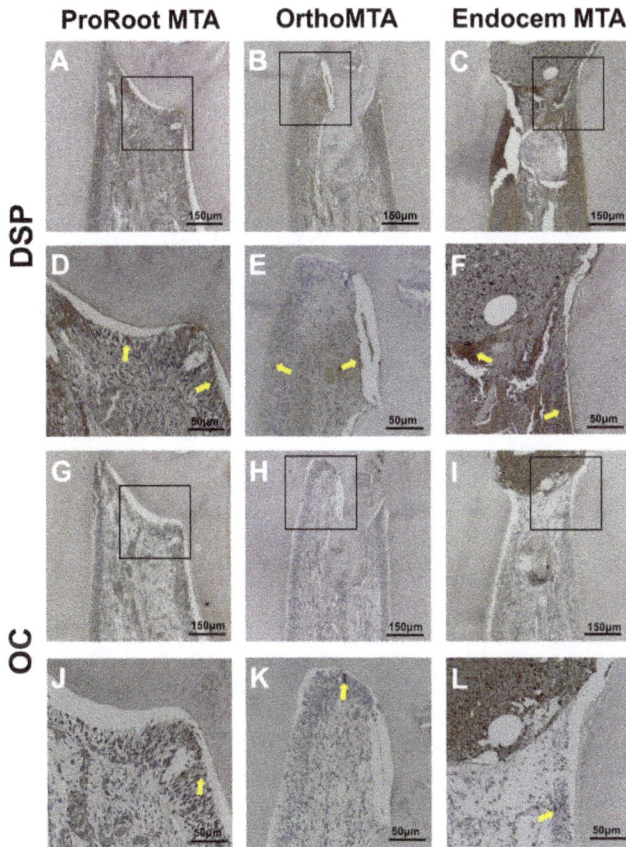

Figure 4. Immunohistochemical staining of dentin sialoprotein (DSP) and osteocalcin (OC). The DSP is highly expressed in all three groups. Although OC is expressed most clearly in odontoblast-like cells in the PMTA group, its expression was reduced in the EMTA group. Yellow arrows indicate cells with a positive signal. ((**A–C,G–I**): scale bars = 150 μm; (**D–F,J–L**): scale bars = 50 μm).

4. Discussion

Many attempts have been made to improve the clinical application of MTA by modifying chemical components or adding a setting accelerator; however, it is important to note that changes in MTA components may cause adverse effects with respect to physical and biological properties [24]. We recommend that in vivo studies explore the pulpal response when using MTA for vital pulp therapy. We recently reported the pulpal responses to pulpotomy with RetroMTA, TheraCal (Bisco Inc., Schamburg, IL, USA), and PMTA [22], and compared the biological efficacies of Endocem Zr (Maruchi, Wonju, Korea) and PMTA at 4 weeks in an in vivo study [23]. We performed this study to compare and evaluate the pulpal response associated with PMTA, OMTA, and EMTA over 8 weeks in a beagle dog pulpotomy model. To our knowledge, this is the first study to examine the biological response from OMTA compared with those from EMTA and PMTA simultaneously in an in vivo model.

Overall, both PMTA and OMTA showed favorable pulpal reactions and formed an almost complete calcific barrier. EMTA induced a slightly greater inflammatory reaction than the other two MTAs and had a lower amount of calcific barrier formation with tunnel defects. These results suggested that EMTA had relatively less biocompatibility than PMTA or OMTA. In addition, PMTA and OMTA formed favorable palisading patterns in odontoblast cells, which demonstrate that these materials had stronger odontogenic differentiation potential.

We suggest several reasons for the low biocompatibility of EMTA observed in this study. A study reported that EMTA resulted in lower cell viability than PMTA immediately after mixing because of its high pH and the heat from the cement surface [25]. This initial cytotoxic effect might contribute to denaturation of adjacent cells [26]. EMTA was also known to express a lower level of osteopontin, a specific bone mineralization marker, compared with other MTAs [27]. Reduced osteopontin production could be a result of lower hard tissue formation. However, these results were not in accordance with the results of a previous in vitro study, which reported that OMTA was significantly more cytotoxic than PMTA and EMTA [27]. The difference may result from interactions with living cells; chemical reactions in the dentin-MTA interfacial layer might have caused the differences in results between the in vivo and in vitro study models. In addition, while PMTA and OMTA exhibit similar components, EMTA contains different chemicals, such as small particles of pozzolan. A pozzolan is a siliceous and aluminous material that reacts chemically with calcium hydroxide to form calcium silicate hydrate in the presence of water [28]. Unlike the initially high pH levels after mixing, EMTA showed a significantly lower pH value than PMTA and OMTA [29]. A previous study found that an acidic environment adversely affected the physical properties and hydration behavior of MTA [30]. We recommend further study because EMTA had inferior cell viability and a low pH level after setting, but was not significantly different from the other two MTAs.

Despite the small number of reports regarding OMTA and EMTA, various in vivo and clinical studies have demonstrated successful formation of calcific barriers in vital pulp therapy with PMTA [16,17,22,23,31,32]. It is still controversial whether calcific barrier formation at the interface between the pulp and material indicates the success of the treatment. Therefore, to judge the efficacy of different MTAs, it is important to analyze the presence or absence of inflammation (the type and severity) and the continuity, structure, and tubule formation of the formed calcific barrier [33]. This study interpreted the calcific barrier as a sign of healing and a positive reaction to stimulation. In the three MTA groups, the calcific barrier interfaces had columnar cells projecting into the bridge with polarized nuclei, which indicates the formation of odontoblast cells and reparative dentin synthesis. This study also confirmed odontogenic properties with IHC by using OC and DSP. OC is a specific marker of the late odontoblastic development pathway and DSP is a marker of odontoblast regulation in reparative dentin mineralization [34]. In regards to several molecular mechanisms, in vitro studies with PMTA showed continuously increasing transcription of mRNA for DSP [35]. Further, upregulation of OC mRNA confirmed the odontoblastic pathway of cell differentiation, indicating that cells enter a quiescent phase in another calcium silicate cement, Biodentine [36]. IHC revealed that PMTA had the potential to induce greater odontoblastic differentiation than OMTA or EMTA. In the analysis of the mean calcific barrier areas, the areas of newly formed calcific barrier were significantly greater in the PMTA group than in the EMTA group. When comparing the area of newly formed calcific barrier in this study statistically, the size of the pulp would be different according to the type of tooth. Although we tried to distribute the tooth types evenly among the experimental groups, we excluded some specimens during tooth removal and specimen production, especially in the EMTA group. Further, the PMTA was regarded as a positive control, but there was no negative control group in this experiment. We recommend further study with a larger sample and the inclusion of a negative control.

In the present study, we evaluated the pulpal response in pulpotomy model for a period of 8 weeks. Other studies used the same time interval [37,38]. In one study, osteodentin matrix formation occurred during the first 2 weeks; after 3 weeks, a complete layer of reparative dentin was formed at the capping site [39]. Another study reported the presence of a calcified bridge in all specimens at 5 weeks after capping with MTA [34]. Notably, we designed the study to compare pulpal responses in a previous study with a shorter time interval (4 weeks) to determine whether there was any difference in response, compared with an 8-week interval. At 8 weeks, hard tissue formation at the exposure site was thicker with less inflammation than in our previous in vivo study at 4 weeks [22,23].

The results of the animal model may not correspond with those of human teeth. A complete hard tissue barrier appeared 1 week after pulp capping in canine teeth [33], whereas the initiation of hard tissue formation has been reported to start as early as 2 weeks after pulp capping in humans [40]. Most studies reported that it took 30 to 42 days to form a hard tissue barrier in humans [41,42]. In addition, we performed the evaluation of the pulpal response on healthy, intact teeth from dogs. Therefore, these results do not necessarily reflect the effects of newly developed MTAs on inflamed pulps. We advise clinicians to place a wet cotton pellet over the MTA in the first visit, followed by replacement with a permanent restoration at the second visit. In this experiment under G/A, the long setting time of PMTA and OMTA was a limitation. Accordingly, careful consideration is necessary when adapting these results to clinical situations.

In conclusion, the results of the present in vivo study demonstrated that PMTA and OMTA had favorable outcomes when used as pulpotomy materials. Both materials showed biocompatibility and induced high-quality calcific barrier formation at the interface with the pulp tissue. However, EMTA has appropriate characteristics and clinical advantages of a shorter setting time and no tooth discoloration. Regarding the application of these results on pulpotomy in healthy canine pulps with no inflammation, we recommend further clinical studies using human teeth for an evaluation of the biological efficacy of these materials.

Author Contributions: C.-M.K. performed data curation, formal analysis and contributed in writing original draft; J.H. contributed in formal analysis and writing original draft; J.S.S. performed conceptualization, project administration, and visualization; J.-H.L. contributed in supervision and validation. H.-J.C. contributed in project administration and supervision; Y.S. performed conceptualization, funding acquisition, and resources; C.-M.K. and J.H. contributed equally as first authors.

Acknowledgments: This study was supported by a faculty research grant of Yonsei university new faculty research fund (6-2017-0009).

Conflicts of Interest: The authors declare no conflict of interest.

References

1. Dahlkemper, P.; Dan, B.; Ang, D.; Goldberg, R.; Rubin, R.; Schultz, G.; Sheridan, B.; Slingbaum, J.; Stevens, M.; Powell, W. *Guide to Clinical Endodontics*; American Association of Endodontics: Chicago, IL, USA, 2013.
2. Zhang, W.; Yelick, P.C. Vital pulp therapy—Current progress of dental pulp regeneration and revascularization. *Inter. J. Dent.* **2010**, *2010*, 856087. [CrossRef] [PubMed]
3. Asgary, S.; Hassanizadeh, R.; Torabzadeh, H.; Eghbal, M.J. Treatment Outcomes of 4 Vital Pulp Therapies in Mature Molars. *J. Endod.* **2018**, *44*, 529–535. [CrossRef] [PubMed]
4. Parirokh, M.; Torabinejad, M.; Dummer, P. Mineral trioxide aggregate and other bioactive endodontic cements: An updated overview—Part I: Vital pulp therapy. *Inter. Endod. J.* **2018**, *51*, 177–205. [CrossRef] [PubMed]
5. Endodontology, E.S.O. Quality guidelines for endodontic treatment: Consensus report of the European Society of Endodontology. *Inter. Endod. J.* **2006**, *39*, 921–930. [CrossRef]
6. Torabinejad, M.; Parirokh, M. Mineral trioxide aggregate: A comprehensive literature review—Part II: Leakage and biocompatibility investigations. *J. Endod.* **2010**, *36*, 190–202. [CrossRef] [PubMed]
7. Barrieshi-Nusair, K.M.; Hammad, H.M. Intracoronal sealing comparison of mineral trioxide aggregate and glass ionomer. *Quintessence Int.* **2005**, *36*, 539–545. [PubMed]
8. Iwamoto, C.E.; Adachi, E.; Pameijer, C.H.; Barnes, D.; Romberg, E.E.; Jefferies, S. Clinical and histological evaluation of white ProRoot MTA in direct pulp capping. *Am. J. Dent.* **2006**, *19*, 85–90. [PubMed]
9. Parirokh, M.; Torabinejad, M. Mineral trioxide aggregate: A comprehensive literature review—Part I: Chemical, physical, and antibacterial properties. *J. Endod.* **2010**, *36*, 16–27. [CrossRef] [PubMed]
10. Schmalz, G.; Widbiller, M.; Galler, K. Material tissue interaction—From toxicity to tissue regeneration. *Oper. Dent.* **2016**, *41*, 117–131. [CrossRef] [PubMed]

11. Parirokh, M.; Torabinejad, M. Mineral trioxide aggregate: A comprehensive literature review—Part III: Clinical applications, drawbacks, and mechanism of action. *J. Endod.* **2010**, *36*, 400–413. [CrossRef] [PubMed]

12. Kum, K.Y.; Zhu, Q.; Safavi, K.; Gu, Y.; Bae, K.S.; Chang, S.W. Analysis of six heavy metals in Ortho mineral trioxide aggregate and ProRoot mineral trioxide aggregate by inductively coupled plasma-optical emission spectrometry. *Aust. Endod J.* **2013**, *39*, 126–130. [CrossRef] [PubMed]

13. Chang, S.-W.; Baek, S.-H.; Yang, H.-C.; Seo, D.-G.; Hong, S.-T.; Han, S.-H.; Lee, Y.; Gu, Y.; Kwon, H.-B.; Lee, W.; et al. Heavy Metal Analysis of Ortho MTA and ProRoot MTA. *J. Endod.* **2011**, *37*, 1673–1676. [CrossRef] [PubMed]

14. Physicochemical Analysis. Available online: www.biomta.com (accessed on 26 May 2018).

15. Yoo, J.S.; Chang, S.W.; Oh, S.R.; Perinpanayagam, H.; Lim, S.M.; Yoo, Y.J.; Oh, Y.R.; Woo, S.B.; Han, S.H.; Zhu, Q.; et al. Bacterial entombment by intratubular mineralization following orthograde mineral trioxide aggregate obturation: A scanning electron microscopy study. *Int. J. Oral Sci.* **2014**, *6*, 227–232. [CrossRef] [PubMed]

16. Kang, C.M.; Kim, S.H.; Shin, Y.; Lee, H.S.; Lee, J.H.; Kim, G.T.; Song, J.S. A randomized controlled trial of ProRoot MTA, OrthoMTA and RetroMTA for pulpotomy in primary molars. *Oral Dis.* **2015**, *21*, 785–791. [CrossRef] [PubMed]

17. Kang, C.M.; Sun, Y.; Song, J.S.; Pang, N.S.; Roh, B.D.; Lee, C.Y.; Shin, Y. A randomized controlled trial of various MTA materials for partial pulpotomy in permanent teeth. *J. Dent.* **2017**, *60*, 8–13. [CrossRef] [PubMed]

18. Lee, B.N.; Son, H.J.; Noh, H.J.; Koh, J.T.; Chang, H.S.; Hwang, I.N.; Hwang, Y.C.; Oh, W.M. Cytotoxicity of newly developed ortho MTA root-end filling materials. *J. Endod.* **2012**, *38*, 1627–1630. [CrossRef] [PubMed]

19. Park, S.J.; Heo, S.M.; Hong, S.O.; Hwang, Y.C.; Lee, K.W.; Min, K.S. Odontogenic effect of a fast-setting pozzolan-based pulp capping material. *J. Endod.* **2014**, *40*, 1124–1131. [CrossRef] [PubMed]

20. Choi, Y.; Park, S.-J.; Lee, S.-H.; Hwang, Y.-C.; Yu, M.-K.; Min, K.-S. Biological effects and washout resistance of a newly developed fast-setting pozzolan cement. *J. Endod.* **2013**, *39*, 467–472. [CrossRef] [PubMed]

21. Nowicka, A.; Lipski, M.; Parafiniuk, M.; Sporniak-Tutak, K.; Lichota, D.; Kosierkiewicz, A.; Kaczmarek, W.; Buczkowska-Radlinska, J. Response of human dental pulp capped with biodentine and mineral trioxide aggregate. *J. Endod.* **2013**, *39*, 743–747. [CrossRef] [PubMed]

22. Lee, H.; Shin, Y.; Kim, S.O.; Lee, H.S.; Choi, H.J.; Song, J.S. Comparative Study of Pulpal Responses to Pulpotomy with ProRoot MTA, RetroMTA, and TheraCal in Dogs' Teeth. *J. Endod.* **2015**, *41*, 1317–1324. [CrossRef] [PubMed]

23. Lee, M.; Kang, C.M.; Song, J.S.; Shin, Y.; Kim, S.; Kim, S.O.; Choi, H.J. Biological efficacy of two mineral trioxide aggregate (MTA)-based materials in a canine model of pulpotomy. *Dent. Mater. J.* **2017**, *36*, 41–47. [CrossRef] [PubMed]

24. Kogan, P.; He, J.; Glickman, G.N.; Watanabe, I. The effects of various additives on setting properties of MTA. *J. Endod.* **2006**, *32*, 569–572. [CrossRef] [PubMed]

25. Song, M.; Yoon, T.S.; Kim, S.Y.; Kim, E. Cytotoxicity of newly developed pozzolan cement and other root-end filling materials on human periodontal ligament cell. *Restor. Dent. Endod.* **2014**, *39*, 39–44. [CrossRef] [PubMed]

26. De Deus, G.; Ximenes, R.; Gurgel-Filho, E.D.; Plotkowski, M.C.; Coutinho-Filho, T. Cytotoxicity of MTA and Portland cement on human ECV 304 endothelial cells. *Int. Endod. J.* **2005**, *38*, 604–609. [CrossRef] [PubMed]

27. Kim, M.; Yang, W.; Kim, H.; Ko, H. Comparison of the biological properties of ProRoot MTA, OrthoMTA, and Endocem MTA cements. *J. Endod.* **2014**, *40*, 1649–1653. [CrossRef] [PubMed]

28. Koseoglu, S.; Pekbagr Yan, K.T.; Kucukyilmaz, E.; Saglam, M.; Enhos, S.; Akgun, A. Biological response of commercially available different tricalcium silicate-based cements and pozzolan cement. *Microsc. Res. Tech.* **2017**, *80*, 994–999. [CrossRef] [PubMed]

29. Che, J.-L.; Kim, J.-H.; Kim, S.-M.; Choi, N.-k.; Moon, H.-J.; Hwang, M.-J.; Song, H.-J.; Park, Y.-J. Comparison of setting time, compressive strength, solubility, and pH of four kinds of MTA. *Korean J. Dent. Mater.* **2016**, *43*, 61–71. [CrossRef]

30. Lee, Y.-L.; Lee, B.-S.; Lin, F.-H.; Lin, A.Y.; Lan, W.-H.; Lin, C.-P. Effects of physiological environments on the hydration behavior of mineral trioxide aggregate. *Biomaterials* **2004**, *25*, 787–793. [CrossRef]

31. De Rossi, A.; Silva, L.A.; Gaton-Hernandez, P.; Sousa-Neto, M.D.; Nelson-Filho, P.; Silva, R.A.; de Queiroz, A.M. Comparison of pulpal responses to pulpotomy and pulp capping with biodentine and mineral trioxide aggregate in dogs. *J. Endod.* **2014**, *40*, 1362–1369. [CrossRef] [PubMed]

32. Nair, P.N.; Duncan, H.F.; Pitt Ford, T.R.; Luder, H.U. Histological, ultrastructural and quantitative investigations on the response of healthy human pulps to experimental capping with mineral trioxide aggregate: A randomized controlled trial. *Int. Endod. J.* **2008**, *41*, 128–150. [PubMed]

33. Parirokh, M.; Asgary, S.; Eghbal, M.J.; Stowe, S.; Eslami, B.; Eskandarizade, A.; Shabahang, S. A comparative study of white and grey mineral trioxide aggregate as pulp capping agents in dog's teeth. *Dent. Traumatol.* **2005**, *21*, 150–154. [CrossRef] [PubMed]

34. Bronckers, A.L.; Farach-Carson, M.C.; Van Waveren, E.; Butler, W.T. Immunolocalization of osteopontin, osteocalcin, and dentin sialoprotein during dental root formation and early cementogenesis in the rat. *J. Bone Miner. Res.* **1994**, *9*, 833–841. [CrossRef] [PubMed]

35. Widbiller, M.; Lindner, S.; Buchalla, W.; Eidt, A.; Hiller, K.-A.; Schmalz, G.; Galler, K. Three-dimensional culture of dental pulp stem cells in direct contact to tricalcium silicate cements. *Clin. Oral Investig.* **2016**, *20*, 237–246. [CrossRef] [PubMed]

36. Zanini, M.; Sautier, J.M.; Berdal, A.; Simon, S. Biodentine induces immortalized murine pulp cell differentiation into odontoblast-like cells and stimulates biomineralization. *J. Endod.* **2012**, *38*, 1220–1226. [CrossRef] [PubMed]

37. Zarrabi, M.H.; Javidi, M.; Jafarian, A.H.; Joushan, B. Histologic Assessment of Human Pulp Response to Capping with Mineral Trioxide Aggregate and a Novel Endodontic Cement. *J. Endod.* **2010**, *36*, 1778–1781. [CrossRef] [PubMed]

38. Asgary, S.; Parirokh, M.; Eghbal, M.J.; Ghoddusi, J. SEM evaluation of pulp reaction to different pulp capping materials in dog's teeth. *Iran. Endod. J.* **2006**, *1*, 117–123. [PubMed]

39. Simon, S.; Cooper, P.; Smith, A.; Picard, B.; Naulin Ifi, C.; Berdal, A. Evaluation of a new laboratory model for pulp healing: Preliminary study. *Inter. Endod. J.* **2008**, *41*, 781–790. [CrossRef] [PubMed]

40. AlShwaimi, E.; Majeed, A.; Ali, A.A. Pulpal Responses to Direct Capping with Betamethasone/Gentamicin Cream and Mineral Trioxide Aggregate: Histologic and Micro-Computed Tomography Assessments. *J. Endod.* **2016**, *42*, 30–35. [CrossRef] [PubMed]

41. Eskandarizadeh, A.; Shahpasandzadeh, M.H.; Shahpasandzadeh, M.; Torabi, M.; Parirokh, M. A comparative study on dental pulp response to calcium hydroxide, white and grey mineral trioxide aggregate as pulp capping agents. *J. Conserv. Dent.* **2011**, *14*, 351–355. [PubMed]

42. Shahravan, A.; Jalali, S.P.; Torabi, M.; Haghdoost, A.A.; Gorjestani, H. A histological study of pulp reaction to various water/powder ratios of white mineral trioxide aggregate as pulp-capping material in human teeth: A double-blinded, randomized controlled trial. *Int. Endod. J.* **2011**, *44*, 1029–1033. [CrossRef] [PubMed]

materials

MDPI

Article

Antibacterial, Hydrophilic Effect and Mechanical Properties of Orthodontic Resin Coated with UV-Responsive Photocatalyst

Akira Kuroiwa [1], Yoshiaki Nomura [2,*], Tsuyoshi Ochiai [3,4,5], Tomomi Sudo [1], Rie Nomoto [6], Tohru Hayakawa [6], Hiroyuki Kanzaki [1], Yoshiki Nakamura [1] and Nobuhiro Hanada [2]

[1] Department of Orthodontics, Tsurumi University School of Dental Medicine, Yokohama 230-8501, Japan; 2711005@stu.tsurumi-u.ac.jp (A.K.); sudo-tomomi@tsurumi-u.ac.jp (T.S.); kanzaki-h@tsurumi-u.ac.jp (H.K.); nakamura-ys@tsurumi-u.ac.jp (Y.N.)

[2] Department of Translational Research, Tsurumi University School of Dental Medicine, Yokohama 230-8501, Japan; hanada-n@tsurumi-u.ac.jp

[3] Photocatalyst Group, Research and Development Department, Local Independent Administrative Agency Kanagawa Institute of industrial Science and TEChnology (KISTEC), 407 East Wing, Innovation Center Building, KSP, 3-2-1 Sakado, Takatsu-ku, Kawasaki, Kanagawa 213-0012, Japan; pg-ochiai@newkast.or.jp

[4] Materials Analysis Group, Kawasaki Technical Support Department, KISTEC, Ground Floor East Wing, Innovation Center Building, KSP, 3-2-1 Sakado, Takatsu-ku, Kawasaki, Kanagawa 213-0012, Japan

[5] Photocatalysis International Research Center, Tokyo University of Science, 2641 Yamazaki, Noda, Chiba 278-8510, Japan

[6] Department of Dental Engineering, Tsurumi University School of Dental Medicine, Yokohama 230-8501, Japan; nomoto-r@tsurumi-u.ac.jp (R.N.); hayakawa-t@tsurumi-u.ac.jp (T.H.)

* Correspondence: nomura-y@tsurumi-u.ac.jp; Tel.: +81-045-580-8462

Received: 18 April 2018; Accepted: 21 May 2018; Published: 25 May 2018

Abstract: Photocatalysts have multiple applications in air purifiers, paints, and self-cleaning coatings for medical devices such as catheters, as well as in the elimination of xenobiotics. In this study, a coating of a UV-responsive photocatalyst, titanium dioxide (TiO_2), was applied to an orthodontic resin. The antibacterial activity on oral bacteria as well as hydrophilic properties and mechanical properties of the TiO_2-coated resin were investigated. ultraviolet A (UVA) (352 nm) light was used as the light source. Antibacterial activity was examined with or without irradiation. Measurements of early colonizers and cariogenic bacterial count, i.e., colony forming units (CFU), were performed after irradiation for different time durations. Hydrophilic properties were evaluated by water contact angle measurements. While, for the assessment of mechanical properties, flexural strength was measured by the three-point bending test. In the coat(+)light(+) samples the CFU were markedly decreased compared to the control samples. Water contact angle of the coat(+)light(+) samples was decreased after irradiation. The flexural strength of the specimen irradiated for long time showed a higher value than the required standard value, indicating that the effect of irradiation was weak. We suggest that coating with the ultraviolet responsive photocatalyst TiO_2 is useful for the development of orthodontic resin with antimicrobial properties.

Keywords: orthodontic resin; photocatalyst TiO_2; antibacterial; cariogenic; early colonizer; hydrophilic properties; irradiation

1. Introduction

Maintenance of good oral hygiene after an active orthodontic treatment is one of the most important procedures. In general, after an active orthodontic treatment, moved teeth and jawbone are retained by acrylic resin based retainer [1]. The retainer is typically used for at least two years.

The longer the retainer is used, the more unhygienic it becomes. Gradually, micro-organisms colonize on the surfaces of retainers, just as on dentures, and they often cause stomatitis, dental caries, periodontal disease, chronic atrophic or candidiasis [2–11]. Porosities on the outer and inner surfaces of retainer also provide favorable conditions for microbial colonization [12]. Therefore, prevention of micro-organism colonization is essential for the maintenance of good oral hygiene and prevention of oral diseases.

Management of oral biofilm is important to maintain a good oral status. Common oral diseases, dental caries and periodontitis, are caused by an imbalance between biofilms and host defenses [13]. The initial process in the oral biofilm formation starts with pellicle, covering the tooth surface within a few minutes after mechanical cleaning. The pellicle plays a major role in the development and maintenance of bacterial communities. Then, the complex bacterial communities develop on the pellicle within a few days, and their components can be divided into two categories—early colonizers, and late colonizers [14]. Early colonizers that directly adhere to the pellicles are predominantly streptococci [15]. Streptococci constitute 60% to 90% of the bacteria that colonize on the teeth in the first 4 h after professional cleaning [16]. These species are mainly Gram-positive and have minor pathogenic effects on periodontal tissue. Late colonizers, such as *Fusobacterium nucleatum*, *Porphyromonas gingivalis*, *Tannerella forsythia*, *Treponema denticola*, and *Aggregatibacter actinomycetemcomitans* tend to be more pathogenic than the early colonizers. The late colonizers alone cannot form a biofilm on the tooth surface, but they form a biofilm by their parasitical adherence to the early colonizers [14].

In the formation of a biofilm it is inevitable that colonizing bacteria primarily adhere to the surface of the retainer [17]. Therefore, it is critical to prevent the bacteria from adhering to the retainers. As the orthodontic acrylic resin and denture base acrylic resin have similar requirements for clinical use (ISO 20795), the results for the denture base resin are also applicable for orthodontic acrylic resin. Various approaches have been employed for the prevention of microbial biofilms. These include dental disinfection, denture cleaner, mixing resin with antibacterial agents, and coating resin with antiseptic [6,18–32]. However, the biofilms are resistant to antibacterial and antifungals agents [33,34]. Moreover, a long-term use of denture cleanser corrodes metals such as clasps [35,36]. Denture cleanser affects the color stability of the denture base acrylic resin [37]. Mechanical cleaning with the adjunctive use of antimicrobial solutions is helpful in reducing biofilm growth or preventing its formation. However, such approaches rely primarily on a patient's compliance, and may be compromised in pediatric, geriatric, and handicapped individuals. Silver nanoparticles impregnated in acrylic resin make the appliance antibacterial, but releasing silver nanoparticles from the resin is a limiting factor [24,25,27,31]. Incorporating fluorine and silver ions into resin elutes antimicrobials but it is available only for the first few weeks [27]. Moreover, elution of antimicrobial agents may result in deterioration of the mechanical properties of the retainer over time. This reduction renders the appliance more susceptible to fracture, due to its low resistance to impact, low flexural strength, or low fatigue strength [38]. Hence, novel and alternative methods to prevent the micro-organism colonization are required.

Application of a photocatalyst is one of the effective and safe approaches to remove biofilm from the dentures or retainer [39,40]. Photocatalyst reaction is defined as a photocatalyst promoted reaction on a solid surface, usually a semiconductor [41]. Titanium dioxide (TiO_2) is the most studied photocatalyst [39,41–50]. TiO_2 is biocompatible, nontoxic, and inexpensive. TiO_2 generates reactive oxygen species (ROSs) upon ultraviolet A (UVA) irradiation, and its strong oxidative power decomposes micro-organisms and organic materials [33,42–44,46,50]. Further, the photocatalyst reaction can obtain superhydrophilic properties by UV irradiation, and superhydrophilicity prevents dirt accumulation on the device.

The purpose of the present study was to test the clinical applications TiO_2 coated orthodontic resin based retainer. To this end, we investigated the antibacterial effects as well as mechanical, and hydrophilic properties of acrylic based orthodontic resin coated with the photocatalyst TiO_2 after irradiation with UVA. Thus, we determined its clinical suitability for use as an orthodontic retainer material. First, we examined the effect of the orthodontic resin coated with TiO_2 on bacteria for various irradiation durations. Further, we investigated the antimicrobial effect against early colonizers, the bacteria which are first attached to the appliances. The effects on *S. mutans* and *S. sobrinus*, which are the most well-known cariogenic bacteria, were investigated. Second, mechanical properties of the orthodontic resin are evaluated by irradiating for about 2 years, which is the recommended usage period of the orthodontic retainer. Third, the hydrophilic properties were investigated as one of the photocatalytic effects. Acquiring hydrophilic properties by decreasing the contact angle could lead to the prevention of bacterial adhesion and a self-cleaning function.

2. Materials and Methods

Autopolymerizable orthodontic acrylic resin (Ortho Crystal, Nissin Co., Tokyo, Japan), which consisted of a liquid component and a powder component, was used for this study. In the following sections, autopolymerizable orthodontic acrylic resin are referred to as 'resin'.

2.1. Sample Preparation

According to the manufacturer's instructions, a powder-to-liquid ratio of 10 g:4.5 mL was used. Powder and liquid components were mixed under vibration for homogenization and removal of the trapped air. The slurry was poured into an aluminum open mold and pressed using a pair of glass plates to fabricate the specimens of different dimensions: 50 mm × 50 mm × 3.0 mm (n = 60), 50 mm × 50 mm × 3.0 mm (n = 20), and 64 mm × 10 mm × 3.5 mm (n = 35) for the antibacterial properties tests, photoinduced hydrophilic tests, and mechanical properties tests, respectively. The slurry resin was immediately transferred in the polymerization equipment for a dental technique at 40 °C (manufacturer's recommendation: 30–40 °C), and 0.25 MPa for 30 min to enhance curing (Fit Resin Multicure, Shofu Inc., Kyoto, Japan). Following preparation, each specimen was kept at room temperature in water for 12 h to eliminate the residual monomer. All test specimens were gradually grinded with waterproof polishing paper, having a grain size of approximately 30 μm (P500), 18 μm (P1000), and 15 μm (P1200). The specimens were then divided into four test groups for the assessment of the antibacterial properties and water contact angle measurement of base resin coated with thin film of photocatalytic TiO_2. Uncoated resin and non-lighted resin were used as control groups for their respective experimental group.

A spin-coating methods was used to apply ultraviolet-light-responsive photocatalytic titanium dioxide (UV-TiO_2) to the surface of the materials. The surface modification of the specimen with commercialized photocatalytic TiO_2 (NRC 350A and 360C, Nippon Soda Co., Ltd., Tokyo, Japan) was carried out by a sol-gel thin film spin-coating method according to the manufacturer's instructions. NRC 350A was coated to the surface of the materials. After coating, they were dried in a desiccator under 30 °C for 48 h. Then NRC 360Cwas coated, and dried in a desiccator in the same way. After the coating, the surface conditions of the materials were observed by a scanning electron microscope (SEM; JSM-5600LV, JEOL, Tokyo, Japan) at an accelerating voltage of 15 kV. Specimens were sputter-coated with Au prior to the SEM observations.

We covered samples with a glass to prevent drying, and irradiation was carried out from above. UVA from a black light source (wavelength: 352 nm, FL15BLB, Toshiba Co., Tokyo, Japan) was selected as the light source for catalytic excitation. The irradiation was performed at a distance of 10 cm (1.0 mW/cm^2 under the glass).

2.2. Bacterial Strains

Streptococcus mutans ATCC 25175 (*S. mutans*), *Streptococcus sobrinus* ATCC33478 (*S. sobrinus*), *Streptococcus gordonii* ATCC 10558 (*S. gordonii*), *Streptococcus oralis* ATCC 35037 (*S. oralis* ATCC), *Streptococcus oralis* GTC 276 (*S. oralis* GTC), *Streptococcus sanguinis* ATCC 10556 (*S. sanguinis*), and *Streptococcus mitis* MRS 08-31 (*S. mitis*) were used in this study. *S. mutans* and *S. sobrinus* are cariogenic bacteria. *S. gordonii*, *S. oralis* ATCC, *S. oralis* GTC, *S. sanguinis*, and *S. mitis* are the early colonizers. These bacterial species were inoculated into 4 mL of Tryptic Soy (TS) broth (Becton, Dickinson and Company, Sparks, MD, USA) and were cultured aerobically at 37 °C for 16 h. They were harvested by centrifugation at 3000× rpm for 15 min and then suspended with phosphate buffered saline (PBS) resulting in an optical density at 540 nm (OD_{540} equal to 1.0).

2.3. Antibacterial Test

A micro-organism suspension (500 μL) adjusted to $OD_{540} = 1.0$ was dropped directly on the surface of the coated and non-coated specimen on ice, and UV irradiation was performed for 0, 15, 30, 60, 90, 120, 150, and 180 min. After irradiation, each bacterial cell pellet was suspended in 1 mL PBS and subjected to serial 10-fold dilutions in PBS. The dilutions of each bacteria were inoculated on MS agar (Difico Mitis Salivarius Agar (semi-selective medium for streptococci); BD Biosciences, Flanklin Lakes, NJ, USA) in petri dishes with spiral plating equipment (Eddy Jet, IUL SA, Barcelona, Spain), and petri dishes were incubated under anaerobic conditions in an AnaeroPack-Anaero box (AnaeroPack System, Mitsubishi Gas Chemical Co., Inc., Tokyo, Japan) at 37 °C for 48 h. The number of colonies was counted in accordance with the spiral plater instruction manual. Each measurement was repeated three times.

2.4. Bending Test

The rectangular plates (64 mm × 10 mm × 3.5 mm) were immersed in distilled water at 37 °C for 0, 200, 400, 600, 800, 1000, 1200 h under UVA irradiation from a distance of 10 cm. The three-point bending test was conducted using a universal testing machine (EZ Test 500 N, Shimadzu Co., Kyoto, Japan) at a crosshead speed of 5 mm/min and a span length of 50.0 mm ($n = 5$). Then, load-displacement curves were plotted to measure bending strength, elastic modulus, and toughness. The test was conducted according to ISO20795-2 standard.

2.5. Water Contact Angle Measurement

The rectangular plates (50 mm × 50 mm × 3.0 mm) were prepared as indicated above and spin-coated with 125 μL of the experimental coating materials on each sample. The water contact angles were measured using a contact angle device (FTA125, First Ten Ångstroms, Portsmouth, VA, USA) at 25 °C. For surface analysis of the hydrophilic characteristics, 3.5 μL of deionized water (Milli-Q Plus system, Japan Millipore, Tokyo, Japan) was dropped on the surface, and video images were taken. Video images were automatically inputted to an attached computer in which the contact angles were measured using an image analysis program (FTA32 video, First Ten Ångstroms). Water contact angles were measured every 30 min for a period of 20 s at 25 °C.

2.6. Statistical Analysis

Antibacterial effects were analyzed by three factors: TiO_2 coating, presence of UVA, and irradiated time ($n = 3$). The significance of differences in the antibacterial effects was examined using three-way ANOVA or two-way ANOVA. Mechanical properties were examined using one-way ANOVA, with irradiated time included as a factor ($n = 5$). Tukey's HSD test was then used to determine the positions of significance. All statistical analyses were performed using IBM SPSS Statistics version 22.0 (IBM, Tokyo, Japan). A significance level of $p < 0.05$ was used.

3. Results

Antibacterial Test

The antimicrobial activity of the resin coated with TiO_2 was examined by bacterial count of the early colonizers and cariogenic bacteria under UV irradiation. Figure 1 shows the antibacterial activity of the TiO_2-coated resin surfaces against *S. gordonii* ATCC 10558, and *S. oralis* ATCC 35037 after irradiation. The coat(−)light(−) group was used as control group. The coat(+)light(−) group showed no significant differences compared with the control group. Hence, there was no effect with the coat alone. We also examined the activity with UV alone. Coat(−)light(+) induced a significant reduction in the number of colony. The number of *S. gordonii* was reduced from 1.6×10^7 colony-forming units/mL (CFU/mL) to 1.6×10^6 CFU/mL (after 120 min of UV irradiation), and that of *S. oralis* ATCC was reduced from 6.5×10^5 CFU/mL to 3.1×10^4 CFU/mL (after 90 min of UV irradiation). Consequently, about 90%, 95% colonies of *S. gordonii* and *S. oralis* ATCC were not formed on the coated plates upon irradiation (Figure 1A,B). On the other hand, coat(+)light(+) group showed a significant reduction in the CFU. The number of *S. gordonii* was reduced from 1.6×10^7 CFU/mL to 2.7×10^4 CFU/mL (after 120 min of UV irradiation) and that of *S. oralis* ATCC from 6.5×10^5 CFU/mL to 5.2×10^3 CFU/mL (after 90 min of UV irradiation). Thus, about 99.9% or more colonies of both *S. gordonii* and *S. oralis* ATCC were not formed on the coated plates upon irradiation (Figure 1A,B). Hence, coat(+)light(+) group showed a hundred times higher antibacterial effect, when compared with coat(−)light(+).

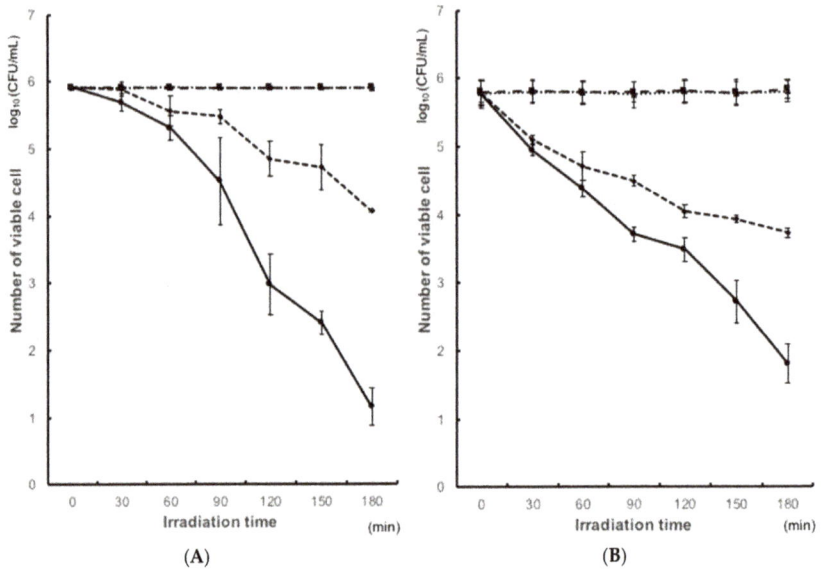

Figure 1. Antibacterial effects of TiO$_2$ photocatalysis against (**A**) *Streptococcus gordonii*; (**B**) *Streptococcus oralis* ATCC. ● coat(+)light(+): experiment group containing powdered TiO$_2$ with irradiation. ◆ coat(−)light(−): control group without both TiO$_2$ and irradiation. ■ coat(+)light(−): experiment group in the presence of powdered TiO$_2$ without irradiation. ▲ coat(−)light(−): experiment group without TiO$_2$, but with irradiation.

Compared with the TiO$_2$-noncoated samples, TiO$_2$-coated samples showed a rapid decrease in the level of CFU of *Streptococcus gordonii*, particularly in the early stages (90 min and 120 min) of irradiation. Therefore, other strains were also examined at 90 and 120 min (Figure 2A,B). The cell viability on each sample before UV irradiation was set as 100%. The coat(+)light(+) clearly showed a great reduction in the cell viability. Cell viability was reduced to 0.2% for *S. sobrinus*, 0.9% for *S. oralis* GTC, 5.4% for *S. mutans*, and 9.9% for *S. sanguinis*. Clearly, the coat(+)light(+) samples exhibited the best antimicrobial performance in all microbes after 90 min UV irradiation. Similar results were observed even after 120 min of irradiation.

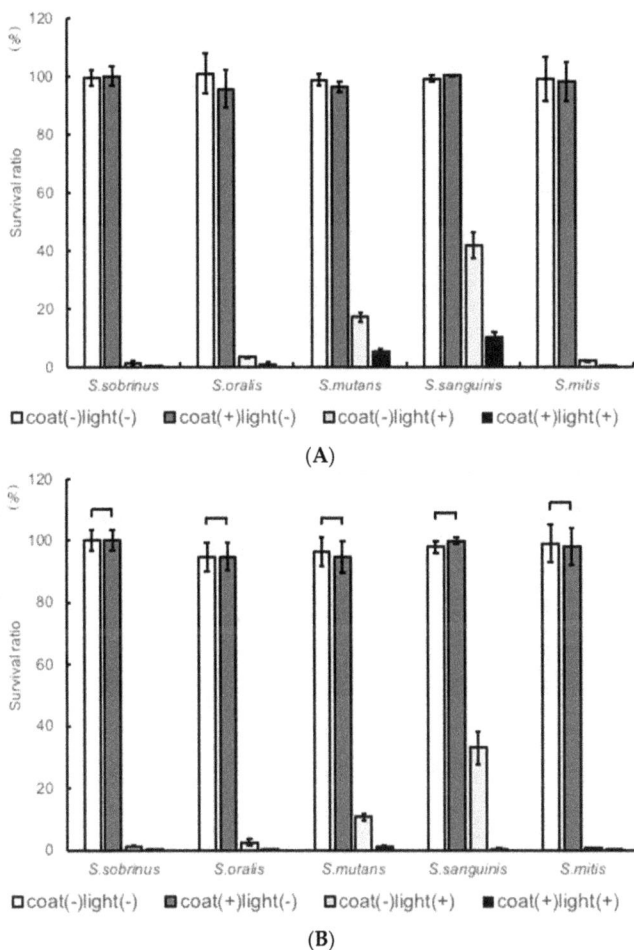

Figure 2. Antibacterial effects on various specimens of *Streptococcus sobrinus*, *Streptococcus oralis* GTC, *Streptococcus mutans*, and *Streptococcus sanguinis* at (**A**) 90 min, and (**B**) 120 min of ultraviolet A (UVA) irradiation.

Table 1 shows explanatory variables related to bacterial counts by three way ANOVA. The coefficients of bacterial counts in light(+) and interaction of light(+) and coat(+) were all negative. The coefficients of light(+) for *S. oralis* and *S. sobrinus* were higher than those by interaction of light(+) and coat(+). In contrast, the coefficients of light(+) of *S. gordonii*, *S. mutans*, and *S. sanguinis* were higher than those by interaction of light(+) and coat(+). The coefficient of both light(+) and interaction of light(+) and coat(+) of *S. mutans* was the highest in all the bacteria. The coefficient for irradiation time of 120 min were significantly different in all bacterial species. In all the bacteria, the coefficient of coat(+) was almost 0. On the other hand, the coefficient of light(+) and interaction of light(+) and coat(+) were statistically significant. These results indicated that the antibacterial effect of the photocatalyst was exerted by UVA irradiation. It also showed differences in susceptibility of oral bacteria to UVA.

Table 1. Models of three way ANOVA for changes in bacterial counts. p-values less than 0.05 were considered statistically significant.

	S. gordonii ATCC10558			S. mitis			S. oralis ATCC35037			S. oralis GTC276		
	Coefficient	95% CI	p-Value	Coefficient	95% CI	p-Value	Coefficient	95% CI	p-Value	Coefficient	95% CI	p-Value
Intercept	6.563	6.04–7.08	0.999<	8.267	7.57–8.96	0.999<	6.584	6.18–6.98	0.999<	8.840	8.30–9.37	0.999<
time 30 min	−0.061	−0.67–0.55	0.845	−0.306	−1.13–0.52	0.462	−0.372	−0.84–0.10	0.123	−0.498	−1.13–0.13	0.122
60 min	−0.238	−0.85–0.37	0.444	−0.753	−1.57–0.07	0.073	−0.616	−1.09–−0.14	0.012	−0.700	−1.33–−0.06	0.031
90 min	−0.458	−1.07–0.15	0.143	−1.246	−2.07–−0.42	<0.001	−0.846	−1.32–−0.37	<0.001	−0.877	−1.51–−0.24	<0.001
120 min	−1.001	−1.61–−0.38	<0.001	−1.671	−2.49–−0.84	<0.001	−1.003	−1.47–−0.52	<0.001	−1.260	−1.89–−0.62	<0.001
150 min	−1.175	−1.79–−0.55	<0.001	−1.935	−2.75–−1.10	<0.001	−1.231	−1.70–−0.75	<0.001	−1.576	−2.21–−0.94	<0.001
180 min	−1.654	−2.27–−1.03	<0.001	−2.432	−3.25–−1.60	<0.001	−1.491	−1.96–−1.01	<0.001	−1.820	−2.45–−1.18	<0.001
coat(+)	−0.001	−0.46–0.46	0.996	−0.015	−0.63–−0.60	0.963	0.017	−0.34–0.37	0.925	−0.008	−0.48–0.47	0.975
light(+)	−0.695	−1.16–−0.22	<0.001	−1.457	−2.08–−0.83	<0.001	−1.247	−1.60–−0.88	<0.001	−1.338	−1.81–−0.85	<0.001
interaction of light and coat	−1.216	−1.87–−0.55	<0.001	−1.845	−2.72–−0.96	<0.001	−0.727	−1.23–−0.21	0.006	−1.206	−1.88–−0.52	<0.001

	S. mutans			S. sobrinus			S. sanguinis		
	Coefficient	95% CI	p-Value	Coefficient	95% CI	p-Value	Coefficient	95% CI	p-Value
Intercept	5.609	5.38–5.83	0.999<	11.066	10.68–11.44	0.999<	9.280	8.90–9.65	0.999<
time 30 min	−0.119	−0.38–−0.14	0.376	−0.585	−1.03–−0.13	0.011	−0.105	−0.54–0.33	0.638
60 min	−0.272	−0.53–−0.01	0.044	−0.938	−1.38–−0.49	<0.001	−0.231	−0.67–0.20	0.299
90 min	−0.347	−0.61–−0.82	0.011	−1.121	−1.56–−0.67	<0.001	−0.348	−0.78–0.09	0.120
120 min	−0.512	−0.77–−0.24	<0.001	−1.146	−1.59–−0.70	<0.001	−0.718	−1.15–−0.27	0.002
150 min	−0.512	−0.77–−0.24	<0.001	−1.404	−1.85–−0.95	<0.001	−0.848	−1.28–−0.40	<0.001
180 min	−0.908	−1.17–−0.64	<0.001	−1.555	−2.00–−1.10	<0.001	−1.239	−1.67–−0.79	<0.001
coat(+)	<0.001	−0.20–0.20	0.998	0.001	−0.33–0.33	0.995	<0.001	−0.33–0.33	0.999<
light(+)	−0.480	−0.67–−0.27	<0.001	−1.587	−1.92–−1.25	<0.001	−0.514	−0.84–−0.18	0.003
interaction of light and coat	−0.557	−0.84–−0.27	<0.001	−0.680	−1.15–−0.20	0.006	−0.948	−1.41–−0.47	<0.001

SEM was used to observe the cross section of coating; cross sectional photographs are shown in Figure 3.

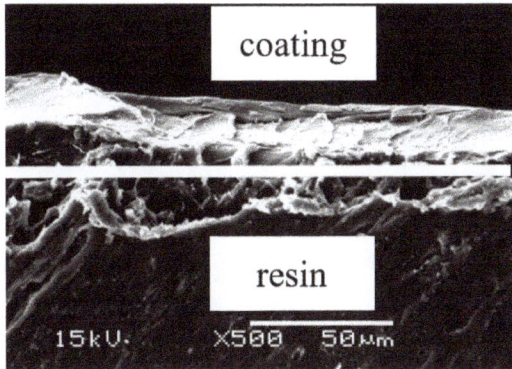

Figure 3. Scanning electron microscopy of cross sectional photographs of the TiO_2 coating.

The Flexural strength (Fs) and flexural modulus (Fm) of resin plates after UV irradiation are shown in Figure 4A,B, respectively. There was no difference between the irradiated test pieces, and all irradiated specimens fulfilled the requirements of the ISO 20795-2:2010 standard for Fs testing after 1200 h of UV irradiation (>65 MPa) (Figure 4A).

In the same way, all irradiated TiO_2-coated specimens fulfilled the requirements of the ISO 20795-2:2010 standard for Fm testing after 1200 h of UV irradiation (>2000 MPa) (Figure 4B).

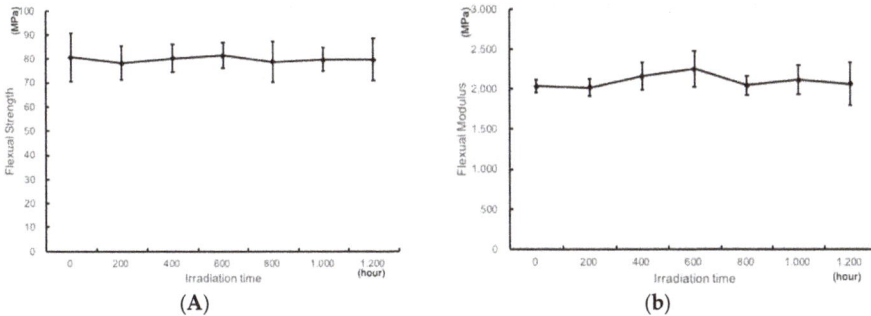

Figure 4. Flexural strength (**A**) and Flexural modulus (**B**) of the TiO_2-coated resin plates upon UV irradiation.

Figure 5A shows the water contact angle for TiO_2-noncoated groups and TiO_2-coated groups. Figure 5B shows the images illustrating the wettability of water on TiO_2-coated specimens or TiO_2-noncoated specimens after 120 min of UV irradiation.

The water contact angle of TiO_2-coated specimen was very small from the start of experiment, compared to that of TiO_2-noncoated and it gradually decreased with time.

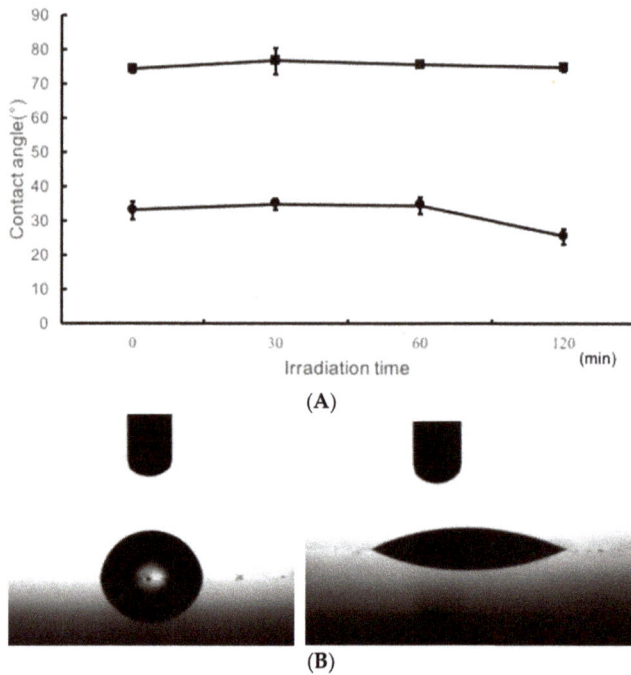

Figure 5. (A) Water contact angle of the resin plates with TiO$_2$-coating upon UV irradiation for 0, 30, 60, and 120 min. **(B)** Image shows contact angle of a water droplet on non-coated (**left**) and TiO$_2$-coated resin plate (**right**) after 120 min of UV irradiation.

4. Discussion

This study demonstrates the antibacterial effects of TiO$_2$-coating on *S. mutans*, *S. sobrinus*, and early colonizers upon UVA irradiation. When TiO$_2$-coating and UVA were used together, a significant reduction in the microbial count was observed. In a previous study, TiO$_2$ coated orthodontic arch wires showed the photocatalytic antibacterial effects on *S. mutans*, and its reduction rate was more than 99.99% by bacterial count after 1 h irradiation [51]. Also, a tissue conditioner containing a TiO$_2$ photocatalyst decreased bacterial counts for *Escherichia coli* (about 90%), *Staphylococcus aureus* (>99.99%), and *S. mutans* (about 90%) after 2 h of irradiation [52]. Moreover, the photocatalytic antibacterial effects of metal specimens coated with two crystalline forms of TiO$_2$ by thermal and anodic oxidation decreased bacterial counts for *S. mutans* (about 90%) after 60 min irradiation [43]. Furthermore, titanium disks coated with anatase-rich titanium dioxide (TiO$_2$) reduced amount of viable cells of *S. oralis* by 40% after 24 h UVA exposure [53]. Interestingly, anodized titanium (AO) decreased survival ratio of *S. sanguinis* (70%) upon 2 h of UV irradiation [50]. Also, after 20 min of UV exposure to TiO$_2$ surfaces, viabilities of *S. mutans* were reduced by 65% [54]. There were a few reports on *S. gordonii* and *S. sobrinus*. Finally, our results showed the bacterial count reduction rate of *S. gordonii* (>90%) (Figure 1), *S. mutans* (87%), *S. sobrinus* (>99%), *S. oralis* (98%), *S. sanguinis* (90%), and *S. mitis* (>99%) (Figure 2) after 90 min irradiation. The reduction rates of *S. oralis*, and *S. sanguinis* were better than those of past study [50,53]. The reduction rate of *S. mutans* was similar as previous studies [43,52–54]. However, a greater reduction rates of bacteria than those in our study were reported [51]. Although factors that contribute to the difference in the reduction rate are unknown, this may be due to the difference in the components of the photocatalyst used. This may also be due to the difference in the surface properties of the coating. Similar to the previous reports, the photocatalytic reaction induced a relatively mild decrease in the bacterial counts after the first 20 min of irradiation and showed a rapid decrease upon subsequent irradiation (Figure 1A) [55]. There is a difference in reaction by bacteria, but only upon irradiation for at least 90 min. Also, compared to conventional cleaning methods, cleaning is facilitated by coating TiO_2. This may lead to improvement in patient's compliance, reduced cost for equipment cleaning, and prevention of unpleasant odors.

UVA irradiation alone showed a decrease in bacterial counts. Furthermore, the photocatalytic activities of TiO_2 coating decreased significantly in bacterial counts of *S. gordonii*, *S. oralis* ATCC 35037 (Figure 1), *S. sobrinus*, *S. mutans*, *S. oralis* GTC, *S. sanguinis*, and *S. mitis* (Figure 2). When uncoated resins were irradiated with UVA light, all the bacteria reduced in counts. These reductions in the light(+)coat(−) groups were expected outcomes and followed a similar trend as a previous study, which reported that the viability of *S. mutans* decreased significantly after 60 min of UVA irradiation as compared to the control which was not irradiated [43]. The decrease may be associated with the cell-damaging effect of UVA. Coat(+)light(+) group showed a higher antibacterial effect, as compared to the coat(−)light(+) group. The difference of antibacterial effect between these group was explained in the Table 1.

The primary step in photocatalytic decomposition consists of hydroxyl radical attack on the bacterial cell wall [48]. This leads to increased permeability which allows radicals to reach and damage the cytoplasmic membrane causing lipid peroxidation and thereby causing membrane disorder [48]. The antibacterial effect of TiO_2 is associated to this disorder of cytoplasmic membrane [48,56]. We propose that the observed bacterial type-dependent variation in the antimicrobial effects may be due to differential effects of hydroxyl radicals on distinct bacterium species [57,58]. The antibacterial effect of the TiO_2 coating for various organisms is determined primarily by the complexity and density of the cell walls, as well as by the types of micro-organisms [59].

In this study, we observed that UVA irradiation has antibacterial effect. In cariogenic bacteria, *S. mutans* was more resistant to UVA than *S. sobrinus*. This may be because of the higher GC content of *S. sobrinus* than *S. mutans*. It has been proposed that species with genomes exhibiting a high GC content are more susceptible to UV-induced mutagenesis [60]. Also, our results showed that the coefficients for *S. sanguinis* and *S. gordonii* in light(+) treatment were low. Consistently, it has been demonstrated that *S. sanguinis* and *S. gordonii* have higher resistance to H_2O_2 [61].

In all bacteria, *S. mutans* was the most resistant to UVA. Conversely, *S. sobrinus* and *S. oralis* were highly susceptible to UVA. Several factors may contribute to the cause of these variable responses to UVA among species. For instance, the production of various ROSs involved in inducing UV damage may vary among species. In addition, the method of defending and repairing DNA damage differs among bacteria [61–63]. Also, the oxidative damage to biomolecules and counteracting protective mechanisms underlie the variability in UVA sensitivity among different bacterial species [64]. However, the reason of higher sensitivity of *S. oralis* and *S. sobrinus* to UV irradiation is unknown (Table 1). Further experiments are needed to understand the mechanistic basis for the variable susceptibility of oral bacteria to UVA.

In general, orthodontic patients are young, and oral care is a major problem for these patients. *S. mutans* and *S. sobrinus* are the most harmful cariogenic bacteria and TiO_2 coating have shown to be effective against them.

Bacteria form biofilms on the surface of the device. During biofilm formation, adherent bacteria produce a polymer matrix in which the community becomes embedded and biofilm bacteria are notoriously resistant to antimicrobial substances [65]. Once they are established on the exposed surface of a dental device, removal of biofilms can be extremely difficult. Effective methodology for cleaning of dental device is not well established. Application of photocatalyst is considered one of the potential strategy to overcome these problems. Therefore, further study is necessary for understanding the effects of TiO_2 on various organisms not only in vitro conditions but also in vivo and intraoral conditions.

With regard to flexural strength, there was no significant difference in bending strength even after irradiation for a long time. UV irradiated specimens fulfilled the requirements of the ISO 20795-2 standard for Fs testing (Figure 4A). Similarly, regarding strength modulus, there was no difference in bending strength even after irradiation for a long time (Figure 4B). These data imply that the resin can withstand irradiation for a long time to ensure a long-term clinical use of orthodontics. Even when irradiating for about 2 years, which is the recommended use period of the retainer, the durability was satisfactory. It was shown that clinical application is achievable.

Moreover, TiO_2-coating improved the hydrophilic properties of the surface of the denture base acrylic resin (Figure 5A,B). A previous study reported that the water contact angle of surfaces in TiO_2-coated resin was 68.1 ± 3.4 degrees [66]. However, TiO_2-coating makes the resin surface more hydrophilic, with a water contact angle of 25.4 ± 2.1 degrees (Figure 5). It has been reported that TiO_2 coating applied to acrylic resin inhibits the adhesion of *S. sanguinis* and *C. albicans* organisms [67,68]. In this study, enhancement in the hydrophilic properties of acrylic resin based orthodontic resin surface suppresses the adhesion of early colonizer, the subsequent adhesion of other microbes, which could reduce the total number of microbes adhering to orthodontic resin. Suppression of early colonizer could reduce further bacterial adhesion thereby reducing the risk of systemic disease. Moreover, improvement in the hydrophilic properties of orthodontic resin surface can suppress adhesion of other dirt such as food debris. Even without irradiation, the coat(+) group showed higher hydrophilicity making it easier to remove dirt.

We would like to establish the novel home care method for orthodontic retainer, with use of TiO_2-coating and UV irradiation. As one of the clinical applications, patients place the retainer under the UV lamp and they can also easily clean it at home. This cleaning methods is very simple. In addition, cleaning up the device can be carried out by a professional at the time of visit to the clinic. Our method can be applied not only to a retainer but also to other orthodontic appliances (expansion plate, functional orthodontic appliance), as well as to denture base and occlusal splint.

One of the important aspect of our method is biocompatibility. It has been reported that in animals the TiO_2-coated resin has no irritation to the oral mucosa, nor does it cause skin sensitization. Any elution of components from the coating has no deleterious effects on the tissues [69].

Overall, we demonstrate that the TiO_2-coated resin exposed to UVA irradiation shows great reduction of microbial counts when compared with uncoated and coated without UVA-exposed samples. In addition, the durability of the specimen showed a higher value than the required standard value, indicating that the effect of irradiation was small. In conclusion, the results of this preliminary study suggest that the antibacterial effect of TiO_2-coated resin can be beneficial in long-lasting orthodontic treatments.

5. Conclusions

The antimicrobial activity of the resin coated with TiO_2 was examined by bacterial count of the early colonizers and cariogenic bacteria under UV irradiation. The results of present study suggest that coating with the ultraviolet responsive photocatalyst TiO_2 is useful for antimicrobial properties of removable orthodontic resin based retainer.

Author Contributions: A.K., Y.N., T.O., T.S., R.N., T.H., H.K., Y.N., and N.H. conceived of and designed the experiments; A.K. and T.S. performed the experiments; A.K. and Y.N. analyzed the data; Y.N., T.O., R.N., and T.H. contributed materials/analysis tools; A.K., Y.N., T.O., and Y.N. wrote the manuscript.

Funding: This research received no external funding.

Acknowledgments: The authors are grateful to Nihon Sotatu Co., for supplying the TiO_2 coating.

Conflicts of Interest: The authors declare no conflict of interest.

References

1. Littlewood, S.J.; Millett, D.T.; Doubleday, B.; Bearn, D.R.; Worthington, H.V. Retention procedures for stabilising tooth position after treatment with orthodontic braces. *Cochrane Database Syst. Rev.* **2006**, *51*, 94–95.
2. Batoni, G.; Pardini, M.; Giannotti, A.; Ota, F.; Giuca, M.R.; Gabriele, M.; Campa, M.; Senesi, S. Effect of removable orthodontic appliances on oral colonisation by mutans streptococci in children. *Eur. J. Oral Sci.* **2001**, *109*, 388–392. [CrossRef] [PubMed]
3. Bjerklin, K.; Gärskog, B.; Rönnerman, A. Proximal caries increment in connection with orthodontic treatment with removable appliances. *Br. J. Orthod.* **1983**, *10*, 21–24. [CrossRef] [PubMed]
4. Hibino, K.; Wong, R.W.; Hagg, U.; Samaranayake, L.P. The effects of orthodontic appliances on candida in the human mouth. *Int. J. Paediatr. Dent.* **2009**, *19*, 301–308. [CrossRef] [PubMed]
5. Kiyoko, T.; Kazuhiko, N.; Sonoko, M.; Atsuko, T.; Takashi, O. Clinical and microbiological evaluations of acute periodontitis in areas of teeth applied with orthodontic bands. *Pediatr. Dent. J.* **2005**, *15*, 212–218.
6. Kuroki, K.; Hayashi, T.; Sato, K.; Asai, T.; Okano, M.; Kominami, Y.; Takahashi, Y.; Kawai, T. Effect of self-cured acrylic resin added with an inorganic antibacterial agent on *Streptococcus mutans*. *Dent. Mater. J.* **2010**, *29*, 277–285. [CrossRef] [PubMed]
7. Madurantakam, P.; Kumar, S. Fixed and removable orthodontic retainers and periodontal health. *Evid. Based Dent.* **2017**, *18*, 103–104. [CrossRef] [PubMed]
8. Ramage, G.; Tomsett, K.; Wickes, B.L.; López-Ribot, J.L.; Redding, S.W. Denture stomatitis: A role for candida biofilms. *Oral Surg. Oral Med. Oral Pathol. Oral Radiol. Endodontol.* **2004**, *98*, 53–59. [CrossRef]
9. Turkoz, C.; Canigur Bavbek, N.; Kale Varlik, S.; Akca, G. Influence of thermoplastic retainers on *Streptococcus mutans* and *Lactobacillus* adhesion. *Am. J. Orthod. Dentofac. Orthop.* **2012**, *141*, 598–603. [CrossRef] [PubMed]
10. Zharmagambetova, A.; Tuleutayeva, S.; Akhmetova, S. Microbiological aspects of the orthodontic treatment. *Georgian Med. News* **2017**, *264*, 39–43.
11. Zingler, S.; Pritsch, M. Association between clinical and salivary microbial parameters during treatment with removable orthodontic appliances with or without use of fluoride mouth rinse. *Eur. J. Paediatr. Dent.* **2016**, *17*, 181–187. [PubMed]
12. Shay, K. Denture hygiene: A review and update. *J. Contemp. Dent. Pract.* **2000**, *1*, 28–41. [PubMed]
13. Song, W.S.; Lee, J.K.; Park, S.H.; Um, H.S.; Lee, S.Y.; Chang, B.S. Comparison of periodontitis-associated oral biofilm formation under dynamic and static conditions. *J. Periodontal Implant Sci.* **2017**, *47*, 219–230. [CrossRef] [PubMed]
14. Kolenbrander, P.E.; Andersen, R.N.; Blehert, D.S.; Egland, P.G.; Foster, J.S.; Palmer, R.J. Communication among oral bacteria. *Microbiol. Mol. Biol. Rev.* **2002**, *66*, 486–505. [CrossRef] [PubMed]
15. Teles, F.R.; Teles, R.P.; Sachdeo, A.; Uzel, N.G.; Song, X.Q.; Torresyap, G.; Singh, M.; Papas, A.; Haffajee, A.D.; Socransky, S.S. Comparison of microbial changes in early redeveloping biofilms on natural teeth and dentures. *J. Periodontol.* **2012**, *83*, 1139–1148. [CrossRef] [PubMed]
16. Nyvad, B.; Kilian, M. Microbiology of the early colonization of human enamel and root surfaces in vivo. *Scand. J. Dent. Res.* **1987**, *95*, 369–380. [CrossRef] [PubMed]
17. Verran, J.; Motteram, K.L. The effect of adherent oral streptococci on the subsequent adherence of candida albicans to acrylic in vitro. *J. Dent.* **1987**, *15*, 73–76. [CrossRef]
18. Da Silva, P.M.; Acosta, E.J.; Pinto Lde, R.; Graeff, M.; Spolidorio, D.M.; Almeida, R.S.; Porto, V.C. Microscopical analysis of candida albicans biofilms on heat-polymerised acrylic resin after chlorhexidine gluconate and sodium hypochlorite treatments. *Mycoses* **2011**, *54*, e712–e717. [CrossRef] [PubMed]
19. De Andrade, I.M.; Cruz, P.C.; da Silva, C.H.; de Souza, R.F.; Paranhos Hde, F.; Candido, R.C.; Marin, J.M.; de Souza-Gugelmin, M.C. Effervescent tablets and ultrasonic devices against candida and mutans streptococci in denture biofilm. *Gerodontology* **2011**, *28*, 264–270. [CrossRef] [PubMed]
20. De Freitas Fernandes, F.S.; Pereira-Cenci, T.; da Silva, W.J.; Ricomini Filho, A.P.; Straioto, F.G.; Cury, A.A.D.B. Efficacy of denture cleansers on *Candida* spp. Biofilm formed on polyamide and polymethyl methacrylate resins. *J. Prosthet. Dent.* **2011**, *105*, 51–58. [CrossRef]
21. De Souza, R.F.; de Freitas Oliveira Paranhos, H.; Lovato da Silva, C.H.; Abu-Naba'a, L.; Fedorowicz, Z.; Gurgan, C.A. Interventions for cleaning dentures in adults. *Cochrane Database Syst. Rev.* **2009**, CD007395. [CrossRef] [PubMed]

22. Dhamande, M.M.; Pakhan, A.J.; Thombare, R.U.; Ghodpage, S.L. Evaluation of efficacy of commercial denture cleansing agents to reduce the fungal biofilm activity from heat polymerized denture acrylic resin: An in vitro study. *Contemp. Clin. Dent.* **2012**, *3*, 168–172. [CrossRef] [PubMed]

23. Farhadian, N.; Usefi Mashoof, R.; Khanizadeh, S.; Ghaderi, E.; Farhadian, M.; Miresmaeili, A. *Streptococcus mutans* counts in patients wearing removable retainers with silver nanoparticles vs. those wearing conventional retainers: A randomized clinical trial. *Am. J. Orthod. Dentofac. Orthop.* **2016**, *149*, 155–160. [CrossRef] [PubMed]

24. Monteiro, D.R.; Gorup, L.F.; Takamiya, A.S.; de Camargo, E.R.; Filho, A.C.; Barbosa, D.B. Silver distribution and release from an antimicrobial denture base resin containing silver colloidal nanoparticles. *J. Prosthodont.* **2012**, *21*, 7–15. [CrossRef] [PubMed]

25. Oei, J.D.; Zhao, W.W.; Chu, L.; DeSilva, M.N.; Ghimire, A.; Rawls, H.R.; Whang, K. Antimicrobial acrylic materials with in situ generated silver nanoparticles. *J. Biomed. Mater. Res. B Appl. Biomater.* **2012**, *100*, 409–415. [CrossRef] [PubMed]

26. Regis, R.R.; Zanini, A.P.; Della Vecchia, M.P.; Silva-Lovato, C.H.; Oliveira Paranhos, H.F.; de Souza, R.F. Physical properties of an acrylic resin after incorporation of an antimicrobial monomer. *J. Prosthodont.* **2011**, *20*, 372–379. [CrossRef] [PubMed]

27. Shinonaga, Y.; Arita, K. Antibacterial effect of acrylic dental devices after surface modification by fluorine and silver dual-ion implantation. *Acta Biomater.* **2012**, *8*, 1388–1393. [CrossRef] [PubMed]

28. Sodagar, A.; Kassaee, M.Z.; Akhavan, A.; Javadi, N.; Arab, S.; Kharazifard, M.J. Effect of silver nano particles on flexural strength of acrylic resins. *J. Prosthodont. Res.* **2012**, *56*, 120–124. [CrossRef] [PubMed]

29. Sousa, F.A.; Paradella, T.C.; Koga-Ito, C.Y.; Jorge, A.O. Effect of sodium bicarbonate on candida albicans adherence to thermally activated acrylic resin. *Braz. Oral Res.* **2009**, *23*, 381–385. [CrossRef] [PubMed]

30. Vieira, A.P.; Senna, P.M.; Silva, W.J.; Del Bel Cury, A.A. Long-term efficacy of denture cleansers in preventing *Candida* spp. Biofilm recolonization on liner surface. *Braz. Oral Res.* **2010**, *24*, 342–348. [CrossRef] [PubMed]

31. Wady, A.F.; Machado, A.L.; Zucolotto, V.; Zamperini, C.A.; Berni, E.; Vergani, C.E. Evaluation of candida albicans adhesion and biofilm formation on a denture base acrylic resin containing silver nanoparticles. *J. Appl. Microbiol.* **2012**, *112*, 1163–1172. [CrossRef] [PubMed]

32. Wolff, M.S.; Larson, C. The cariogenic dental biofilm—Good, bad or just something to control? *Braz. Oral Res.* **2009**, *23*, 31–38. [CrossRef] [PubMed]

33. Chandra, J.; Mukherjee, P.K.; Leidich, S.D.; Faddoul, F.F.; Hoyer, L.L.; Douglas, L.J.; Ghannoum, M.A. Antifungal resistance of candidal biofilms formed on denture acrylic in vitro. *J. Dent. Res.* **2001**, *80*, 903–908. [CrossRef] [PubMed]

34. Jagger, D.C.; Harrison, A. Denture cleaning the best approach. *Br. Dent. J.* **1995**, *178*, 413–417. [CrossRef] [PubMed]

35. Budtz-Jørgensen, E. Materials and methods for cleaning dentures. *J. Prosthet. Dent.* **1979**, *42*, 619–623. [CrossRef]

36. Fitjer, L.C.; Jonas, I.E.; Kappert, H.F. Corrosion susceptibility of lingual wire extensions in removable appliances. An in vitro study. *J. Orofac. Orthop.* **2002**, *63*, 212–226. [CrossRef] [PubMed]

37. Hong, G.; Murata, H.; Li, Y.; Sadamori, S.; Hamada, T. Influence of denture cleansers on the color stability of three types of denture base acrylic resin. *J. Prosthet. Dent.* **2009**, *10*, 205–213. [CrossRef]

38. Rantala, L.I.; Lastumäki, T.M.; Peltomäki, T.; Vallittu, P.K. Fatigue resistance of removable orthodontic appliancereinforced with glass fibre weave. *J. Oral Rehabil.* **2003**, *30*, 501–506. [CrossRef] [PubMed]

39. Yamada, Y.; Yamada, M.; Ueda, T.; Sakurai, K. Reduction of biofilm formation on titanium surface with ultraviolet-c pre-irradiation. *J. Biomater. Appl.* **2014**, *29*, 161–171. [CrossRef] [PubMed]

40. Tomomi, S.; Susumu, I.; Tsuyoshi, O.; Hiroyuki, K.; Yusuke, M.; Nobuhiro, H.; Yoshiki, N. Evaluation of antibacterial activity of visible light-responsive TiO_2-based photocatalyst coating on orthodontic materials against cariogenic bacteria. *Asian Pacfic. J. Dent.* **2016**, *16*, 15–22.

41. Fujishima, A.; Honda, K. Electrochemical photolysis of water at a semiconductor electrode. *Nature* **1972**, *238*, 37–38. [CrossRef] [PubMed]

42. Cho, M.; Chung, H.; Choi, W.; Yoon, J. Linear correlation between inactivation of *E. coli* and oh radical concentration in TiO_2 photocatalytic disinfection. *Water Res.* **2004**, *38*, 1069–1077. [CrossRef] [PubMed]

43. Choi, J.Y.; Chung, C.J.; Oh, K.T.; Choi, Y.J.; Kim, K.H. Photocatalytic antibacterial effect of TiO_2 film of tiag on streptococcus mutans. *Angle Orthod.* **2009**, *79*, 528–532. [CrossRef]

44. Choi, J.Y.; Kim, K.H.; Choy, K.C.; Oh, K.T.; Kim, K.N. Photocatalytic antibacterial effect of TiO$_2$ film formed on ti and tiag exposed to lactobacillus acidophilus. *J. Biomed. Mater. Res. B Appl. Biomater.* **2007**, *80*, 353–359. [CrossRef] [PubMed]

45. Ochiai, T.; Fujishima, A. Photoelectrochemical properties of TiO$_2$ photocatalyst and its applications for environmental purification. *J. Photochem. Photobiol. C Photochem. Rev.* **2012**, *13*, 247–262. [CrossRef]

46. Hoffmann, M.R.; Martin, S.T.; Choi, W. Environmental applications of semiconductor photocatalysis. *Chem. Rev.* **1995**, *95*, 69–96. [CrossRef]

47. Hosseinpour, S.; Tang, F.; Wang, F.; Livingstone, R.A.; Schlegel, S.J.; Ohto, T.; Bonn, M.; Nagata, Y.; Backus, E. Chemisorbed and physisorbed water at the TiO$_2$/water interface. *J. Phys. Chem. Lett.* **2017**, *8*, 2195–2199. [CrossRef] [PubMed]

48. Sunada, K.; Watanabe, T.; Hashimoto, K. Studies on photokilling of bacteria on TiO$_2$ thin film. *J. Photochem. Photobiol. A Chem.* **2003**, *156*, 227–233. [CrossRef]

49. Takeuchi, M.; Sakamoto, K.; Martra, G.; Coluccia, S.; Anpo, M. Mechanism of photoinduced superhydrophilicity on the TiO$_2$ photocatalyst surface. *J. Phys. Chem. B* **2005**, *109*, 15422–15428. [CrossRef] [PubMed]

50. Unosson, E.; Tsekoura, E.K.; Engqvist, H.; Welch, K. Synergetic inactivation of staphylococcus epidermidis and streptococcus mutans in a TiO$_2$/H$_2$O$_2$/uv system. *Biomatter* **2013**, *3*, e26727. [CrossRef] [PubMed]

51. ÖZyildiz, F.; Uzel, A.; Hazar, A.S.; GÜDen, M.; ÖLmez, S.; Aras, I.; Karaboz, İ. Photocatalytic antimicrobial effect of TiO$_2$ anatase thin-film–coated orthodontic arch wires on 3 oral pathogens. *Turkish J. Biol.* **2014**, *38*, 289–295. [CrossRef]

52. Uchimaru, M.; Sakai, T.; Moroi, R.; Shiota, S.; Shibata, Y.; Deguchi, M.; Sakai, H.; Yamashita, Y.; Terada, Y. Antimicrobial and antifungal effects of tissue conditioners containing a photocatalyst. *Dent. Mater. J.* **2011**, *30*, 691–699. [CrossRef] [PubMed]

53. Westas, E.; Hayashi, M.; Cecchinato, F.; Wennerberg, A.; Andersson, M.; Jimbo, R.; Davies, J.R. Bactericidal effect of photocatalytically-active nanostructured TiO$_2$ surfaces on biofilms of the early oral colonizer, streptococcus oralis. *J. Biomed. Mater. Res. A* **2017**, *105*, 2321–2328. [CrossRef] [PubMed]

54. Ahn, S.; Han, J.; Lim, B.; Lim, Y. Comparison of ultraviolet light-induced photocatalytic bactericidal effect on modified titanium implant surfaces. *Int. J. Oral Maxillofac. Implants* **2011**, *26*, 39–44. [PubMed]

55. Kayano, S.; Yoshihiko, K.; Kazuhito, H.; Akira, F. Batericidal and detoxification effects of TiO$_2$ film photocatalysts. *Environ. Sci. Technol.* **1998**, *32*, 726–728.

56. Saito, T.; Iwase, T.; Horie, J.; Morioka, T. Mode of photocatalytic bactericidal action of powdered semiconductor TiO$_2$ on mutans streptococci. *J. Photochem. Photobiol. B* **1992**, *14*, 369–379. [CrossRef]

57. Dongari, A.I.; Miyasaki, K.T. Sensitivity of actinobaciiius actinomycetemcomitans and haemophiius apiiropiiiius to oxicjative killing. *Oral Microbiol. Immunol.* **1991**, *6*, 363–372. [CrossRef] [PubMed]

58. Jagger, J. Near-uv radiation effects on microorganisms. *Photochem. Photobiol.* **1981**, *34*, 761–768. [CrossRef] [PubMed]

59. Maness, P.; Smolinski, S.; Blake, D.; Huang, Z.; Wolfrum, E.; Jacoby, W. Bactericidal activity of photocatalytic TiO$_2$ reaction: Toward an understanding of its killing mechanism. *Appl. Environ. Microbiol.* **1999**, *65*, 4094–4098. [PubMed]

60. Matallana-Surget, S.; Meador, J.; Joux, F.; Douki, T. Effect of the gc content of DNA on the distribution of uvb-induced bipyrimidinephotoproducts. *Photochem. Photobiol. Sci.* **2008**, *7*, 794–801. [CrossRef] [PubMed]

61. Arrieta, J.M.; Weinbauer, M.G.; Herndl, G.J. Interspecific variability in sensitivity to uv radiation and subsequent recovery in selected isolates of marine bacteria. *Appl. Environ. Microbiol.* **2000**, *66*, 1468–1473. [CrossRef] [PubMed]

62. Matallana-Surget, S.; Douki, T.; Cavicchioli, R.; Joux, F. Remarkable resistance to uvb of the marine bacterium photobacterium angustum explained by an unexpected role of photolyase. *Photoch. Photobio. Sci.* **2009**, *8*, 1313–1320. [CrossRef] [PubMed]

63. Santos, A.L.; Lopes, S.; Baptista, I.; Henriques, I.; Gomes, N.C.; Almeida, A.; Correia, A.; Cunha, A. Diversity in uv sensitivity and recovery potential among bacterioneuston and bacterioplankton isolates. *Lett. Appl. Microbiol.* **2011**, *52*, 360–366. [CrossRef] [PubMed]

64. Santos, A.L.; Oliveira, V.; Baptista, I.; Henriques, I.; Gomes, N.C.; Almeida, A.; Correia, A.; Cunha, A. Wavelength dependence of biological damage induced by uv radiation on bacteria. *Arch. Microbiol.* **2013**, *195*, 63–74. [CrossRef] [PubMed]

65. Gilbert, P.; Das, J.; Foley, I. Biofilm susceptibility to antimicrobials. *Adv. Dent. Res.* **1997**, *11*, 160–167. [CrossRef] [PubMed]
66. Kado, D.; Sakurai, K.; Sugiyama, T.; Ueda, T. Evaluation of cleanability of a titanium dioxide (TiO$_2$)-coated acrylic resin denture base. *Prosthodont. Res. Pract.* **2005**, *4*, 69–76. [CrossRef]
67. Arai, T.; Ueda, T.; Sugiyama, T.; Sakurai, K. Inhibiting microbial adhesion to denture base acrylic resin by titanium dioxide coating. *J. Oral Rehabil.* **2009**, *36*, 902–908. [CrossRef] [PubMed]
68. Obata, T.; Ueda, T.; Sakurai, K. Inhibition of denture plaque by TiO$_2$ coating on denture base resins in the mouth. *J. Prosthet. Dent.* **2017**, *118*, 759–764. [CrossRef] [PubMed]
69. Tsuji, M.; Ueda, T.; Sawaki, K.; Kawaguchi, M.; Sakurai, K. Biocompatibility of a titanium dioxide-coating method for denture base acrylic resin. *Gerodontology* **2016**, *33*, 539–544. [CrossRef] [PubMed]

materials

MDPI

Correction

Correction: Stencel, R., et al. Properties of Experimental Dental Composites Containing Antibacterial Silver-Releasing Filler. *Materials* 2018, *11*, 1031

Robert Stencel [1], Jacek Kasperski [2], Wojciech Pakieła [3], Anna Mertas [4], Elżbieta Bobela [4], Izabela Barszczewska-Rybarek [5] and Grzegorz Chladek [3,*]

[1] Private Practice, Center of Dentistry and Implantology, ul. Karpińskiego 3, 41-500 Chorzów, Poland; robert.stencel@op.pl
[2] Department of Prosthetic Dentistry, School of Medicine with the Division of Dentistry in Zabrze, Medical University of Silesia, pl. Akademicki 17, 41-902 Bytom, Poland; kroczek91@interia.pl
[3] Faculty of Mechanical Engineering, Institute of Engineering Materials and Biomaterials, Silesian University of Technology, ul. Konarskiego 18a, 44-100 Gliwice, Poland; wojciech.pakiela@polsl.pl
[4] Chair and Department of Microbiology and Immunology, School of Medicine with the Division of Dentistry in Zabrze, Medical University of Silesia in Katowice, ul. Jordana 19, 41-808 Zabrze, Poland; amertas@sum.edu.pl (A.M.); ebobela@sum.edu.pl (E.B.)
[5] Department of Physical Chemistry and Technology of Polymers, Silesian University of Technology, 44-100 Gliwice, Poland; Izabela.Barszczewska-Rybarek@polsl.pl
* Correspondence: grzegorz.chladek@polsl.pl; Tel.: +48-32-237-2907

Received: 21 September 2018; Accepted: 25 September 2018; Published: 2 November 2018

In the published article, "Properties of Experimental Dental Composites Containing Antibacterial Silver-Releasing Filler" [1], we found two editing errors. The Vickers hardness was calculated according to equation 7 of reference [1], but HV should be in the place of E in reference [1]:

$$HV = \frac{1.8544 \times F}{d^2} \qquad (7)$$

Moreover, F was the load in kgf, not in N, as it was previously described in reference [1].

We also found an editing error in Figure 6 of reference [1]. The axis should be described as "Vickers microhardness, kgf/mm^2", not "Vickers microhardness, MPa".

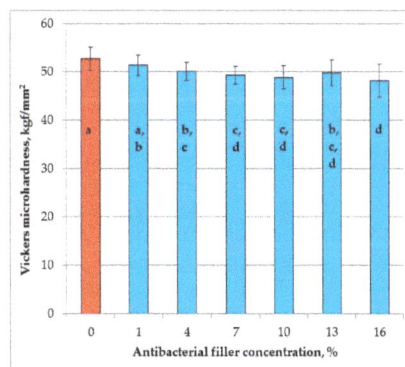

Figure 6. Mean Vickers microhardness values with standard deviations; different lowercase letters show significantly different results at the $p < 0.05$ level.

The changes do not affect the results. The values were correct. We apologize for the inconvenience this has caused and we would like to thank the editorial office for publishing the correction. The manuscript will be updated and the original will remain online on the article webpage, with a reference to this Correction.

References

1. Stencel, R.; Kasperski, J.; Pakieła, W.; Mertas, A.; Bobela, E.; Barszczewska-Rybarek, I.; Chladek, G. Properties of experimental dental composites containing antibacterial silver-releasing filler. *Materials* **2018**, *11*, 1031. [CrossRef] [PubMed]

MDPI

St. Alban-Anlage 66

4052 Basel

Switzerland

Tel. +41 61 683 77 34

Fax +41 61 302 89 18

www.mdpi.com

Materials Editorial Office

E-mail: materials@mdpi.com

www.mdpi.com/journal/materials

www.ingramcontent.com/pod-product-compliance
Lightning Source LLC
Chambersburg PA
CBHW051846210326
41597CB00033B/5789